BREAKING THE CODE O

These are critical times.

The dwindling of the Earth's energy, water and food resources, global warming and geopolitical instability pose a threat to humanity on a scale once unimaginable.

David Murrin began his career as a geophysicist with a passion for military history and went on to become a leading market analyst and fund manager. He has developed what might be called a 'grand unifying theory' of the social and political dynamics that have propelled us from the first human civilizations to our present perilous position.

In this prescient book of astonishing breadth – the culmination of decades of personal research across a wide range of disciplines – Murrin offers a conclusion as startling in its simplicity as complex in its workings. Human history, he argues, is not random, but determined by specific, quantifiable and predictable patterns fuelled by our need to survive and prosper. As individuals and collectively, we have been governed for thousands of years by unconscious responses that repeatedly mire us in catastrophe.

According to Murrin, to resolve the difficult issues confronting us today we cannot merely study the past - we also need to understand the precise algorithm of behaviour that has caused us to re-enact the same destructive cycles in ever-greater magnitudes.

Murrin takes the reader on a journey from the first civilizing impulses of early humans to the modern era, offering evidence of how civilizations develop, wax and wane in a five-stage replicating process. These 'Five Stages of Empire' are repeated the world over as regional powers accumulate resources, expand, mature, overextend and finally decline.

Set within a framework informed by analysis of geopolitical strategy and commodity theory, Murrin's masterful thesis is also a work of prophecy. His starting point is the current decline of the American empire juxtaposed with the rapid and accelerating ascent of China. He then shows us a world in which new and surprising alliances will render today's international relations obsolete – with religious polarization, war, disease and climate change acting as additional, dangerously combustible factors. Yet, he argues, we can save ourselves, if we can take the necessary first steps towards consciousness.

The philosopher George Santayana famously wrote: *'Those who cannot remember the past are condemned to repeat it.'* Murrin takes this idea further. We must not only remember the past, he says, but understand it. Only by deciphering the code of history and breaking age-old patterns can we transcend our destructive cycles and rise to the immense challenges facing humankind.

BREAKING THE CODE OF HISTORY

BREAKING THE CODE OF HISTORY

PAST · PRESENT · FUTURE

DAVID MURRIN

Dedicated to my children, Winston, Kai, Horatio and Madelaine,
and to all our children.

First published in 2010 by Apollo Analysis Limited
Copyhold Lane
Fernhurst
West Sussex
GU27 3DZ
United Kingdom

Jacket/book design and art direction: Peter Dawson, www.gradedesign.com
Design and production: Peter Dawson, Paul Palmer-Edwards

A catalogue record of this book is available from the British Library.
ISBN: 978-0-9567175-0-4

Contents

Contents (continued)

Introduction

Some of my earliest recollections are of my mother reading to me. She would frequently express surprise at my choice of a book on warfare – generally the story of a battle, or one where the technology of the clashing armies featured prominently. By the age of six I knew the names of most of the tanks, warships and planes of the twentieth century. What drove me to seek such knowledge and understanding was a growing recognition that human history was dominated by warfare and conflict. Later, as an adult, I would come to believe that seeking mastery of the subject of human conflict would provide not only a structural context for history, but a predictive model of potential outcomes for our future as well.

The sixteenth-century French seer Nostradamus also fascinated me as a child. Was his work truly clairvoyant, I wondered, and if so, how was he able to foretell the future? This scepticism remained with me, but later in life it would occur to me that the cycles that underlie the affairs of mankind could have served as a basis for his prophecies.

My interest in military history continued into my teens, but I was discouraged from formally pursuing history as a discipline by a misguided teacher who mistook my dyslexia for sloppiness. Nor could this passion be eclipsed by my eventual choice to obtain a physics degree. I continued to ponder World War Two and the Cold War as a light distraction. In physics, meanwhile, I had discovered a means of analysing the universe based on equations and models. My first two years were difficult, as I sought a way to understand the basics of each theory. In my third year, I experienced a paradigm shift: concepts that had eluded me now became clear. I also discovered geophysics, where theories could be equated to visible outcomes in the analysis of the Earth's processes. I developed a love and continuing fascination for my chosen specialisation.

After graduating, I became a seismologist with a company specialising in forming seismic cross-sections used in the search for oil deposits. The company had twenty crews at the time, stationed in sixteen countries across the globe, and I waited expectantly for news of my first posting. One night in October 1984, I was informed by telephone that I would be catching a plane the next day to join a new crew in the Sepik River Basin in Papua New Guinea, then as now one of the remotest jungle environments left in the world. My encounter with the population of this faraway land would eventually change my view of the nature of humanity and history, leading directly to the writing of this book.

The local population where I was posted had been scarcely touched by outside civilisation. The people lived in small tribes of sixty to a hundred, in a malaria-ridden swamp enveloping one of the world's greatest river systems. On the first day I found myself, groggy from jet lag, standing on a jungle helipad at 7 am in a drizzle 100 miles away from base camp. Before me were sixty Papua New Guineans, armed with bows, arrows and machetes. I had a small labour-relations problem. My interpreter, who spoke Pidgin English, informed me that the men were not prepared to work in the rain. The rain was only light, and I felt the crew was testing my authority as their new boss. I replied that in my country, women and even young children worked in such conditions. (This approach is not generally recommended in the manual on labour relations with Papua New Guineans, which was then unwritten.) First, the chief's son, whose name was Augustine, became apoplectic with rage at hearing my riposte. Next, as though Augustine was the proverbial pebble thrown into the pond, ripples of anger spread to the other men, swiftly generating a level of hostility so strong it was almost palpable; I believed they were about to kill me there and then.

I have no doubt these warrior tribesmen would have responded with fatal violence had I wavered. I walked through the baying mob, holding my head high and locking eyes with the most threatening man ahead of me, as though my will was overpowering his. I retired to my tent, ignoring the continuous stream of physical and verbal abuse. After three hours, during which time I amused myself by writing what I believed to be my last thoughts to my family, I emerged to find that the scene had transformed to one of general bewilderment and calm. Six hours after that, it was as though nothing had happened: miraculously, I was still alive. This kind of scene was played out many times over the next six months across the survey area, and after witnessing numerous such incidents, I began to reflect upon and develop a theory of collective tribal consciousness based on the following three observations:

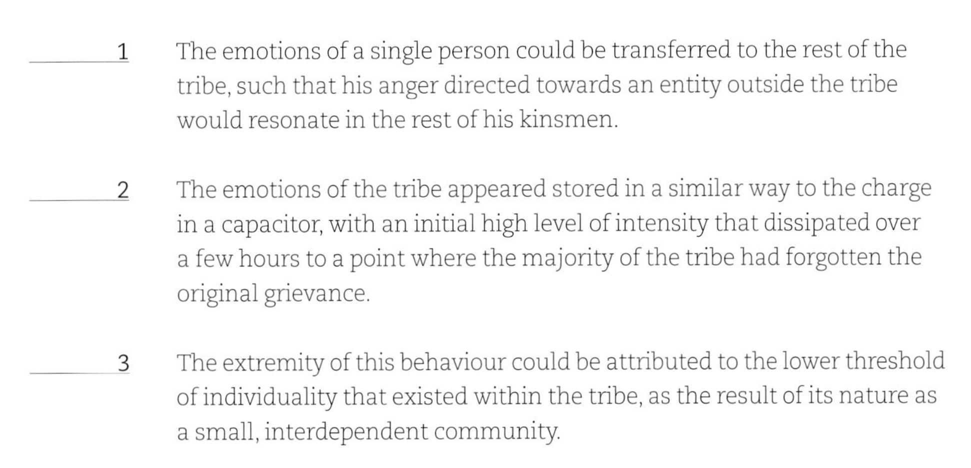

1. The emotions of a single person could be transferred to the rest of the tribe, such that his anger directed towards an entity outside the tribe would resonate in the rest of his kinsmen.

2. The emotions of the tribe appeared stored in a similar way to the charge in a capacitor, with an initial high level of intensity that dissipated over a few hours to a point where the majority of the tribe had forgotten the original grievance.

3. The extremity of this behaviour could be attributed to the lower threshold of individuality that existed within the tribe, as the result of its nature as a small, interdependent community.

I felt privileged to be witnessing what I believed was the human state as it once had been, thousands of years in the past. This imperfect conclusion was only an illusion, however, and it would be shattered years later.

In 1986, after two years in the jungle and significant deterioration in my health, I decided it was time for a change of career. Via one of those strange twists that occur in a life, I ended up on the trading floor at J. P. Morgan. Without a jot of financial training,

RIGHT **A Wara Tamba Masalai elder from Papua New Guinea wears a headdress made of moss.**

I approached my indoctrination like a good scientist, observing the criteria that would produce success. I noticed that people in senior positions, with economics and finance backgrounds, often did not perform highly; on the other hand, some of the boys on the floor from London's rough East End made substantial sums for the bank. Above all, I observed levels of emotion at work that seemed disproportionately high, especially to one who had been in a few tight squeezes where lives were at stake. It dawned on me that I had seen such behaviour like this before: among the tribes in Papua New Guinea. Suddenly, there it was – the shocking realisation that today's human beings had not developed as far as I had believed. We were, in fact, still tribal creatures. It was just that the boundaries of the tribe had become so much larger and more complex that we did not readily perceive them anymore.

I came to realise that prices moved with emotions, and I began to quantify the effects by recording price fluctuations. Soon, I appreciated that by doing so I was also registering and measuring collective human behaviour. (Needless to say, subscribing to such ideas at J. P. Morgan at that time was the equivalent of coming to work on a broomstick!) Later, I was surprised to find an enormous body of work based on the study of price patterns in markets, with the object of predicting future moves. I was especially attracted to the Elliott wave principle, which was developed by the prescient American accountant R. N. Elliott in 1928.

Elliott's principle (conceived during his early retirement as a result of a tropical illness from having worked for years in Central America) was based on Fibonacci numbers and on the methodical analysis of seventy-five years' worth of stock market indices. Many modern scientific concepts were embedded in his work long before becoming established, such as the fractal nature of the universe. (Fractal theory holds that a complex process can be understood by identifying the smaller, simpler processes it contains, each of which is identical to the whole, only smaller. A common example of the many fractals found in nature is broccoli: each floret of the vegetable echoes the whole.) Elliott identified repetitive cycles in market upturns and downturns alike (measured in 'waves'), and was further able to classify the emotional character of each distinctive sub-wave. Encountering Elliott's work was a revelation: to my mind, here was a model for the growth and contraction of the cycles of human endeavour. I believe his work would have been recognised with a Nobel Prize, had he not been so far ahead of his time.

If markets followed such predictable patterns as Elliott had demonstrated, my training as an amateur historian and professional physicist told me that similar underlying mechanisms should be at work in human evolution – thus conferring the potential to predict global events. I was convinced there existed a mathematics of human behaviour and history.

It occurred to me that it should be possible to identify the theories that underpin historical cycles, learn the lessons and apply them to today's changing world. Studying the ebb and flow of empires throughout history, in particular, can enable us to pinpoint the mechanisms that cause civilisations to rise and fall. These principles apply equally to regional powers, which are in effect smaller fractals engaged in the same processes – to scale – as empires (and, in some cases, less well documented).

This book attempts to understand not only the calculus of empire but also the dynamic of religious belief that has influenced it throughout history to the present day. Spiritual beliefs and religions are related, but defined differently: a religion erects a formal collective structure around spiritual beliefs, bringing with it the politics of human relations. Religions are products of the societies that have created them, reflecting the needs of the people at the moment; as they cease to be effective for their respective societies, they evolve – sometimes radically. Religion is but one of the tools a society uses to shape and define itself, especially with regard to how it orders the behaviour of its people – including motivating them to commit violence against other societies with different beliefs.

There is a tradition that history is about the detail, but I have always believed instead that it is determined on a vast scale, by a specific set of dynamics. Moreover, its 'randomness' is nothing more than an illusion: once the sequence of events that we call 'history' is shown to be governed by certain behavioural algorithms, we can then discern, with clarity, the degree to which our lives are bound up in numerous interrelationships. We can understand our world in addition to merely knowing it. The application of these essential principles can enable us to create a model of modern geopolitics to describe the future that awaits us.

There is an important element to writing about history that is generally neglected (but which, from the point of view of a market analyst, is instructive): the experience of evaluating the past – be it ten days ago or ten years – is drastically different from how that given period feels as it is lived out in real time. I believe this is so because the 'now' is the domain of collective consciousness; only afterwards do the prefrontal cortices of our brains analyse the details in an attempt to generate a logical sequence of events. However reassuring that process might be to our need for order, it cannot predict the future with any reasonable degree of accuracy.

This book embarks on a journey to unravel the strands of the past and determine how they have shaped our consciousness on many levels – particularly in the formation of nation-states and religious beliefs. Once the pattern has been deciphered, we shall then be in a position not only to examine our geopolitical present with precision, but also to make predictions about the future of the world.

Following this logic, this book has been organised into three major sections: **PAST**, **PRESENT** and **FUTURE**. The section titled 'The Five Stages of Empire', found in the **PAST** (Chapter Two), is the underlying basis upon which this entire book has been constructed – the key, in fact, to breaking the code referred to in the title. All subsequent chapters are informed by this theory, and are best understood within the context it provides.

PAST reviews the history of humanity, from the first appearance of the genus *Homo* to the present day. It charts the rise and fall of empires – the largest social organisational structures we know – and introduces the Five Stages of Empire as an endlessly repeating cycle of behaviour that, once understood, can allow us to best assess the future of civilisation.

PRESENT contextualises world events in such a way that they may be seen as the product of six main drivers:

1 contemporary geopolitics as understood within the framework of the Five Stages of Empire;

2 the commodity cycle, and its effects on consumers and producers;

3 the polarisation process – the road to war;

4 the global balance of military power;

5 disease;

6 climate change.

FUTURE proposes solutions to prevent these six drivers from condemning us all to catastrophe over the next two decades and beyond.

This book is designed to help the reader understand the patterns inherent in human behaviour since the beginning of our time on Earth, and to show how, if we continue along this unconscious path, we risk all-out disaster. It might be called a self-help book for the planet: it proposes the first steps towards awareness of the patterns that control us and that we reproduce, and initiates a process whereby we can alter those patterns – changing ourselves and hence the very course of our future.

For change can only take place with consciousness. My hope is that every reader who assimilates the ideas in this book will contribute their own small part towards achieving the critical mass of change necessary for us to find new ways to coexist, both as individuals and as nations.

David Murrin, 2010

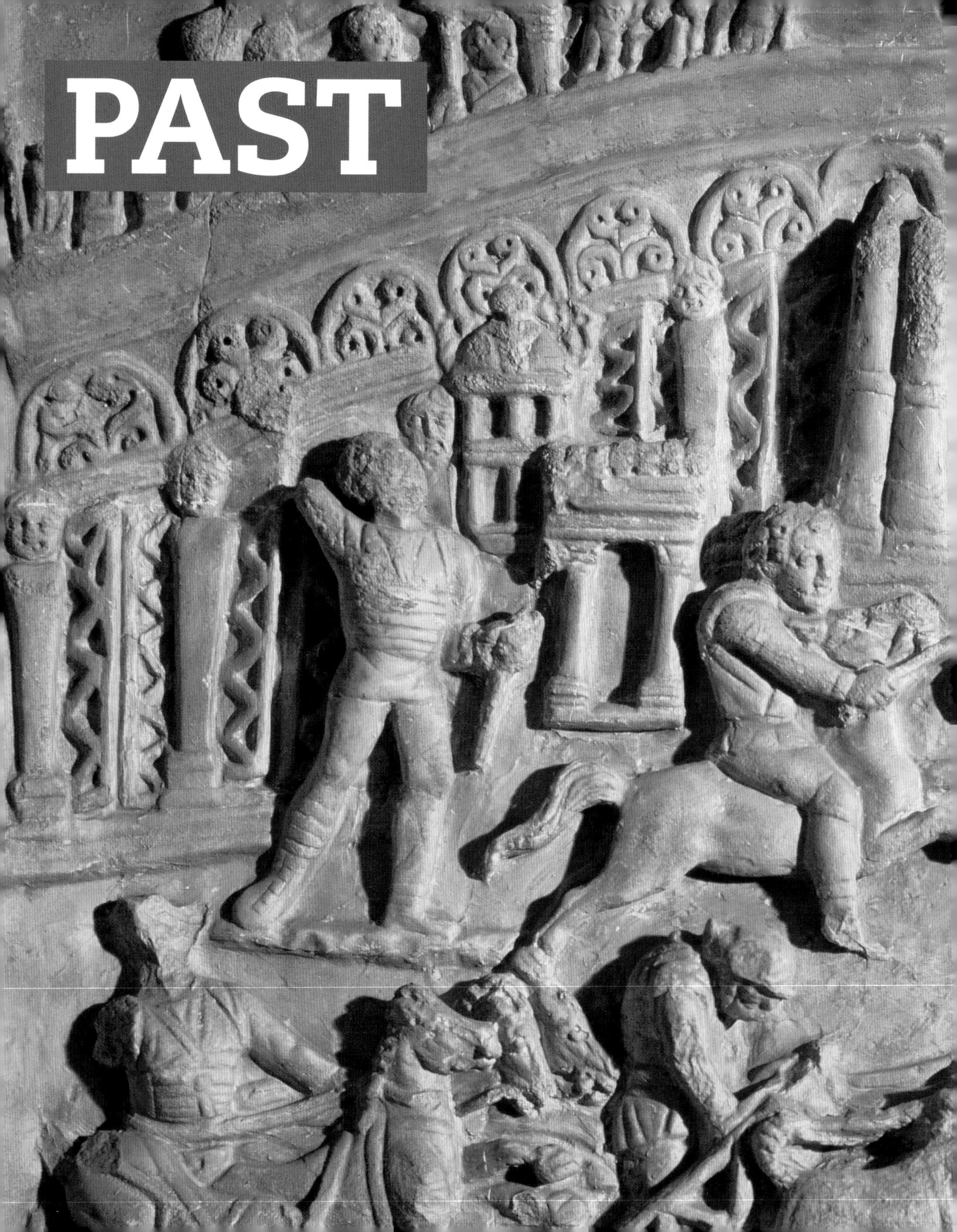
PAST

Chapter One

From Hunter-Gatherer to Industrialist

Overpopulation

The overriding problem in the world today is simply that there are too many people. *Homo sapiens* has been too successful as a species, judging by the degree to which we have populated the planet. As a consequence of our numbers and hence the rate of industrialisation, we have changed and are changing our delicate balance with our environment at an accelerating pace. We are beset by a hubris that blinds us to the fragility of our current position.

The danger stems from the fact that the Earth is a complex closed system, and closed systems seek mechanisms by which to redress imbalances and find a new equilibrium. The mechanism, in this case, may be far from ideal for us: a new equilibrium implies our near-extinction.

By way of a simple analogy, if the Earth is the human body and we are a viral infection, the outcome will unfold according to one of three possibilities:

1. The body fights off the infection and returns to normal, i.e. the host lives and the virus dies.

2. Both the virus and the body are unable to claim victory in an outright battle; they are thus both forced to live with each other in symbiosis. The virus finds a home and prolongs its life cycle, and the body limps along.

3. The virus is so successful that it runs rampant throughout the body, killing its host; then, with nothing to sustain itself, it dies too – both the body and the infection lose the struggle for survival.

The second possibility best describes our relationship with the biosphere until the Industrial Age. Now, some 200 years after the First (Western) Industrial Revolution, we are staring the third possibility in the face as the Second (Eastern) Industrial Revolution gathers increasing momentum. The Earth itself will not die, of course: over its 4.5 billion years or so, it has birthed numerous ecosystems. It is most animal life – including human beings – that faces annihilation.

Populations have historically been reduced by natural catastrophes, climate change, disease and war. War is the result of intense competition for resources, which in

PREVIOUS **Stone relief depicting games at the *Circus Maximus*, first century AD. (Museo della Civiltà Romana, Rome, Italy.)**

RIGHT **Devout Hindus throng the streets of Mumbai as a huge idol of the elephant-headed god, Ganesha, is carried through a busy thoroughfare so as to be immersed in the Arabian Sea.**

turn generates social stress. Weakening of the immune systems of a traumatised population then provides opportunities for disease to manifest as epidemics. (One example, as I shall discuss in greater detail in Chapter Nine, is the Spanish flu epidemic following World War One.) As humans proliferate within the closed system that is our planet, we can expect this destructive cycle and its consequences to become more extreme – unless we attain a new collective awareness of the need to neutralise it.

With potential disaster so close at hand, the problem of population is the primary issue we must confront.

THE EXPANSION OF THE HUMAN POPULATION OVER TIME

From the Earth's perspective, the human race resembles a plague that (along with other mammals) has been manifest over the course of the last 0.04 percent of its existence. To understand the progression of the 'disease of humanity', it is worth reviewing our race's relatively brief history within the context of the Earth's aeons-old existence. (See Figures 1–3 for three different diagrammatic views of world population growth.)

Homo habilis ('handy man') arose some 2 million years ago in East Africa, developed stone tools and hunted for meat. The larger and more intelligent *Homo erectus* either followed or coexisted with *Homo habilis* (recent findings suggest the latter), and migrated from Africa, first to Asia and then to Europe. To survive the harsh climate of that age, *Homo erectus* devised shelter and clothing, and mastered fire. These hunter-gatherers were resourceful enough to prosper even in the cold, frost-covered areas of the Northern Hemisphere for over 1 million years.

Somewhere between 500,000 and 300,000 BC, *Homo erectus* evolved into two separate species: *Homo sapiens* ('wise/knowing human') and *Homo neanderthalensis* (whom we generally know as 'Neanderthals'). These two species were, in effect, competitors for control of the Earth's surface.

Homo sapiens sapiens, the genus, species and sub-species, respectively, to which modern human beings belong, evolved in Africa around 200,000–130,000 BC. Our direct forebears proved to be a considerable upgrade on their ancestors, with bigger brains and enhanced capabilities that, it is thought, included language by 100,000–50,000 BC. They too began migrating from Africa, and wherever they went they displaced all other descendants of *Homo erectus*.

One can only imagine the competitive process that led to the demise of the Neanderthals and to *Homo sapiens sapiens*' domination of the planet's surface by 30,000 BC. It is logical to suppose that it was ruthless. Humans inhabited Asia by 50,000 BC, Europe by 40,000 BC and the Americas by approximately 12,000 BC, just at the peak of the last Ice Age. Then, by 10,000 BC, as the climate warmed and sea levels rose by 70 m, many of these human communities became isolated from each other, precipitating the diversification of the human race into the racial subsets with which we are now familiar, all living as hunter-gatherers in small, nomadic groups.

By 8000 BC, based independently around rivers and stimulated by the climatic changes that warmed the planet's surface, humans began to develop agriculture and made the transition from hunter-gatherers to agrarians. This Neolithic revolution led to the Bronze and Iron Ages, beginning in approximately 3000 BC. Advanced metal technology combined with irrigation further enhanced productivity and promoted a massive population expansion. The agricultural boom increased the population spectacularly, by seventeen times in 4,000 years, and spawned the first riverine empires and the birth of recorded history.

The next big leap occurred around 1800 BC, as cities and towns combined to form substantial urbanised civilisations that facilitated the acceleration of population growth over the next two millennia. By the beginning of the Christian era, there were some 300 million humans on the planet, and this remained constant for 1,000 years.

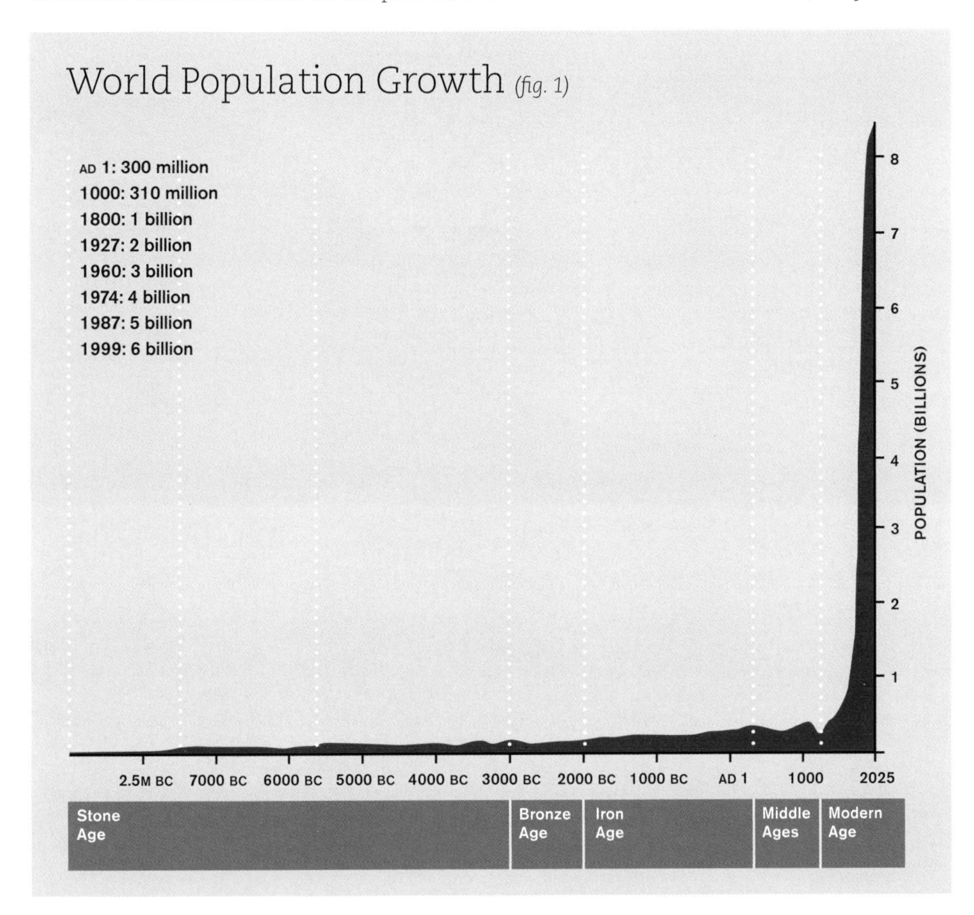

Yet the expansion of the world's population is an algorithmic mechanism that has only really impacted the planet as a whole since the 1800s, when the total exceeded 1 billion. This period coincided with the gathering momentum of the First Industrial Revolution. Substantial improvements in agricultural technology led to surging food production, and human numbers continued to grow. The resulting population explosion then triggered mass migration from the developed countries – with the help of bigger, safer ships – to the 'new frontiers', where fresh opportunities awaited.

World Population over Time *(fig. 2)*

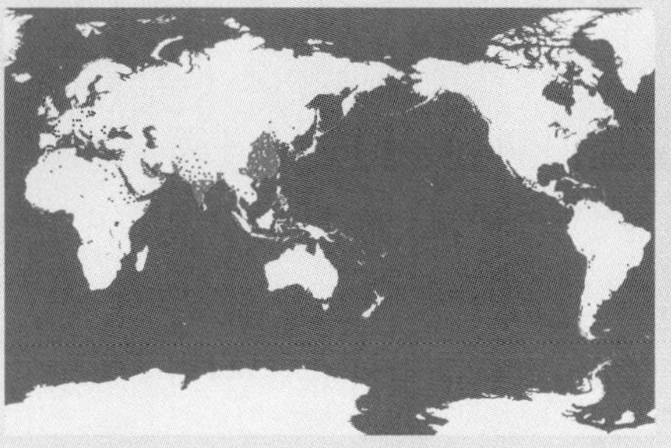

AD 1000: population = 310 million

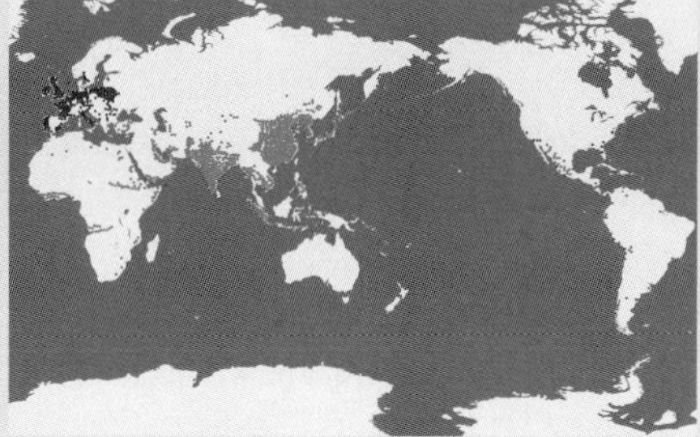

1800: population = 1 billion

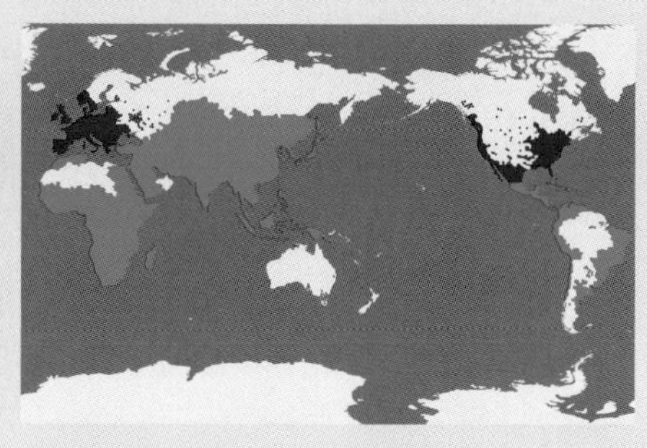

2050: estimated population = 9 billion

World Population and Development *(fig. 3)*

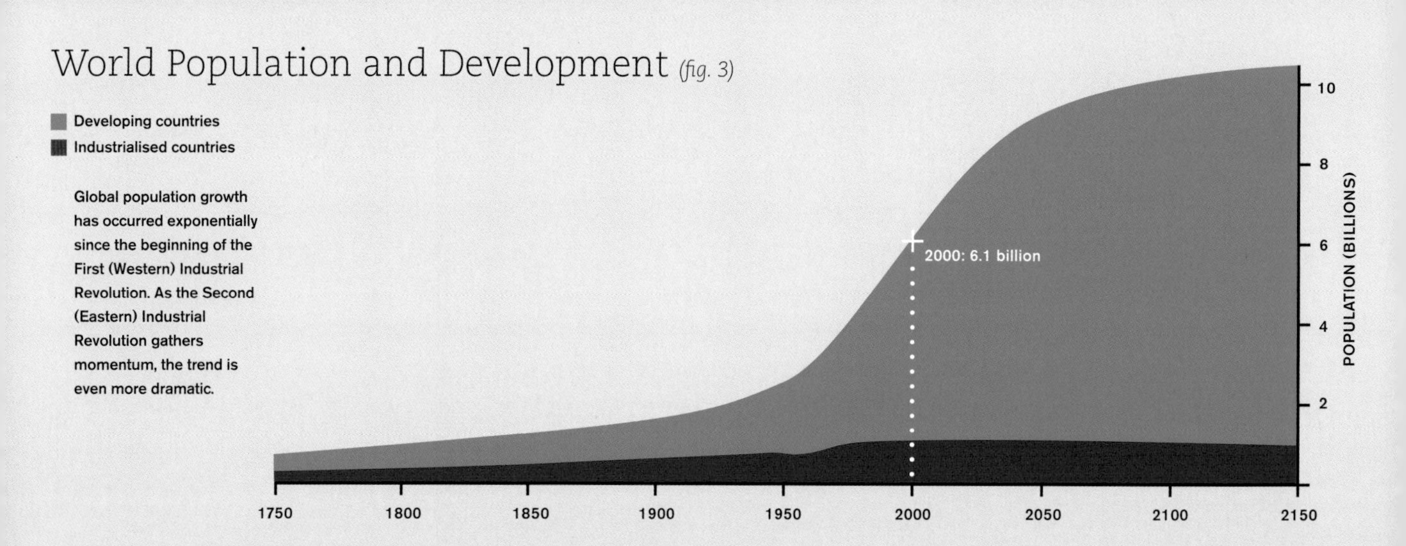

Global population growth has occurred exponentially since the beginning of the First (Western) Industrial Revolution. As the Second (Eastern) Industrial Revolution gathers momentum, the trend is even more dramatic.

Today, expanding industrialisation around the developing world is compounding the damage to the environmental balance of the planet begun by the West's own Industrial Revolution. The resulting climate change can be viewed as the Earth's attempt to rebalance itself, in order to rid itself of a parasitic human race. Unless we take active measures to control our population, the planet's natural 'defence mechanism' will do it for us.

BELOW LEFT **Densely populated high-rise apartment blocks are widespread in Shanghai, China.**

OLD AND NEW WORLD POPULATIONS

There are two distinct global population dynamics at work today: the 'old population' of the so-called developed world and the 'new population' of the developing world. The developed world's population – approximately 1.2 billion – has been stable since the 1970s. (The net effect of this is foreign policy focused on maintaining the *status quo*.) On the other hand, the explosive population growth of developing countries in the context of the Second Industrial Revolution has resulted in a hugely competitive demand for limited resources. In the case of China, which is not that far behind the West on the industrialisation timeline, this demand is manifesting itself as competition for access to the traditional suppliers of resources to the West. The bellwether for this process is the urbanisation of the Chinese population. By 2025, China's urban population is expected to have risen to 926 million from 572 million in 2005. Some 80–90 million rural residents per year will become urban residents. This massive population transfer is at the heart of the current relentless squeeze on the world's commodity resources.

African states, which are the world's least developed and least structured, cannot exercise an expansive response to population growth; they are hemmed in by neighbours in similar situations. A turning inward has resulted, instead, in civil war, ethnic conflict and genocide. These are not one-off events, but a reflex mechanism caused by overcrowding. Examples include Rwanda (1 million killed in 1994), the Democratic Republic of Congo (over 5 million dead since 1999) and Sudan/Darfur (at least 300,000 dead since 2003).

The consequences of overpopulation are already observable, then, and will dominate the next twenty-five years, in which wars are likely to be waged over access to decreasing resources.

KEY ELEMENTS OF POPULATION GROWTH

There are six interrelated factors affecting the expansion and contraction of our species:

1 **CLIMATE.** Ancient hunter-gatherers, responding to changes in their ecosystem over which they had no control, migrated in barren times to more amenable climes. As civilisations grew in size and trading complexity, underpinned by diversified agrarian foundations, humans became less reliant on local climate. This process was extended further during the Industrial Age across the world.

2 **FOOD SUPPLY.** As hunter-gatherers gave way to agrarians, new food cultivation technologies emerged to support increased populations. Over millennia, crop yields improved vastly with the introduction of the plough, irrigation and new cultivars, and later the steam plough, natural and chemical fertilisers and, finally, plough-less farming and genetically modified crops. Each jump in production has been associated with a corresponding leap in population size.

3 **WAR.** There is an argument that humans are inherently warlike, driven by a need since prehistory to conquer and kill. Yet no creature on the planet is programmed to risk life without potential gain. War occurs in order to expand resource bases, population and power. The absorption of a conquered foe rather than outright eradication has tended to be the natural preference, enabling a conquering civilisation to increase its power base. Notably, as agricultural technologies improved and the population expanded, so did the destructive technology of war, enabling the slaughter of ever-greater numbers of people in a gruesome equation.

4 **DISEASE.** Epidemics have always acted as a limitation on population growth. While human beings have proliferated over the past century, in part on account of the success of modern medicine, the risk is nevertheless always present that the very drug and antibiotic technology that has protected us could result in resistant strains of viruses with population-blighting properties. (In addition, the globalised context today has imposed heavy challenges on the containment of epidemics.)

5 **MIGRATION.** Relocation has always been the human response to environments rendered inhospitable, and it served us well for millennia. However, we have now reached the point where there are relatively few uninhabited places left to which to migrate: we are reaching the limits of space on Earth. The only other option is migration to the stars, which will not solve our population issues in the short or medium term but may provide hope for future generations.

6 **COMPULSORY POPULATION CONTROL.** This topic is controversial, although we have some perspective on such measures from their having been attempted in China. It goes against most human instincts to not procreate; our biological imperative is to replace two parents with at least two children and maintain the population. The problem with controls applied to reproduction is social inequality: the powerful always find ways to evade the proscription, while the poor are forced to abide by the rules. Whatever the political problems associated with population control, the world's states may well need to confront this difficult issue as we rapidly approach resource capacity. This is less the case in the developed world, where populations are now relatively stable, but must be faced in developing countries witnessing surging populations.

THE SHADOW OF EXTINCTION

Throughout the history of the human race, all manner of prophets and mystics have warned of the world's end. For our ancestors, their tribe, town or city would for the most part have been the full extent of their horizons, as long-distance travel was (relatively) limited. Their society might have been subjected to natural disasters, such as earthquakes, tidal waves, volcanic eruptions or even meteor impacts, but the diversity of humanity across the globe ensured its survival as a species.

However, with the industrialisation and globalisation that began in the 1800s, the world has become smaller and more interconnected; humanity has become increasingly vulnerable to major catastrophic events determining its fate.

The twentieth-century world wars inaugurated a new and effective technology of mass destruction. World War Two, in particular, spurred the creation of weaponry that was capable of eradicating human existence. Consequently, since 1945, we have been living in a new phase of human history, contending with the prospect of extinction not just from the technology of warfare (which increases in destructive power daily) but also from disease and the huge pressure on the food supply played out on a global scale. (One 'Malthusian catastrophe' has been forecast. This term describes the projected inability of the food supply system to provide for an exponentially expanding human population).

In the past, diversification was our survival mechanism against the failures in judgement of any one society. Now we are a single global community, a multi-polar world interlinked in such a way that humanity's survival depends on the decisions and actions of all nations. Our interrelationship means we simply do not have the ability to withstand massive-scale catastrophes the way we once did. It is imperative that we raise collective awareness if we are not to drift deeper into the shadow of extinction.

Human Social Structures

THE EVOLUTION OF COLLECTIVE BEHAVIOUR

RIGHT **Traders signal offers in the euro pit of the Chicago Mercantile Exchange in 2004, following an announcement that the US Federal Reserve would raise interest rates.**

History and the collective behaviour patterns that determine its course are products of the human decision-making process, and therefore of the human mind and its engine, the brain. The earliest section of the human brain to develop was the stem, which governed the majority of the body's basic functions. During the early stages of human existence, it was also instrumental in assessing the place of an individual within the tribe. The acceptability of every action, and the continued inclusion of a single member of the tribe, were gauged within the collective. Rejection by the tribe meant a severe reduction in an individual's ability to survive. Those who did survive would, by definition, have developed a keen sense of what was required to preserve their place. As a result, the tribe developed a collective consciousness to which all its members were connected by various degrees.

Only relatively late in human development have we acquired sophisticated functions in the middle part of the brain, where emotions are processed, and in the frontal lobes, where logic and reasoning are enabled, allowing for a more comprehensive thinking process. The recently evolved frontal lobes have not yet had time to assume complete control of human thought and action; the older sections of our brains still predominate, producing highly emotional behavioural patterns that can, at times, override the brain's higher centres.

The inescapable conclusion is that logic has not governed human history, which has instead been influenced by collective emotional responses. Although individuality is valued in many societies, we are all to some extent deeply linked to each other via our lower-brain functions. Few people are able to maintain their independent-mindedness in the face of strong group response – such individuals are the ones most suited to the task of leadership and invention. For the purposes of historical analysis, we may conclude that human social constructs, such as a city-state, regional power, empire or religion, all manifest a collective consciousness that processes information and then responds on a predominantly emotional basis, reflecting, for example, fear and greed. It is this group engine that drives both short- and long-term patterns; individuals may not recognise their part in this dynamic, but, when observed from a distance, these responses can be perceived as existing within a cycle.

It is these cycles that I shall examine in this book. Once the algorithm of a natural cycle is understood and recognised, its characteristics can be used to discern where a society or societies are situated within it and where they might be heading.

The hierarchy of needs is also an important factor in the way societies cooperate, focusing their energies on an ascending pyramid – the developmental sequence of human social organisation. Food, water, shelter and security take top priority. Once these conditions for survival are met, they are transcended by the need for emotional fulfilment, procreation and spiritual connection.

GLOBEX
LIBOR
5704 53.7 27.5 06.5B EMH EMJ EMK EMM5
9688 9853 27.5 15.5 00.5 83.5 9675
15.5 9697A 83.5 9675
1324.37 PK +13 40 EDU7 EDM8
00.5
9700 15.5B 00.5 83.5B 75B
15.5 9697A 83.5 9675
EYH.P9975 11
LYH.P9975 11 12
EYH.P9975 11 1- 11 12
EYH.P9975 11 1- 9 11
BLUE +13 40 1324
3-YR +12 100 1326
GLBX GLBX GLBX GLBX
9700 53.7 27.5 06.5B
TTAB
008

THE ORIGINS OF CIVILISATION

Early humans were as scattered seeds: those tribes that fell on barren ground stayed weak or perished, while those that settled in fertile lands grew stronger. These different fates are clearly illustrated in Papua New Guinea. In the Sepik Northern Basin, with humidity at almost 100 percent and malaria endemic, and where the population density of crocodiles vastly outnumbers that of humans, tribes rarely exceed 100 in number. However, in the highlands, where the climate is more temperate, a tribe might number in the thousands.

RIGHT Aerial views of five great river systems. Clockwise from top left are the Nile; the Tigris and the Euphrates; the Yellow River; and the Indus.

The genesis of civilisation occurred in areas exhibiting two factors: a temperate to subtropical climate and fertile river plains.

FIVE RIVERS, FOUR POWERS

The history of civilisation is the history of four distinct power blocs separated by geography, which might be thought of as the pistons in a four-cylinder engine: separate from one another but collectively driving mankind forward:

1. **THE NILE**. This would serve as the crucible not only for indigenous Egyptian culture but, after millennia, for Greek and Roman culture as well, leading ultimately to what I shall later term the Western Christian Super-Empire (WCSE).

2. **THE TIGRIS AND THE EUPHRATES**. These rivers define the region known as Mesopotamia, birthplace of the Sumerians, Akkadians, Babylonians and Assyrians, and later the powerhouse Persian Empire. The entire Middle East was shaped in greater or lesser part by this region.

3. **THE INDUS**. From approximately 1800 BC, the Indo-Aryan civilisation began establishing itself in this region, drawing from local cultures to develop what would be its enduring legacy, Hinduism. Much later, from the Middle East, the Islamic caliphates extended their power into this area, now known as India and Pakistan.

4. **THE YELLOW RIVER**. First irrigated around 2200 BC, this is the cradle of Northern Chinese civilisation and its subsequent great dynasties.

When one of these blocs evolved into an exceptionally powerful empire, it would infringe on the borders of a neighbouring bloc for a time; but for many millennia, these regions remained separate on account of the mountains and deserts between them that prevented cross-communication.

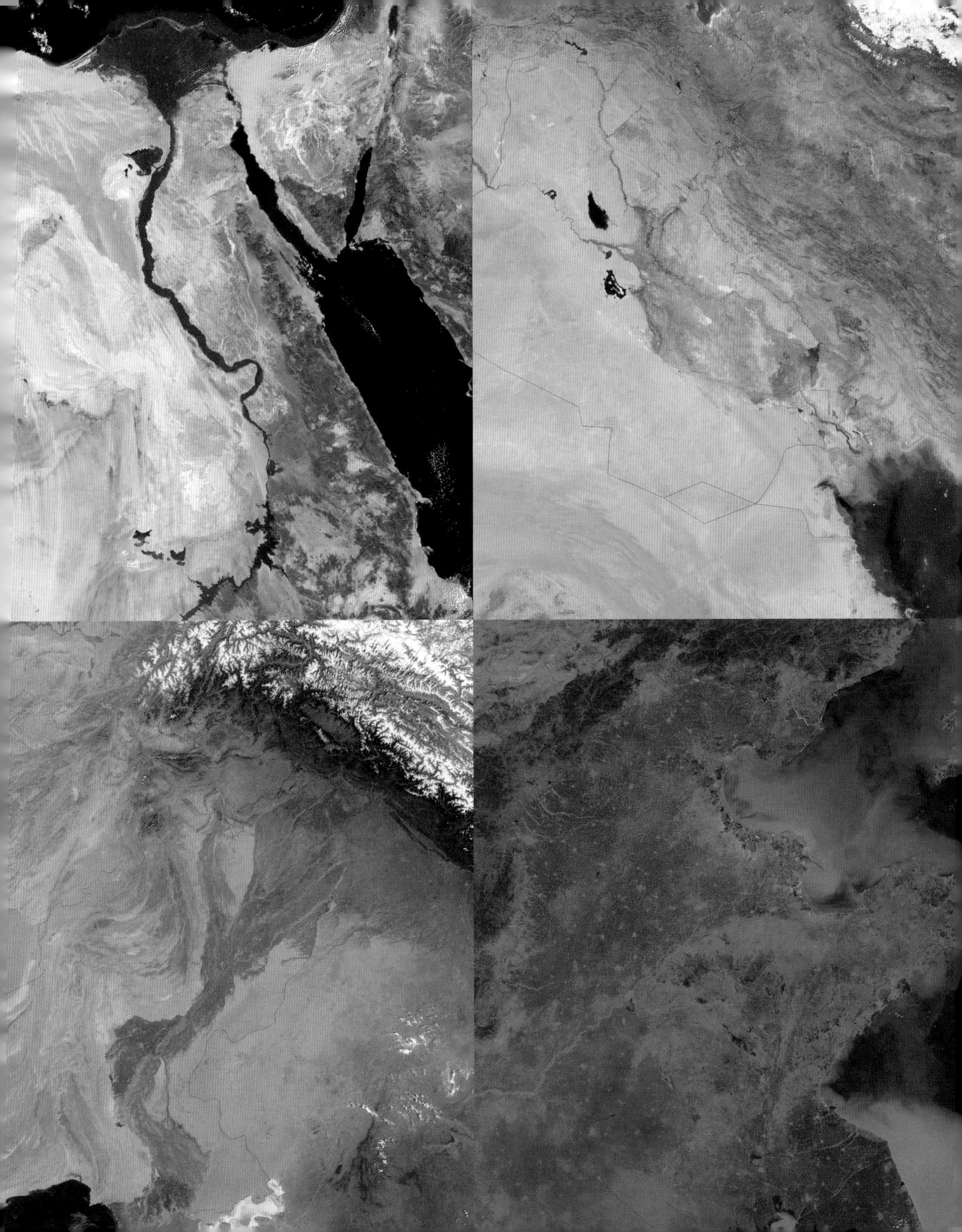

ABOVE **Detail of a Corinthian vase which shows a battle between hoplites, dating from *circa* 600 BC.**

ANCIENT WARFARE

Civilisations began as a number of small tribes aggregating (usually by conquest) to become a larger entity as they competed for resources, with the spoils going to the larger, better-organised forces. Initially such conflict could have been predominantly ritualistic, but, as human density and competition for resources increased, the nature of warfare became more vicious.

During my stay in Papua New Guinea, I managed to learn Tok Pisin or Pidgin English, the *lingua franca* of this country of extreme tribal diversity (6 million people speaking some 860 languages). In my time there, I befriended a number of local tribes and heard many stories, one of which stands out clearly. One night, probably in the 1950s, the Hauna people invited a neighbouring tribe to a feast that included plenty of betel nut and lime (the local stimulant, producing a mild euphoria). The night was a great success, ending with the tribespeople falling asleep in the main meeting hut. But all was not as it appeared; the hosts had not consumed the betel nut at the same rate as their guests, remaining alert for a signal from their chief. When it came, the Hauna rose up and butchered their guests in their sleep, engaged in the ritual cannibalism still common at that time and appropriated the other tribe's land and crops. (Such stories are repellent, but one must ask if the genocides perpetrated in Germany during the 1940s or in the Balkans during the 1990s were ultimately motivated by drives any less basic. Modern warfare, for all its sophisticated weaponry, is still conducted with the same motives that drove the earliest civilisations: the acquisition of power, resources and status.)

The conquered peoples of the past would either be killed or be converted to the cause of the victors, increasing the size of the latter's armies. The Spartans of the Peloponnese enslaved the Greeks they conquered and set them to work the land as helots, thus freeing all Spartan men for combat. The women were used both as workers and to breed the offspring of the Spartans to enhance the next generation. The lack of utility represented by young children would often result in their deaths.

As one tribe assimilated the next, each conquest became easier as newly accumulated resources were absorbed into the society. This algorithm underlay the growth of ancient civilisations to the point where, assuming all the tribes were of equal size, the rate of expansion would have been exponential until the natural borders of a region were reached. This point would generally mark a period of consolidation as a regional power for a growing civilisation. The process would subsequently be re-enacted with other regional powers – slowly at first, as the warring sides would be more or less equal in strength, and more rapidly later on as the accumulated resources of two or three regional powers were matched against a lone power. On it would continue, until empire status was attained at a balance point between natural borders and resources. Our definition of an empire, therefore, is a civilisation that at its peak has no competition within its geographic sphere of influence.

History is replete with tales of nomadic, 'barbarian' tribes possessing low levels of civilisation but high levels of aggression, descending and preying on advanced cultures that had become complacent in their comfortable lifestyles. Great civilisations are often cast initially as 'barbarians'. (The Greeks referred to the Persians in this way, although the culture and technology of the latter at least matched, and arguably exceeded, that of the Greeks.) Mongolian nomads became the Timurids and then the glorious Mughals; the Hyksos invaded Egypt (according to the predominant theory) and later begat pharaohs; and so on.

War began as, and remains, a means of imposing the will of 'one tribe' upon another. The objective of warfare is to increase the aggressor's resource base in terms of raw materials, food supplies and, of course, the human stock necessary for growing its economy as well as expanding its military manpower.

Stone Age implements were likely the first instruments of war. There must have been many battles pre-dating recorded history whereby clubs, slings, rocks, spears and axes were used as weapons, countered by wooden shields and leather garments. The advent of copper- and bronze-working sometime between 4000 and 3000 BC ushered in a period of increased sophistication, and the way wars were fought reflected this. By 2500 BC, Bronze Age technology had spread around the Nile in Egypt, the Tigris and Euphrates in Mesopotamia and around the Indus Valley and Yellow River. By 1000 BC, the use of iron had replaced bronze in the Middle East and Europe, followed a few centuries later by the Indus Valley and Yellow River cultures.

Raising armies to defend agrarian Bronze and Iron Age societies took men away from the fields, thereby limiting cultivation and harvesting. This, in turn, limited the percentage of the male population that could be sent on long campaigns. It thus produced the need for more decisive action in conflicts, and so the ritualistic battlefield came of age, where nations' differences could be resolved in a day of bloody combat.

RIGHT Ancient Theban mural of Anubis, the jackal-headed god of mummification and the afterlife, at the tomb of Horemheb in the Valley of the Kings, Egypt.

The fundamental drivers of conflict have not changed over millennia, only the social structures that prosecute war and respond to it. By examining the history of empires through the prism of technological development, and the tactics and strategy of warfare, we can gain an understanding of the military edge enjoyed by successful empires as well as the themes that governed their rise and fall.

THE GODS

One other key element to understanding the evolution of civilisations is the structure of their religious beliefs. To primitive humans, the natural world would have been full of wonders. Ascribing these inexplicable natural phenomena to spirits – embodied by the sun, thunder, etc – led, in time, to the development of deities. (In Papua New Guinea, every tree is perceived as being inhabited by a spirit.)

To a society like Ancient Egypt, reliance on the flood patterns of the Nile to break its banks and fertilise the surrounding floodplain would have been without equal in terms of priorities. They believed they were dependent on their gods for this natural occurrence. Only when a regional state expanded outside its borders into other ecosystems with different flood and crop cycles would the role of religion have transformed significantly. At some stage in a society's development cycle, its many gods would have been reduced to a few, governing the areas of life that human technology and expansion had not yet tamed. At a later stage, a single god whose domain was perceived to be the most important influence would become the supreme figure.

The empires of the Babylonians, Medes, Persians, Greeks and Romans all had their gods, but they were effectively secular societies. The shift from polytheism to monotheism took many centuries to unfold, and went hand in hand with the growth of science and technology that enabled humans to control their environment. Indeed, the trend has one further progression: the cult of atheism, with no place for gods – or God. In Europe, atheism as a publicly espoused, credible conviction first stirred faintly in Ancient Greece and much later during the Renaissance, but it would become more boldly clarified in the eighteenth century, when the French Revolution ushered in the world's first atheist political regime. In the twentieth century, the ideological forces of Fascism and Communism replaced the central role of a supreme deity with the State, differentiating themselves from the rest of a crumbling Europe from which they had split and which they were now challenging for power.

In sum, the gods we do or do not worship define the largest tribe to which our loyalties are affiliated, within the larger social structure of the world.

Chapter Two

The Elements of Empire

RIGHT Life-size terracotta warriors in the tomb of 'First Emperor' Qin Shi Huang, Shaanxi, Xi'an, China.

Human beings organise their loyalties in a series of concentric circles, beginning with their immediate family and extending outwards to their tribe, town, region, country and religion. All things being equal, humans will defend their inner circle against a threat originating from anywhere in the outer rings. I shall examine this concept in greater detail in Chapter Seven, but suffice it to say here that all human societies are fractal in nature: patterns at the smallest levels of social organisation are replicated at larger levels, with the same internal dynamics at work.

The empire is the largest known social structure to which an individual can owe allegiance; it is therefore vital, in geopolitics, to take account of empire dynamics past, present and future. Over the ages, the mechanics of empire have grown more sophisticated, evolving from the simpler structures of the ancient world into the complex system of the super-empire. Advances in technology – travel, communications and weaponry – have enhanced their effective governance. However, despite this increased complexity, great powers continue to develop along similar lines, because human behavioural patterns (operating for the most part just below the level of consciousness) have remained relatively unchanged. The nine factors described below can be considered essential in the rise and success of empires.

1. Human Resources

Civilisations bound for empire have, throughout history, risen to power based on the productivity of their citizens and their slaves. Typically, empire building in agriculturally based societies depended on a supply of labour for working the land as well as to build up the armed forces that are necessary for defence and expansion. Equally vital was the role of women for reproduction, in order to increase the population and replenish the stock of labourers, soldiers and future child-bearers. The first civilisations to emerge, beginning with the Sumerians in approximately 5000 BC, were greatly advantaged: the fertile ground they occupied enabled them to multiply their population base rapidly, giving them a demographic edge over their neighbours dwelling on less bountiful soil. With a more numerous and better-structured army, a civilisation so blessed could then conquer smaller tribes, adding their numbers to the incipient empire. This new supply of labour could be channelled into combat, growing surplus food for times of scarcity, or the construction of fortifications, general infrastructure or for places of worship.

The new labour pool would further enhance the growth of the civilisation as it challenged older, established ones.

The organisation of labour to sustain empires has gone hand in hand with social evolution. Slavery was a bulwark of the Egyptian, Greek, Persian and Roman Empires; it is estimated that 50–65 percent of the total population of ancient societies consisted of slaves. Slaves could be conquered peoples, those born into servitude or criminals and debtors sentenced to slavery. They generally served in all professions apart from the army, the priesthood and the law. Slave conditions varied according to the empire and to the phase of its cycle. Being a slave in a wealthy Ancient Roman household was to enjoy a high standard of living relative to other slaves during the mature stage of Rome's empire cycle, when oppressive slave conditions were imposed. (Conditions, however, did vary according to the rule of different emperors during this stage.)

Feudalism was another approach to fulfilling labour demands, with serfs working for their overlord in return for protection against invasion. Although a serf's existence was harsh, it was superior to that of a slave and included a degree of free choice. With time, this system evolved into landlord–tenant-farmer relationships in Central Europe and in the Arab world, and proved very successful owing to the strong motivation and high productivity exhibited by the workforce in return for their freedom.

The final stage in the evolution of bonded labour came in the form of the indentured servant, who would work for a set period of time and then gain freedom, payment and possibly land. This method enabled new settlement of the east coast of the US, and prompted a flood of Indian and Chinese immigration into the British Empire at the height of its maturation phase.

The modern, Capitalist labour model offers freedom and motivation to the worker, who then pays taxes back to local and national governments for shared services.

Whatever system is in operation, societies with fast-growing populations behave expansively, much as people in their late teens and early twenties: they learn rapidly and are prepared to take risks. This very process provides the engine that spurs the rise of an empire. Conversely, demographic contraction is comparable to old age, with conservatism, low energy and divestment – all hallmarks of decline.

2. Natural Resources

Early human organisational structures, from villages to empires, required sufficient food, shelter, manpower, base metals and possibly gold as currencies for survival and expansion. The absence or deficiency of any vital component of the resource pool would give rise to the natural human response – acquiring it from others. The Assyrians, for example, were one of the first cultures to expand based on a need for more resources, in this case into the Indus Valley, seeking copper and tin. (Iron, notably, was more readily available across the Earth's surface: there were very few 'iron wars'.) Expanding demographics went hand in hand with resource acquisition, leading in turn to the further development of the armed forces and/or a military edge that would make

ABOVE Miners work along the walls and in the crater of the open-pit Serra Pelada gold mine in northern Brazil.

conquest successful. The striving for access to resources is at the heart of most conflicts throughout history. To name a few examples:

1 Spain's expansion into the Americas was fuelled by the need to acquire gold and silver, after coming close to bankruptcy as a result of the Spanish Habsburg rulers' profligacy.

2 The British attack on Copenhagen in 1801 was driven largely by concern that the League of Armed Neutrality formed by Denmark, Prussia, Sweden and Russia to isolate Britain threatened the latter's access to Scandinavian timber, essential in the construction of British warships.

3 The Japanese attack on Pearl Harbor in 1941 was conceived in order to weaken the US naval fleet after a US oil embargo against Japan was put in place. The Japanese, greatly dependent on US oil, found themselves in crisis, and subsequently aimed to invade the Dutch East Indies for oil (and rubber). Reckoning on a US declaration of war as a response, Japan acted pre-emptively.

Religion is widely believed to have been, and to be, the underlying cause of many of the world's most destructive wars. In fact, most 'religious wars' are misnomers: religion has tended to mask the main reason motivating human beings to kill each other in organised conflict, namely the acquisition of resources. Wars of colonisation, from Spanish wars against the Aztecs, Incas and other 'heathen' peoples to the subjugation of African and Asian territories by other European powers, invariably transpired under the pretext of Christianisation. Although there was undoubtedly a religious component to colonisation, by far the overriding motive was land and resource control.

In ancient times, wars were predominantly land-based conflicts; but, as civilisations expanded and trade came to be conducted by sea, warfare spread to the oceans. One of the most profound outcomes of this progression was that, in general, a sea power would always beat a land power if it could defend its base of operations: in the long run, it could gather more resources from a wider sphere of influence. The Western empires, beginning with Greece and followed by Rome, Spain, Britain and the US, best exemplified this dynamic. As the modern era took hold, the need for resources to feed the beast of industrialisation only exacerbated this trend.

The demand for resources accelerated exponentially as technology advanced by quantum leaps. First came the British-led Industrial Revolution, followed by the Germans' chemical revolution, the US revolution in mechanisation and – in the present day – the Asian-led Second Industrial Revolution. Each new wave of innovation was accompanied by a boost in economic power, generating wealth to pay for an empire's military expansion, along with an ever-growing need for more resources. (In Chapter Six I shall look at how this process applies to China's current ambitions.)

3. Home, Land, Security

As estate agents are famous for saying, the key factor in purchasing a house is 'location, location, location'; so it is with villages, towns, cities, countries...and empires. As we have seen, river valleys were critical to the development of early empires, endowing them with fertile soil and the production of food to feed population growth. This, in turn, nurtured the development of the human resources required for imperial expansion. Water, then, has played a vital role, whether in the form of rivers for irrigation or oceans for effective commerce. Cities such as Rome were based at the confluence of land routes at a river crossing. Situated next to a ford across the Tiber River, the initial seven hill settlements derived income from the trade that passed across the river through the hills. As the settlements grew, they amalgamated. This was the genesis of Rome, which, as it expanded, took advantage of its proximity to the coast via its harbour at Osti Antiqua to spread across the Mediterranean and become an empire.

Location took on new importance after 500 BC, as the world population mushroomed and the competition for resources intensified. Natural barriers at the frontiers of a civilisation were invaluable, making defence more cost-effective. The Spanish Empire is a good example of a great power protected by sound natural

ABOVE **This engraving from 1628, illustrating William Gilbert's book on magnetism, *De Magnete* (first published twenty-eight years earlier), depicts the North Sea. Britain's natural defence – the sea barrier – is clearly evident.**

defences, with a coastline and the Pyrenees as a northern border. Probably the best example is Britain: its sea barrier provided formidable protection once the island's population had exceeded 1–2 million people, as enough forces were available to adequately man the defences. The Low Countries, on the other hand, had few borders with few natural defences and were the victims of regular invasions from east and west.

The Asian empires, on the whole, remained active mostly on land, maintaining large armies and secondary navies. In the West, beginning with the Mediterranean empires of Greece and Rome, the combination of land and sea power was always believed to be important. Ocean access, as noted previously, played a crucial part in the development of the Spanish and British Empires; the former used the Atlantic as a conveyor belt, sailing its westbound currents to reach the New World, and its eastbound currents to return just off Spanish shores. Imperial Spain was thus able to plunder South America with ease. Britain, with a small island population, lacked the demographics to compete as a continental power; instead, it expanded its fleets and competed for control of the oceans.

Once again, location came into play, as without the luxury of proximal west- and east-flowing currents, British seafarers had to master the science of navigation and sail with levels of skill as yet unseen. Their application would become the foundation of

the biggest naval empire in history. Britain's location at one end of the Western European landmass also prevented other European civilisations from attaining the status of sea powers. This advantage, along with control of the Strait of Gibraltar, corked Europe's naval aspirations until the end of World War Two. Since 1945, the US has supplanted Britain as the West's dominant sea power. As globalisation took hold, its isolated continental position provided the US with advantages comparable to those enjoyed by Britain at its height: an 'island' landmass surrounded by water over which resources could flow to feed its industrial growth.

LEFT **The Ancient Egyptian tradition of building enormous, complex pyramids requiring intensive labour and extensive resources was made possible in part by collective empowerment on an equally monumental scale. In this 1893 engraving, artist O. Grosch imagines a pharaoh come to assess the progress of a pyramid under construction *circa* 2560 BC.**

4. Collective Empowerment

The groundwork for the primary identification of a populace with a given empire is laid when individuals within a society bond together in common purpose against those outside it. (The mechanics that underlie this process, polarisation, will be discussed more fully in Chapter 7.) Religion is one of the key factors of any social organisation. The god or gods worshipped, along with their associated values, were probably among the earliest differentiating factors between tribes. Deities provided justification for aggression – after all, if the 'other side' did not recognise the same gods, they did not deserve to live. War became about more than just expansion, taking on the additional onus of carrying out a righteous, purifying act. (As noted, however, given that most wars are resource-driven, religion should not be blamed as their primary cause.) Similarly, nationalism is also a major tribal identifier, empowering and uniting the collective to risk all for ascension to the status of dominant power.

Increased polarisation was always the precursor to conflict. It heightens the perceived differences between two societies, leading in turn to an accommodation with violence in order to attain the goals of the collective. One of the strongest examples was the militant Catholicism of the Spanish Empire, which ruthlessly and forcibly spread the religion across what is now Latin America. Although the drive to colonise the Americas, as noted, was primarily motivated by access to resources, the religious fervour that accompanied it was nevertheless genuine. Civilisations require an unshakeable belief in their own values in order to ascend to empire, a motivating sense of superiority and a justifying set of principles. Religion fills these needs, and Spain certainly made good use of its Christianising mission to colonise the New World so relentlessly.

5. A Military Edge

Without a military advantage, no fledgling regional power has any chance of ascending to empire. Throughout history, various civilisations have enjoyed a strategic advantage in battle; in some cases, combinations of the following factors were, and are, present:

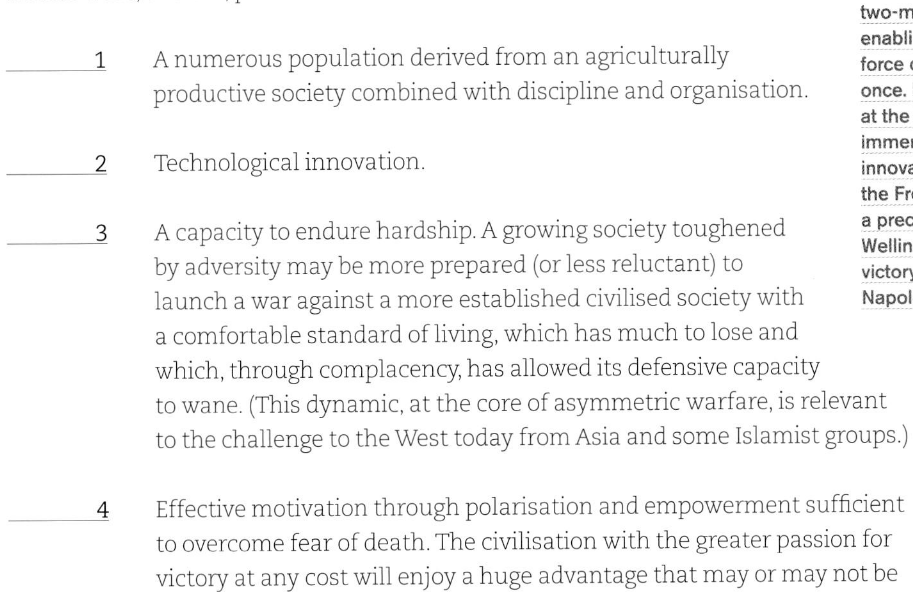

1. A numerous population derived from an agriculturally productive society combined with discipline and organisation.

2. Technological innovation.

3. A capacity to endure hardship. A growing society toughened by adversity may be more prepared (or less reluctant) to launch a war against a more established civilised society with a comfortable standard of living, which has much to lose and which, through complacency, has allowed its defensive capacity to wane. (This dynamic, at the core of asymmetric warfare, is relevant to the challenge to the West today from Asia and some Islamist groups.)

4. Effective motivation through polarisation and empowerment sufficient to overcome fear of death. The civilisation with the greater passion for victory at any cost will enjoy a huge advantage that may or may not be countered by the technology of the 'softer' civilisation.

LEFT ***The Battle of Waterloo*, a painting by the nineteenth-century artist William Heath. Late in the battle, British troops under Wellington reorganised as a two-man-deep 'thin red line', enabling them to fire with the full force of their musket power at once. European military convention at the time favoured charging in immense columns. This British innovation therefore bewildered the French forces and converted a precarious situation for Wellington's army into a decisive victory that spelled the end of Napoléon's French Empire.**

It is of great importance for a growing empire to avoid fighting wars on its own territory. For an expanding power, this is the most likely scenario in any case (e.g. Spain, Britain and the US). From an economic perspective, in the industrialised world, remote military conflicts facilitate growth of a war economy. Conversely, the ravages of war on the homeland of an aspiring power will greatly inhibit its growth; France's development from the end of the Napoleonic era until 1945 was impeded by wars and occupations. The advantage of being a sea power, expanding away from the homeland in expeditionary wars, is once again highlighted in this context.

6. Longevity

Three factors contribute to the duration of an empire:

1 **THE BREADTH OF ITS ADMINISTRATIVE APPARATUS.** The core foundation of the Roman empire was its administrative infrastructure, developed over many centuries. In sharp contrast, the empire launched by Alexander the Great rose and fell quickly without an essential administrative body able to maintain control of it. Similarly, the Mongol Empire rose to great heights under Genghis Khan, but a few generations later had collapsed back to its original state. Given a firmer administrative structure and less divisiveness, it could have continued to ascend to ever-greater power. More durable was the successful long-term conquest of India by the Indo-Aryans, who used the Hindu religion to regulate society (which it still does, to some degree, today). It may be the case that a rapid rise to empire without sophisticated enough infrastructure will most likely produce a short-lived maturity and a rapid decline. Longevity as an empire is also a function of the extent to which conquered peoples can be integrated and managed successfully.

2 **TECHNOLOGICAL INNOVATION.** New technology, and specifically that applied to war, can very quickly change the balance of power. The Ancient Egyptian Empire survived from approximately 3150 to 525 BC, its longevity due in no small part to the relative consistency in military innovation during this period, with perhaps one exception: the introduction of the chariot by the Hyksos, who managed to conquer and rule for 100 years or so before being ousted. (The new technology had by then been integrated into the Egyptian army, returning the empire to a power equilibrium with neighbouring states.)

3 **COMPETITION.** For much of history, a mechanism has appeared to operate akin to the law of conservation in physics, whereby the total number of empires in existence at any one time is kept constant for each major region of the globe. A large empire is eventually replaced by a new one, and it heads into decline around the time when the challenger is ready to make aggressive moves. The impression is that when one empire overextends, other regional powers sense the change and move in to fill the vacuum. Like a young predator, a nascent empire can only mount a successful challenge if and when the established empire has passed its peak; to do so at any other time would be suicidal. Another way of viewing this round of alternating civilisations is to consider that, at any one time, humanity as a collective organism manifests a certain level of social organisation expressed as empires: as one wanes, another waxes.

ABOVE LEFT Sculptured head of Alexander the Great, *circa* 330 BC.

ABOVE RIGHT Stone carving of Genghis Khan, Khenti, Delgerkhaan, Mongolia. Offerings have been left at its base.

7. Vitality and Innovation

The notion of 'national vitality' takes on a different quality depending on where the country in question is situated along its power curve. That in a growth phase enters a virtuous cycle where hard work is well rewarded, further encouraging continued effort; its citizens work harder until they reach a plateau where their desires are fulfilled, and the higher quality of life slows down the motivation factor. (The German workforce in the 1980s, for instance, exhibited the highest productivity in Europe. By the 1990s, however, German workers were enjoying a comfortable lifestyle, and the country lost its pole position.) The other side of this equation is decline, whereupon the cycle becomes non-virtuous: hard work does not prevent failure, and renewed effort is met with further failure. Society atrophies.

These growth and contraction cycles are linked to an empire's creative force. In its regional-power stage, a country will attempt to emulate the best of its neighbours' advancements. Later, as it heads towards empire, it shifts to a phase of innovation. Macedonia looked to Greece; Rome to Etruria and Greece; Britain to Spain; Germany to Britain; and, in our time, China to the US. The process can be seen as mirroring the life cycle of an individual, whose greatest advancements are (generally) made relatively early in life, between the ages of twenty and forty; an individual's rate of innovation begins to slow down with age, later declining further before winding down completely.

Conventional opinion tends to the view that Western society has been more innovative and creative than have its Eastern counterparts. This false assessment is a linear extrapolation from a period when the West was the world's dominant empire. Yet the truth is that the West is now in decline, the East in ascension; the pattern of relative creativity will shift dramatically in the coming decades. In thirty years' time, most of the innovations that drive the world will originate in Eastern countries, such as Japan and China. This is already the case with regard to the innovations that have emerged from Japan (and to a lesser extent from Korea) in electronics, robotics and automotive technology. In the future, we can expect to derive new advances in technology across all fields from China.

RIGHT **Egyptian charioteer, from a carving in the Temple of Ramses II at Abu Simbel. The chariot, a formidable military innovation, had been introduced into Egypt by the Hyksos, Semitic-Asiatic invaders who first appeared as early as the twenty-first century BC and eventually rose to power. By the end of the Hyksos Dynasty several hundred years later, the Egyptians had gained mastery of the chariot.**

BELOW RIGHT **Some 80,000 German political leaders take an oath of allegiance to Adolf Hitler in Berlin *circa* 1935, in a sobering display of the kind of national energy that can galvanise a population during the expansive stage of empire.**

8. National Energy

This concept describes the degree of conviction a society has in its prospects and abilities. During phases of imperial expansion, that energy is positive; in decline, negative. Just as a young person goes forth into the world with an adventurous nature that can outweigh their caution and good judgement, growing empires do likewise. From 1871, the German Second Reich held a robust overestimation of its own abilities, and underestimated its enemies – a formula that would prove disastrous for the country in 1914. Conversely, Britain and France showed considerable negative energy in their response to the mounting threat from Germany during the period from 1931 to 1939, on account of their own declining national energy. National energy is also manifested in the concept of intention – the greater will to win. This is a key component in assessing the risks of war. (In the present day, the US has exhibited early signs of negative national energy by failing, for instance, to initially commit sufficient resources and troops to the wars in Iraq and Afghanistan; to ensuring the peace after claiming victory in the First Gulf War; and to addressing public disillusionment with what are now perceived as hopeless military adventures.)

9. Leaders, Heroes and Champions

One more factor that deserves lengthy discussion when considering the elements of empire is the importance of heroes. Exceptional, inspiring people who are seen as exemplifying noble attributes in the preservation, salvation or progression of a society are essential to any human social structure, and empires are no exception. Heroes can be classed into three categories:

RIGHT **Publius Cornelius Scipio Africanus (237–183 BC) was a Roman general and statesman best known for defeating his Carthaginian counterpart, Hannibal, in the final battle of the Second Punic War. Both generals were among the finest in military history. Hannibal is arguably the more famous, but Scipio proved himself to be the better commander.**

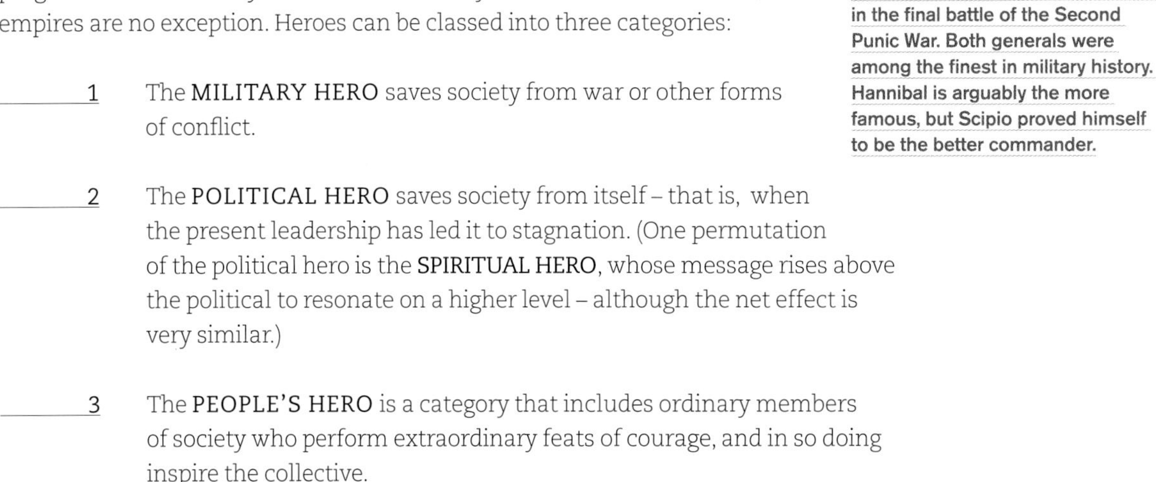

1 The **MILITARY HERO** saves society from war or other forms of conflict.

2 The **POLITICAL HERO** saves society from itself – that is, when the present leadership has led it to stagnation. (One permutation of the political hero is the **SPIRITUAL HERO**, whose message rises above the political to resonate on a higher level – although the net effect is very similar.)

3 The **PEOPLE'S HERO** is a category that includes ordinary members of society who perform extraordinary feats of courage, and in so doing inspire the collective.

The three types of hero are servants of empire, in that they further the collective goals of the dominant social structure. Their appearance is not a random event but firmly linked to the cycle of the empire in which they arise. They appear during expansive periods as well as near the end of contractive phases, but generally not near a peak or during a decline: those are times for antiheroes. Heroes play a subtle role in the human organisational process, analogous to white blood cells, lying dormant in the body until called upon to fight an invasive infection and saving their host in turn.

History is rich with incidences of heroes who appeared to emerge out of the blue at a time when a society was under strain, and in need of inspiration and leadership. If a society at a critical juncture fails to rise to the challenge, it disappears from history along with its would-be hero; however, if successful, the hero's name is passed down through the ages. (Yet history is also fickle. We all remember Hannibal, who enacted a near-catastrophic crossing of the Alps during the Second Punic War to threaten Rome with a huge army complete with elephants. Less well remembered, however, is Scipio Africanus, the Roman general who was responsible for the bold counterattack and the ultimate defeat and collapse of Hannibal's Carthage, giving Rome its first major victory against an established empire.)

In identifying an authentic hero, and understanding that their arrival is not a chance occurrence, we can gain an observation point from which to discern a locus for the growth and contraction cycles of an empire.

THE MILITARY HERO

Military heroes appear to correlate with the growth phase of a regional or imperial civilisation, when conquests abound and they are popularly received. In a military context, the hero rises rapidly through the ranks, particularly following a period of sustained conflict, and they demonstrate exceptional strategic and tactical genius from the earliest age.

Essential to this process is a perceptive mentor who recognises and nurtures the hero's exceptional talents – which can, at times, border on the maverick, contrary to the very military institution they inhabit. The British admiral Lord Barham was just such a man, and he ensured that Admiral Horatio Nelson was in the right place at the right time to lead the British fleet to a decisive battle against the French during the Napoleonic Wars. In the fourth century BC, King Philip II of Macedonia shaped and facilitated the success of his son, Alexander the Great. Very often, military success and hero status precede a move into politics and overall leadership. Julius Caesar is the obvious Roman case; a few more modern examples would include the Duke of Wellington, Mustafa Kemal Atatürk and Dwight D. Eisenhower.

BELOW **General Dwight D. Eisenhower (right), Commander-in-Chief of the Allied invasion forces, confers with British Field Commander General Bernard Montgomery (left) as they review Allied troops. During World War Two, Eisenhower served as Supreme Commander of the Allied forces in Europe, with responsibility for planning and supervising the successful invasion of France and Germany in 1944–45. He was US President from 1953 to 1961.**

ABOVE LEFT British Prime Minister Margaret H. Thatcher makes a speech during her re-election campaign. She served as prime minister from 1979 to 1990, and as leader of the Conservative Party from 1975 to 1990 (the only woman to have held either post).

ABOVE RIGHT Indian leader Mohandas Karamchand ('Mahatma') Gandhi (1869–1948) led India from British rule to independence, while adhering to a strict policy of civil disobedience and non-violence.

THE POLITICAL/SPIRITUAL HERO

Political heroes appear when empires face major challenges, during the growth acceleration phase and/or at the end of a decline or retrenchment period preceding a new growth phase. They arise when a society is at its point of greatest need. Powerful, revolutionary leaders seem to appear from nowhere to lead a civilisation from a widely perceived abyss, when it becomes apparent to the collective that radical changes are necessary.

Therefore, the hero must be not only capable but innovative – the right person in the right place for a society to thrust forward into a position of responsibility. Complicating the scenario is the hero's dichotomous position of requiring a support base, yet at heart being different enough from it to make radical changes. This complex social interaction can only take place at an unconscious level, where the hero's supporters recognise that the individual has the qualities needed to overcome the problems being faced.

In practice, these qualities are probably more or less latent at the start of the hero's career, when they are most vulnerable to the judgement of the old conservative system. Gradually, with the acceptance of the hero's supporters, their actions become bolder and more pronounced. In politics, the hero will often begin as the apparent outsider and suddenly sweep away the opposition. History is replete with such iconic figures. Their presence is an indicator of the position in which their civilisation finds itself relative to the cycle of change and expansion. Twentieth- and twenty-first-century political heroes include Margaret Thatcher, Junichiro Koizumi and Barack

Obama. Thatcher appeared as Britain was wallowing in the after-effects of having lost its empire. Her leadership re-motivated the economy and reinforced British will to protect its territory at a time when the Soviet Union was a serious threat. Most of all, she generated a sense of pride where disillusionment had come to predominate. Koizumi, the former prime minister of Japan, succeeded in changing the economic foundations of his country after a lost decade, increasing the velocity of money via pension funds that will in time raise Japan's fortunes. US President Barack Obama's confident message of change clearly struck a chord with voters, although it remains to be seen how he will meet the challenges confronting the declining US empire both domestically and abroad. Yet the emotional force and longing for an American hero that brought him to power cannot be denied or underestimated.

RIGHT **Russian Prime Minister Vladimir Putin takes part in a judo training session in St Petersburg, December 2009.**

BELOW RIGHT **African National Congress leader Nelson Mandela, later South African President and now the country's elder statesman, rejoices with then-wife Winnie upon his release after twenty-seven years' incarceration.**

The time of heroes should be viewed with caution, as it carries great risks as well as potential rewards. A society can come under the thrall of a powerful figure who leads it to destruction. Adolf Hitler is a classic case in point. Elected Germany's chancellor in 1933, he subsequently dismantled the democratic system that had elected him. A collective, urgently and emotionally demanding a change in political direction, may not be overly discriminating as to the nature of the change it advocates.

Vladimir Putin currently has the status of a hero, being widely perceived as responsible for reclaiming Russia's pride and power after the collapse of the Berlin Wall in 1989. In the process, he has turned away from democracy towards authoritarianism. The future may recall him as Russia's Margaret Thatcher – or as Josef Stalin redux.

One subset of the political hero, the spiritual hero, appears when a country or region seeks freedom from heavy oppression. A spiritual leader arises whose domain, the hearts and minds of the people, cannot be suppressed by force. Examples are Jesus Christ, who appeared when the Jews were subjugated by the Romans; the Prophet Muhammad, who sought to unify and spiritually renew his people; Mahatma Gandhi, whose values and example guided India to a relatively peaceful break from the British Empire; and Nelson Mandela, who, although not associated with a religious movement, emerged from twenty-seven years' imprisonment with the aura of one spiritually and morally advanced, and came to embody South Africa's peaceful liberation.

THE PEOPLE'S HERO

For every acknowledged hero awarded a place in history, there are countless others who enrich society yet go unrecognised. A few of these figures do, however, become popular heroes, hailed for their courage, defiance, compassion, talent or self-sacrifice. These people's heroes incarnate the values of an empire as espoused by the collective. Their recognition is vital to the continued value system and to the moral survival of the empire; hence the existence of battle honours, medals and commendations for public service and other laurels. A system that ignores such collective recognition is fast receding down the slope of decline.

Examples are legion, but include war heroes such as Manfred von Richthofen, aka 'The Red Baron', Germany's most brilliant and successful World War One flying ace,

who became an international legend (worshipped in Germany, feared and hunted by the enemy). The 139 British soldiers under Lord Chelmsford, who in 1879 held off as many as 5,000 Zulu warriors during an intense assault in the battle of Rorke's Drift, were also hailed at the time for their bravery and military discipline, and provided Britain with a welcome distraction from the disastrous massacre at Isandlwana of some 1,200 British soldiers.

LEFT Rosa Parks was an inspirational hero to the American civil rights movement as a result of her defiance in 1955, when she refused to surrender her seat on a bus to a white passenger.

Many people's heroes attain their status based on acts of defiance. Rosa Parks, the black American woman whose courageous refusal in Alabama in 1955 to give up her seat to a white passenger, was transformed into a symbol of the civil rights movement. The lone unidentified Chinese man who stood in front of advancing tanks during the Tian'anmen Square massacre in 1989, clutching nothing more than a shopping bag, forced the tank column to swerve awkwardly in frustration and became a worldwide symbol of resistance. One person's hero can be another's villain, however. Guy Fawkes, who is annually burned in effigy in Britain for his role in plotting to destroy London's Houses of Parliament in 1605, was nevertheless a martyred hero to Catholics under the Protestant reign of James I. More recently, the Iraqi journalist Muntadhar al-Zaidi, who threw his shoes at US President George W. Bush during a press conference in Iraq in 2008, became a hero throughout the Middle East (and elsewhere) for embodying the anger and despair felt by so many Arabs in the face of the US invasion and its unsuccessful aftermath.

Many figures of defiance do double duty as political/spiritual heroes, such as Mandela, who started out as an activist lawyer with the militant African National Congress, and Aung San Suu Kyi, Burma's opposition leader, who was detained by the ruling junta for over fourteen years.

Heroes of compassion would include such figures as Mother Teresa, the Catholic nun of Macedonian Albanian descent who founded hospices in India and worldwide, and cared for the most vulnerable segment of the population for forty-five years, earning herself a Nobel Peace Prize.

Then, too, there are heroes who perform daring acts of rescue, such as American pilot Chesley Sullenberger, who landed a failing commercial jetliner in New York's Hudson River in 2009, performing coolly under pressure and saving the lives of all 155 passengers on board. 'Ordinary people doing extraordinary things' could best serve as a definition of a people's hero.

COMPETITIVE GAMES AND RISING EMPIRES

In humanity's ancient past, survival was the key focus; but we are social animals, and fights would have occurred in order to resolve differences between members of a tribe. However, struggles to the death would weaken a tribe by depleting its members. Thus, we can speculate, the notion of ritualised competition as a safe channel for aggression must have arisen. With the later advent of weapons, the need to train for their use in a non-lethal way would also have been necessary. Mock combat scenarios would have evolved into sports, begetting, in turn, the first champions. The key intention underlying sporting contests would always have been to strengthen both the individual and the tribe, and to harness the competitive emotions of individuals within the tribe for the greater good. Indeed, the Greeks integrated sport into their culture, and resolved differences through competition among champions.

Prior to the industrial age, sports were linked to the development of skills required in combat, so warriors could train and channel their aggressive tendencies in a less destructive way. In the past 100 years or so, however, non-combat-oriented sports have grown exponentially around the world. Sporting activities provide a peaceful outlet for national energy and its expression, and, as such, a country's performance in international competitions is linked to its status in the pecking order of empires. Sports are a microcosm of geopolitical dynamics, and as such they can offer important insights.

BELOW RIGHT **Fireworks light up the sky as paramilitary policemen stand guard outside Beijing's National 'Bird's Nest' Stadium during the opening ceremony of the 2008 Olympic Games.**

The choice of a national sport is often representative of the values of the parent culture. Gladiatorial combat, for example, was an expression of Roman ideals prizing expansion, strength, honour, courage and skill at arms. Similarly, American football is a complex team game fit for a warlike, industrialised country, a sport of violent struggle reflecting the history of a nation forged in battle. (It can, in fact, be seen to follow on from the rugby of its former colonial masters.)

For Britain in the nineteenth century, as the world's greatest maritime power, yachting was unsurprisingly considered the sport of kings. Naturally, the Americans competed in yacht-racing contests with Britain, and, perhaps remarkably – although not when considering sports as geopolitics – Germany was also a keen competitor. The Kaiser funded the German yachting programme with government money, and he viewed the contest as an expression of the growing national pride and expansionism of his country. This was representative of the first stages of Germany's aspirations to challenge Britain's navy by building one of its own.

During the Cold War, military and economic deadlock was echoed by the Olympic Games, with the US and the Soviet Bloc locked in a race for medals to establish moral and symbolic dominance. (See Figure 4 for a tally of medals by country.) Recently, the world has been witness to the ascendance of Chinese athletes. China has even fielded a sailing team in the America's Cup competition, a sign of growing Chinese economic success and imperial ambition.

Countries with broad-based, competitive and successful sports teams may well indicate in this way that they are in the ascendant phase of an empire cycle. The 2008 Beijing Olympics were thus perfectly timed to herald the rise of China; both the magnificence of the venues and the country's results in those Olympic Games seemed to underline the point vividly.

Olympic Medal Totals by Country: Summer Games, 1896–2008 *(fig. 4)*

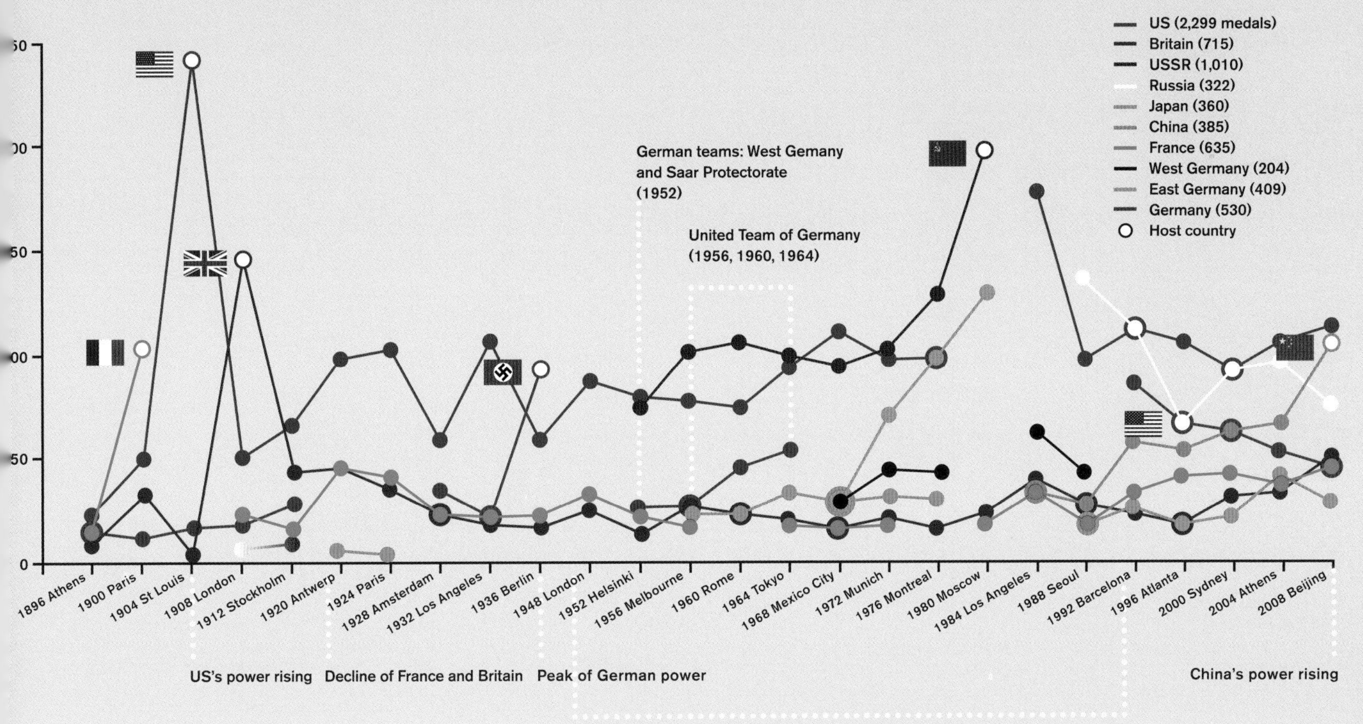

Chapter Three

The Five Stages of Empire

The Five Stages of Empire – a model of the growth and decline of civilisations – can provide a way both to understand history's 'big picture' and to accurately assess current and future geopolitical environments. To illustrate the influence and power projection possessed by an empire, Figure 5 uses a graphical representation of the five stages similar to a Gaussian curve – a statistical probability distribution of data around an average. Empires are not all the same, of course, but the majority of them exhibit a similar distribution, peaking at about 60–70 percent along their life cycles. The Five Stages of Empire, then, are as follows:

1. regionalisation;
2. ascension to empire;
3. maturity;
4. overextension;
5. decline and legacy.

These five stages can be compared to the human life cycle, beginning with birth and a period of nurturing, and followed by independence, self-expression and the manifestation of one's capabilities in the world. A peak is reached after, say, four or five decades; if it could be measured, it would comprise a mixture of wealth, energy, health, contentment, power and creativity. Finally, the decline towards death begins, completing the cycle.

As an empire grows, the world around it tends towards uni-polarity, until at its peak it comes to dominate its surroundings. Then, as it declines, there is a trend towards multi-polarity as it weakens and its neighbours strengthen.

Empires embody a critical dynamic balance between investment, defence, consumption and protectionism. The first two characteristics dominate the growth phases of an empire, as the rising power seeks to build its future and ensure its growth and survival. The third and fourth characteristics become dominant during the contractive phases, when the priority is protecting the *status quo* and maintaining a high standard of living, not investing in the future. (A comparison can be made with an elderly person who generally does not invest in their future but instead consumes during the last years of life.)

The Power Curve of the Five Stages of Empire *(fig. 5)*

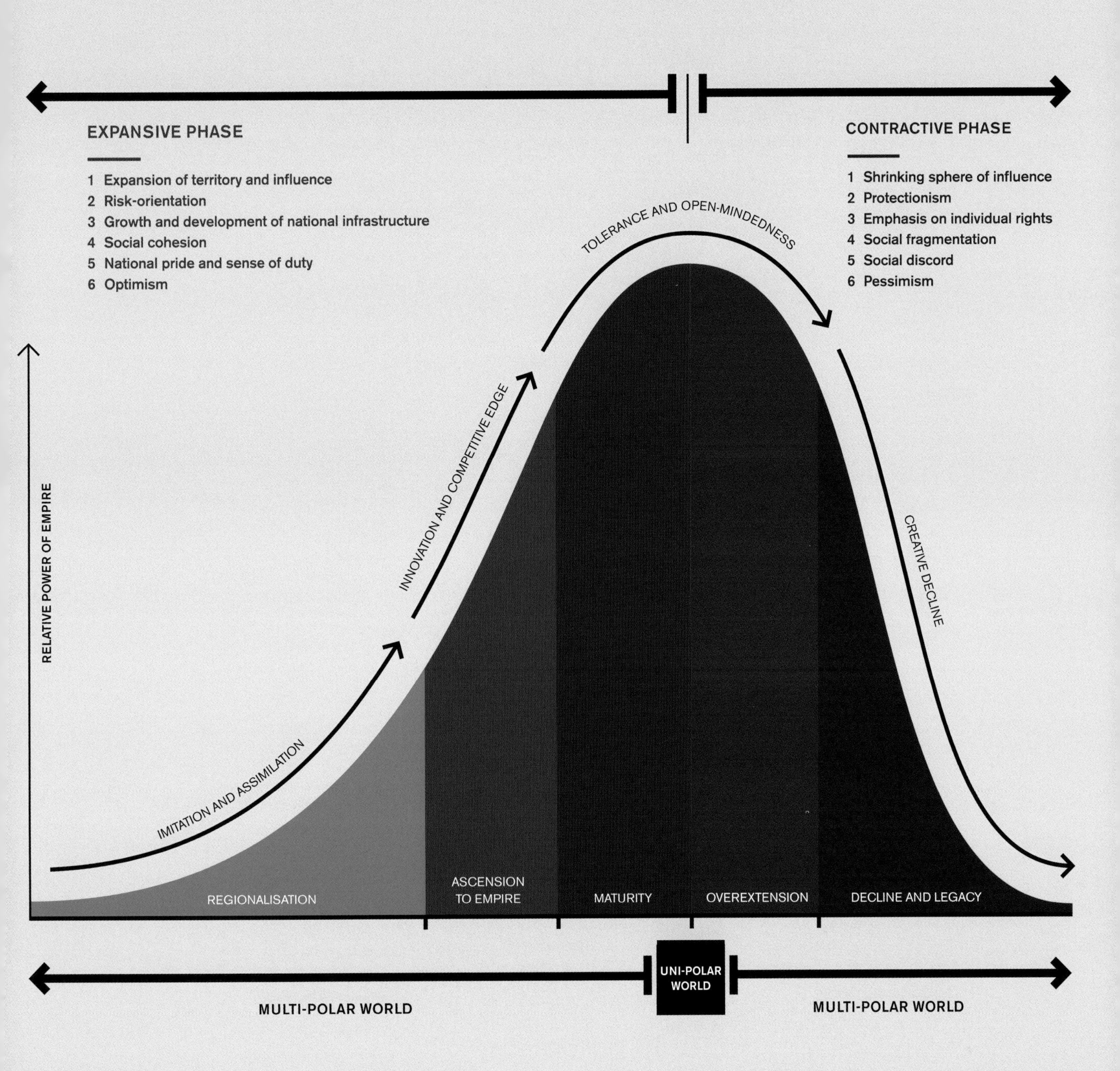

As I have discussed, demographics lie at the heart of an empire's growth, and they provide a measure of its energy, predisposition to risk and value system. Moreover, all empires display a social composition divided between the core population and the workforce that has freed it to focus on expansion. As we have seen, in ancient times, slaves and serfs filled this role. Since the abolition of slavery in the West, they have been replaced by indentured labour, colonial subjects and the working classes. (Figure 6 relates the Five Stages of Empire to the demographics of the population.)

The Expansive and Contractive Phases

The first three stages of an empire – regionalisation, ascension and maturity – are associated with the qualities of expansion; optimism; appetite for both individual as well as collective risk; investment in national infrastructure; a sense of cohesion and national duty; social cooperation; pride in national achievements and values; and, as the limitations of material-world comfort are experienced, the search for individual happiness and spiritual fulfilment.

During the expansive phase, growth is not linear but occurs in spurts interspersed with pauses for consolidation. As the region or empire becomes more economically powerful, it seeks to extend its influence as far and wide as possible. There are no cases in history in which a wealthy country with strong demographics has not chosen to militarise its economic wealth, justifying such action by trade protectionism, territorial control, political influence and domination of the widest possible economic sphere.

With industrialisation, the size and power of empires have increased, along with the destructive potential of their war-making capacity. Nations must now carefully consider the cost/benefit analysis of making war. As a result, they may commit hostile acts that are economic rather than military by nature. The new global economy has helped by creating a common arena in which the players can compete in trade. However, it would be a grave mistake to be lulled into a similar false sense of security as were the nations of the world prior to World War One, who erroneously believed that the close linking of global trade mechanisms would prevent war. All global trade does is to raise the threshold for all-out war; it does not render it obsolete.

The last two stages of empire – overextension and decline – are governed by the process of decline. Its hallmarks include a lack of social cooperation (with a decline in resources every person begins to act in their own interest); an emphasis on the rights of the citizen as opposed to a sense of duty to the nation; protectionism; the inability of the empire to use foresight to invest in vital infrastructure for its future survival; unhappiness and a sense of exclusion; the fracturing of society into social sub-groups; social discord; and pessimism.

The Five Stages of Empire and Demographics *(fig. 6)*

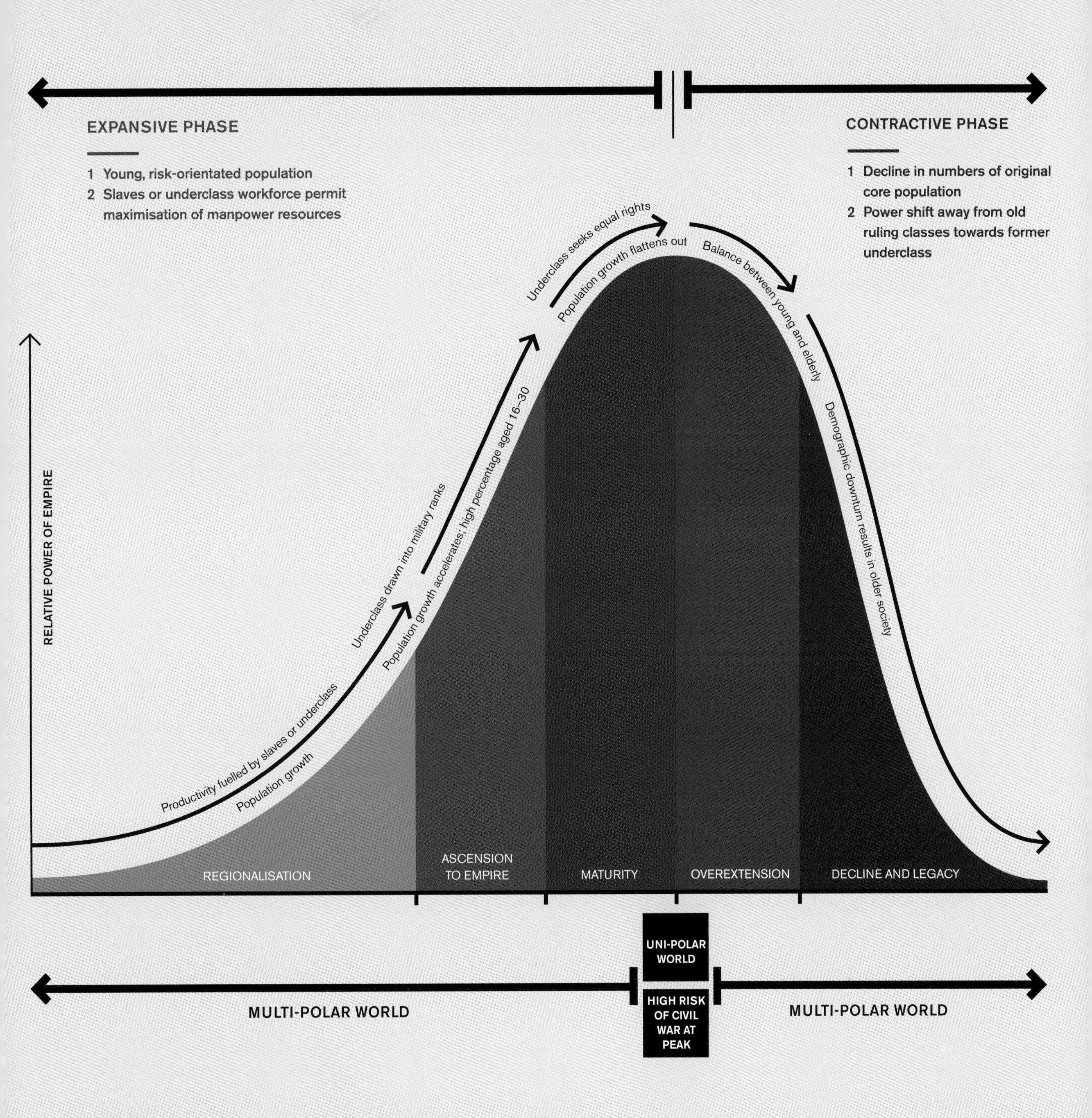

Regionalisation: The First Stage of Empire

Early in the growth of a regional power, a struggle occurs between various states within the same geographical vicinity, with the victor amalgamating all of the others. (Rome, for example, absorbed other city-states before going on to control the whole of the Italian Peninsula.) This enlarged state then becomes a player in the game of nascent empires, aiming to expand further until it attains imperial status. As discussed, the key driver prompting this behaviour is a growing population, which both needs to be fed and provides extra risk capital. The wealth of a new regional power increases through conquest, the development of new trade relationships and the spoils of war (with which the army and society are rewarded).

The regionalised entity's political and military establishments then take root, along with the society's core values. The military would, by this point, have developed a well-honed edge, making it a formidable opponent – although it would still be a long way from becoming the dominant force in its sphere of influence. Plans for expansion would continue to take into account the asymmetry of the regionalised power base in contrast with the local hegemon. The regional power would seek to make gradual and incremental gains until enough strength had been accumulated for a direct confrontation.

The catalyst propelling a nation from regional power to empire is the point at which it can no longer sustain its economy internally, particularly with respect to the acquisition of natural resources, so it is forced to look outside its borders. The crystallising moment comes when its military becomes strong enough to take on the powers around it with a good chance of success.

As an empire's core population increases, so does its demand for an enlarged menial workforce to match its growing economy and to focus the core population's energies on expansion. Traditionally, the army drove this demand: as the need for it to expand became more urgent, manpower was redirected from the maintenance of an agrarian economy. Additional labour was then required in the fields to grow and harvest crops. The Spartan solution was to annex other lands and rededicate the subjugated populations to food production, thereby freeing the Spartans to form one of the first large and permanent armies. (As discussed, slaves satisfied the need for a menial workforce for millennia.)

Civil war often attends rising empires that are approaching the end of their regionalisation stage. For the regional power in question – assuming it survives its civil war intact – the conflict can act as a coalescing agent, preparing the nation for the quantum leap to ascension to empire. Examples include the English Civil War (1641–51), the American Civil War (1861–65) and the Chinese Civil War (1927–37 and 1945–49).

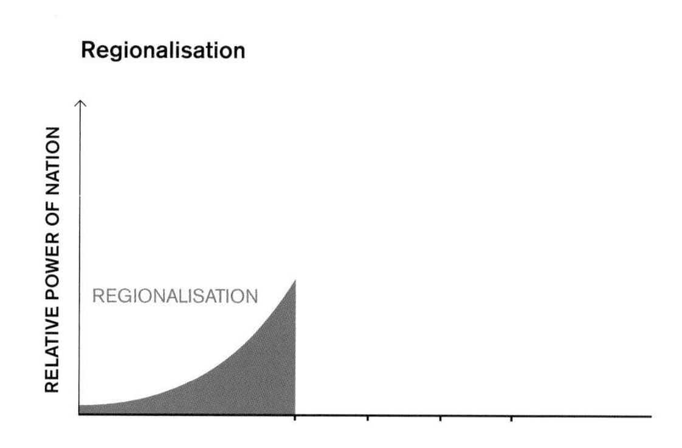

THE FRACTAL OF REGIONALISATION

Understanding the relationship between a regional power and an empire is essential to grasping one of the most critical drivers for change in geopolitics at any one time. At what stage do regional powers become empires? What happens to regional powers that bid for empire but fail?

The regionalisation stage is a fractal of the Five Stages of Empire. There are early and late stages of regionalisation, followed by maturity, overextension and decline. The inflection point – at which a nation either remains a regional power or ascends to empire – occurs at the maturity stage within the regionalisation cycle. If the demographic trend at this point exhibits momentum and continues to grow, it will force the regional power to expand as a result of an ever-pressing need for resources. The more forceful the demographic surge, the more likely the shift from regional power to empire. (Figure 7 is the by-now-familiar Five Stages of Empire graph, but this time showing the fractal of regionalisation and the inflection point.)

However, the success of a regional power in its challenge will directly relate to the status of others situated within or near its domain. If it has no competitors, the regional power will certainly become an empire. However, if other potential empires of the same order exist in the vicinity, the challenger will only have a reasonable chance of success if one of those powers has reached the overextension stage in its own cycle, thereby creating a power vacuum. Timing is, therefore, crucial as to whether or not a new empire comes of age.

Failure by the challenger could result in its absorption by the victor, but might also result in the challenger's rapid overextension and decline. It will then subsequently reorganise itself to repeat the cycle. Thus, Rome's triumph as a regional power was also about Carthage's failure as a mature, overextended state. France built a limited continental empire in the eighteenth century, yet failed to successfully challenge British power in the race towards a global maritime empire. It was therefore limited to the status of regional power in a struggle that lasted for 100 years. France did make two further challenges during this time, as a revolutionary and then later an imperial power, but ultimately failed in both attempts. Similar examples include Germany's challenge to the British Empire in 1914 and 1939, and Japan's offensive against the Asian and Pacific territories of the Western powers in 1941–42. All resulted in the collapse of the regional challenger, which subsequently underwent a new regional cycle.

The Fractal of Regionalisation and the Challenge for Empire *(fig. 7)*

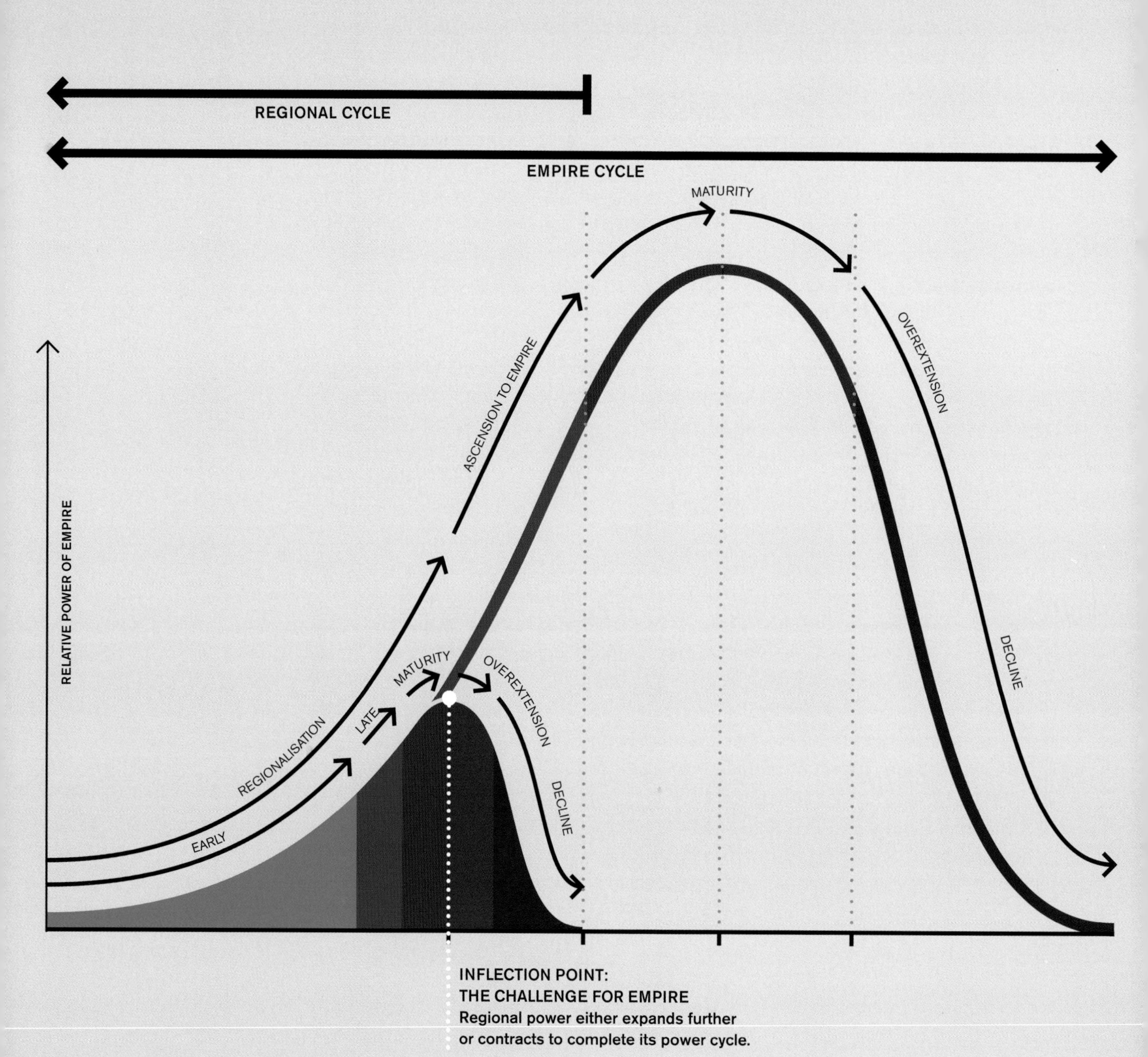

Ascension: The Second Stage of Empire

When a regional power successfully absorbs a number of similar rivals, it then spreads out, projecting its power further. This process marks the ascension stage. Once again, smaller entities are amalgamated and absorbed (e.g. Macedonia and Greece, Rome and Carthage). The algorithm of growth (given a simple model consisting of nothing but regional powers) operates as follows. When one entity conquers another, it becomes twice as powerful as the next entity it takes on; all things being equal, it will therefore succeed in half the time. By the time the ascending empire has assimilated three regional powers and attacks a fourth, it will require only a quarter of the time to succeed as it did for the original conquest. This stage of growth is the most heady and dynamic, as the wealth and power of the new empire increase exponentially. Income from expansion is vastly greater than expenditure.

Demographic expansion is once again the key, pushing regional powers to expand their influence and bring in raw materials to sustain their economies. The population will be achievement- and risk-orientated: a big advantage in confrontations against rivals in a mature, overextended or declining phase of the empire cycle. The ascension phase is characterised by clear, strategic planning and execution, along with an extensive degree of confidence that is expressed as a sense of collective destiny.

Britain entered its ascension stage early in the mid-sixteenth century. It had ample supplies of wood, bronze and iron with which to build ships, but its economy was underdeveloped. It opted to acquire a share of Spain's wealth through privateering and freebooting, and in the process obtained the financial resources (gold and silver) to allow it to ascend to empire. American imperial evolution began with the early twentieth-century oil boom and the motor industry. China began its own ascension in 1996, rather quietly, without anyone paying much attention; it is now aggressively projecting its power into the outside world in a bid to corner resources.

Recalling the law of competition in determining the longevity of an empire, an ascending power seeking to supplant an established, mature one can only do so when the hegemon begins to decline. During the ascension stage, the core population of the ascending civilisation swells with a high concentration of people thirty years old or younger. Imbued with the qualities of youth, it is therefore expansive, risk-orientated, resilient and flexible, ready to embark on wars of accretion.

This risk-positive factor is further pronounced in cultures where males predominate significantly over females. China's population today, for example, is 56 percent male, and therefore has 5 percent extra risk capital to employ during its ascension stage.

Continued population growth also facilitates and supports the manpower drain caused by military expansion and war-making. Demand for a menial workforce is very high as the economy expands.

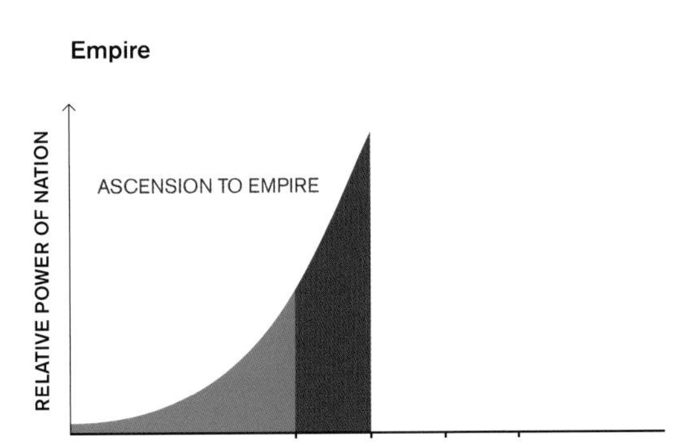

Maturity: The Third Stage of Empire

In the cycle of empire, a phase of equilibrium and stability naturally follows a period of conquest and expansion – assuming the borders of the new empire are well defined and well defended, and the administrative system highly organised: this is the maturity stage of empire. Without gains from conquest, a stable economy is required to generate enough revenue both to sustain a defensive army and to maintain civil harmony. Over time, income and expenditure become balanced during this stage. In its mature stage, Rome restricted the size of its army under Augustus to 300–400,000 men, in order to balance the budget. Even more extreme was the Western Jin Dynasty's attempt to generate a huge peace dividend in China in the third century AD, after attaining supremacy over all of its challengers: the entire army was disbanded. However, neither the Roman nor the Chinese strategy was ultimately successful in forestalling the eventual decline of their respective empires.

The beginning of imperial maturity is often witness to sweeping social changes within the empire. Population growth slows, and the ratio of young to old becomes more balanced. The drive to expand decelerates, and the empire enters a period of unmatched prosperity. The core population grows wealthy, achieving a high standard of living, which blunts the tougher qualities that drove previous generations during the regionalisation and ascension stages. When new wars erupt, the empire finds itself beset by a manpower shortage for the first time that can only be solved by inducting the menial workforce into the military. Once such wars are over, the returning soldiers from this class reject their former lowly status and demand equal rights of citizenship. The core population begins to integrate with the awakening menial workforce, but the reins of power are still held by the former. Significant internal power shifts by the end of the (subsequent) overextension stage usually result from the social changes initiated during maturity.

A frequent characteristic of empires near the end of the maturity stage is the advent of peak civil war. Without an external target to act as a focus for an empire's aggression, its leaders turn inward, and a power struggle results. Such conflicts can weaken the empire significantly. The distraction may also provide a strategic opening for any up-and-coming regional power awaiting the right opportunity to strike. Such critical moments can be particularly explosive in the case of rigid power structures, such as dictatorships and monarchies. On the other hand, democratic empires may witness a more subtle internal struggle between competing groups. In Britain, the decade that followed the Boer War saw a great deal of political upheaval that, according to Winston Churchill, took the empire to the brink of civil war. The US in recent years has seen the radical politics of the Bush-era neo-Conservatives come into play, representing a dramatic swing away from traditional American values and producing a backlash with tumultuous political consequences.

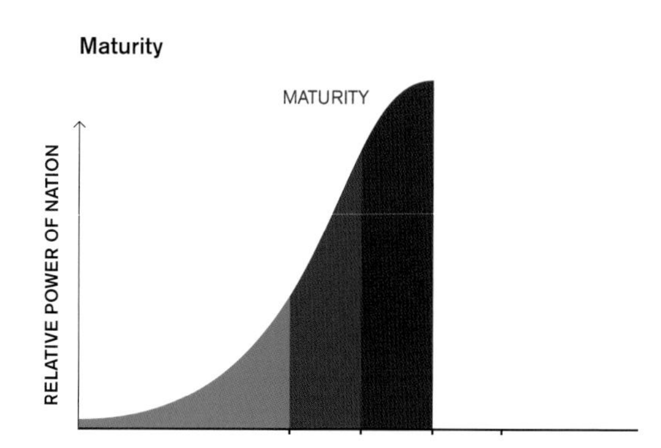

If we accept that empire cycles are unavoidable, and therefore that mature empires will witness significant power struggles, then the value of democracy as a political system can be viewed in a new light. It may well be the solution to prolonging periods of peaceful development. Inherent in the democratic structure is a means of 'blowing off steam', allowing, if necessary, a change in direction without resulting in the sorts of revolutionary, traumatic changes that can leave an empire vulnerable to outside powers. In effect, democracies seem to be better equipped to cope with such stress; unpopular elements can be replaced in elections rather than through the extremes of civil war; the process protects the empire overall.

Overextension: The Fourth Stage of Empire

Overextension signifies the onset of gradual decline, beginning at levels apparent only to the most astute observers and ending in financial disasters and military challenges.

At some point an empire's success induces complacency, arrogance, corruption and other manifestations of decay, as the comforts of civilised society give rise to expectations by the middle classes that the *status quo* will be maintained. The transformation of an empire from 'barbarian' to 'civilised' is now complete, and over time it will become ripe for domination by another aspirant. In the early stages of overextension, the cost to the economy of running an empire is no longer compensated for by revenue. The empire then begins to increase its debt burden, preventing a rise in military expenditure precisely at a time when it is most threatened by new challenges.

The Roman and British Empires each displayed such signs of overextension before declining. As I have posited, the most likely challenge to an established civilisation is always posed by a power that the decaying empire perceives as 'barbarian': one that is riding a growth curve with less to lose and more to gain from open conflict, able to act with greater aggression and fewer inhibitions to achieve its goals. For Rome, the Goths embodied this role; for Britain, it was the US. The 'barbarian' presumption, of course, conveniently masks the reality that the challenger is more developed than the declining power's complacency allows it to discern. The current US perception of both the Islamic world and the People's Republic of China exactly fits this model.

World War Two provides a compelling instance of the problems facing a civilised society fighting a 'barbarian' empire. By the time the Allied forces had left Normandy's beaches and begun the battle for Germany, it had become clear to the Allied generals that they faced problems in motivating and leading a democratic army that placed a high value on individual life. There were critical delays in the prosecution of the land war against Germany as a result. The Russian Red Army, in sharp contrast, was truly barbaric in its advance on Berlin. From the turning point of the Battle of Stalingrad, Russia pushed on relentlessly,

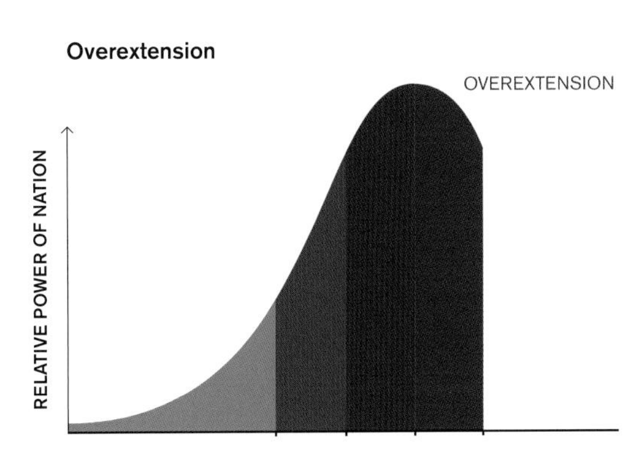

The Social Changes of the American Empire, 1600s–Present *(fig. 8)*

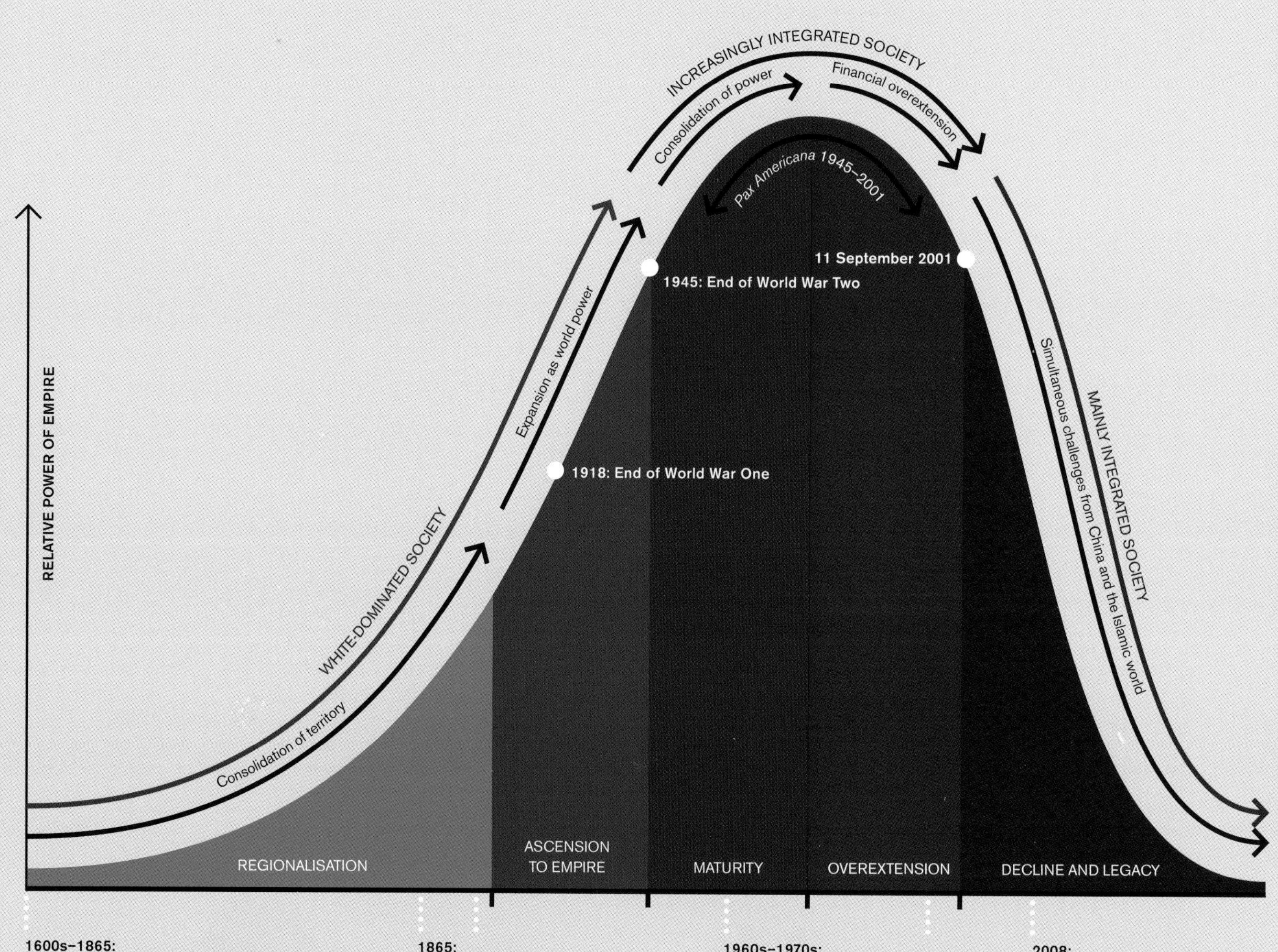

1600s–1865:
Society is completely dominated by population of European descent.

Africans and African Americans are slaves.

1865:
Civil War ends; slaves emancipated.

Black Americans remain second-class citizens.

Late 1800s:
Beginning of successive waves of European immigration.

1960s–1970s:
Black soldiers present in larger numbers than ever during the Vietnam War.

Civil rights movement struggles for equal rights.

Blacks and immigrants begin to gain access to the power core.

Late 1980s–present:
Latin American immigrants begin gaining rights; citizens of Hispanic descent begin to gain power.

2008:
Barack Obama elected President.

despite incurring massive casualties on both sides, right up to the fall of the German capital. Had the Russian army not torn the guts out of the *Wermacht* on the Eastern Front, it is doubtful that the US and Britain could have invaded Europe in 1944; global sea dominance and supreme air power could not by themselves have forced Germany to surrender its continental empire. Of course, the advent of the nuclear bomb would in any case have compelled Germany's capitulation, as it did Japan's. Yet there is a certain irony that, without one ruthless atheist totalitarian empire to counter the other, the outcome of the war might have been very different for the democratic powers.

It is certainly difficult to motivate an overextended empire to fight when its people have a high standard of living. Since 1945, the US has become increasingly reliant on its technology to ensure that American wartime casualty rates remain low. However, in wars such as those in Iraq and Afghanistan, where there is no alternative to soldiers on the ground, considerable public outcry over fatalities occurs.

Wars are expensive, more so as technology bleeds from the hegemon to the challengers, making it very costly to maintain the military edge. In the overextension stage, a dominant empire will not reduce social or defence commitments, nor increase taxation. It moves rather into financial deficit, which saps its strength. Financial-market peaks take place past the pinnacle of the empire cycle, as the system finds ways to increase spending by raising debt. During these periods, there is always talk of reducing costs, but doing so proves consistently unsuccessful. The demise of the USSR as an empire was hastened by military spending against a backdrop of declining revenue.

One way in which an empire can attempt to reduce defence costs is by building new alliances to spread the load – although doing so only delays the inevitable at best. However, it is worth noting that, when an empire overextends and then goes into decline, it is forced to scale back its influence in terms of both military presence and financial holdings. This spells the beginning of its end. (The rally of the US dollar in Autumn 2008 is a perfect example of this process; the fact that the US was forced to repatriate its wealth in order to support its collapsing core signified that its empire is now firmly in decline. The next stage will be the substitution of the dollar as the global currency by a challenger, which is likely to be the *yuan* – probably in as little as a decade.)

The social integration that began during the (preceding) maturity stage with the demand for full rights by disenfranchised citizens now progresses rapidly during the overextension stage. The composition of society by the end of this penultimate phase of the empire cycle will have dramatically altered. Formerly low-status classes can now gain entry to the empire's power core. The barbarian auxiliaries drafted into the Roman legions, for example, gained leverage and ultimately control over Rome. The British Empire's reliance on its colonies in both world wars helped to accelerate its break-up after World War Two as they demanded independence.

In the US, black Americans experienced a transition from slavery in the regionalisation stage to underclass in maturity. Black soldiers have served in US wars since the American Revolution, but, following the Vietnam War – in which they fought in greater numbers and in greater capacity than ever before – they pressed for validation as the civil rights movement intensified, and entered the middle and power classes.

This increasing degree of social integration is perhaps most effectively symbolised by the 2008 election of the first black US president. (Figure 8 summarises these and other social changes within the Five Stage Empire Model.)

As the former underclass rises, a new wave of people from poorer nations fills the roles they have left behind. This process began in the last decade-plus in the US, with hundreds of thousands of mainly Mexican immigrants seeking new lives in the apparently (and certainly relatively) prosperous US, prepared to work on any terms.

The problem of manning the armed forces continues, and the social makeup of the military continues to evolve as a result. The core of the problem lies in the fact that society now has a greater proportion of older citizens than younger ones. As a result, its decision-makers become more conservative and less adaptive in solving the growing challenges of the empire's decline.

Decline and Legacy: The Fifth Stage of Empire

For an empire in the final throes of overextension, the cost of power vastly outweighs its economic benefit. Imperial sustainability becomes increasingly unfeasible, and the system rapidly begins to disintegrate. Although the signs would have been present during the overextension stage, other great powers would, for the most part, not have begun to recognise the waning empire's vulnerability until the final stage of decline and legacy, when external and internal dynamics deteriorate at an alarming rate. Enemies on the periphery then awaken to the progressive ebbing of vitality, and become emboldened by incremental successes that can soon escalate. The old empire is now prey for other regional powers in the ascendant. The rate of decline surprises the world as a formerly iconic empire collapses.

The stage of decline and legacy can be described as the evolution of multi-polarity. The uni-polar world dilutes as the hegemon grows feebler, and challenging nations grow stronger and begin to exert a newfound influence. Characteristic of this stage is the empire's collective denial that it is declining, expressed by the body politic and by the people themselves, who ask, 'How can that which we have built up to be so splendid, so powerful, and which appeared so invincible, simply be erased?'

The illusion of the *status quo*, clung to during the overextension stage, now shatters. Often the catalyst is a key event, like the attacks on the US of 11 September 2001. The response by the leadership is to attempt to more deeply embody the perceived 'original' values of the empire. When their actions fail, a path is opened to new leadership, reflecting the new social order – and often the descendants of the former underclass, now risen, come to power. This process can appear as though hope has been renewed; yet it is, in effect, the beginning of the end of the empire. The election of Barack Obama, widely hailed by Americans and others around the globe as heralding a new age, may nevertheless be seen as conforming precisely to this model. Nor is it without precedent: near the end of the Western Roman Empire's decline, the Germanic chieftain Odoacer (believed to be a Goth) became emperor – although he acknowledged allegiance to the deposed and permanently exiled emperor Julius Nepos. (The Visigoths and Vandals had sacked Rome long before, however, and many historians consider Odoacer's reign as marking the Western Empire's end.)

The final phase of this fifth and ultimate stage of empire is legacy. This endures to some degree after all empires have declined, but is best observed in the case of an empire that has had a rapid military ascension, driven by a strong leader who dies without a clear line of succession or sustainable administrative system (a recurring theme throughout history – the Macedonian and Mongol Empires have it in common, as does the first Islamic caliphate). Fragmentation usually precedes the decline of an empire of this sort, but very often a collective value system – its legacy – persists in the region in which it was formerly active, suffusing its smaller units and remaining until they are, in turn, subsumed by the next invading empire.

The advent of nuclear weapons and the accompanying threat of Mutual Assured Destruction (MAD) may well have altered the dynamics of decline, as borne out by the Cold War. Whereas in the past, empires in decline and legacy were often swallowed up, those wielding the nuclear advantage have the privilege of being protected by it as they re-form into their next incarnations. Europe has certainly benefited from this new model, as the US will, particularly as its missile shield defence system is developed further.

Europe is an example of a region in legacy. Following the collapse of its empires post-World War Two, it has gradually been rebuilding itself into the European Union (EU). The reconstitution is not without pain, as national identities are subsumed into a greater collective. Given enough time, accompanied by the considerable demographic expansion required to propel it out of regionalisation and into empire, a fully integrated EU could see Europe become an essential force in the world once more. (Should this scenario yet materialise, however, it would do so only far in the future.)

Wars are the clocks of empires. Once this simple truth is acknowledged, history can be unravelled, and lessons from the past placed into this context can directly inform both the present and the future. The Five Stages of Empire is an instructive model for today's multi-polar world.

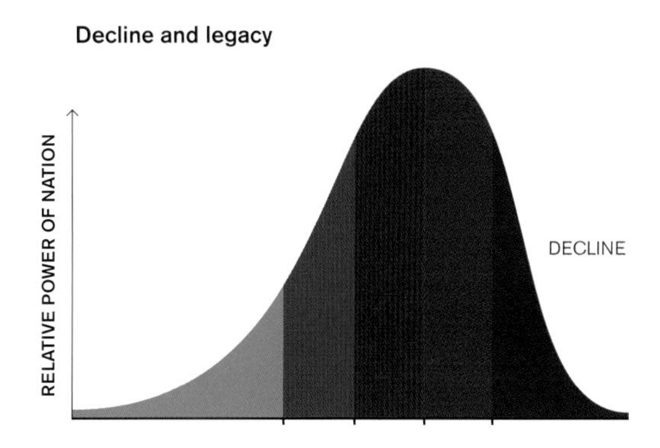

Chapter Four
Empires and Super-Empires

The Roman Empire

REGIONALISATION, EIGHTH CENTURY–202 BC

At the end of the third century BC, as the Macedonian and Greek Empires wound to a close in the eastern Mediterranean, the regional powers of Rome, Carthage and Syracuse were vying for domination across the Ionian Sea.

The kingdom of Rome had been founded just over 500 years earlier, as a group of Iron Age peasant communities near the Tiber River. Etruscan overlords in the sixth century BC initiated a large-scale land engineering programme to promote urban development, and the rise of Rome began. Rome had been a part of the Etruscan system, so much so that it suffered as Etruria went into decline after its war with Carthage in 540 BC. At that time, Greek influence was also widespread in the Italian Peninsula, and provided Rome with a template for its culture and its army. The Romans had achieved supremacy over Etruria by 509 BC, and they began to absorb Etruscan cities. The new Roman Republic would go on to control 350 sq. miles of territory barely a decade later.

The organisation of the original Roman army followed the structure of Roman society. Primacy was awarded to the wealthy patrician equestrian order – essentially Roman knights. Next were 'first-class' soldiers, also wealthy and high-born, equipped with full arms and armour (sword, long spear and breastplate) in the manner of Greek hoplites. Second-class soldiers lacked the breastplate; the third and fourth classes had still less armour, and the fifth comprised skirmishers with slings and stones. Military service was compulsory for all but the poorest (who could not afford to equip themselves). The hierarchical system thus placed the fate of the city in the hands of those with the most to lose. A damaging invasion of Rome by Gaul in the late fourth century BC prompted drastic reforms of the Roman army. The rigid phalanx was abandoned in favour of the legion formation, inspired by the tactics of the Gauls and the tribes of Samnium in the Apennines, who had also defeated the Romans in battle. The legion was tactically flexible enough to overcome armies of lightly armed barbarians. The loss of so many aristocrats in the war against Gaul had also refocused priorities, and armour and weapons became standardised.

By 396 BC, Rome's new army had enabled the state to become the most powerful in the northern Italian Peninsula, controlling 10,000 sq. miles. By 264 BC, Rome had secured the entire Italian Peninsula, leading a confederacy of communities with special privileges for Latin citizens. A further 42,000 sq. miles came under its control, confirming its status as a regional power. (See Figure 9 for additional data.)

The Five Stages of the Roman Empire *(fig. 9)*

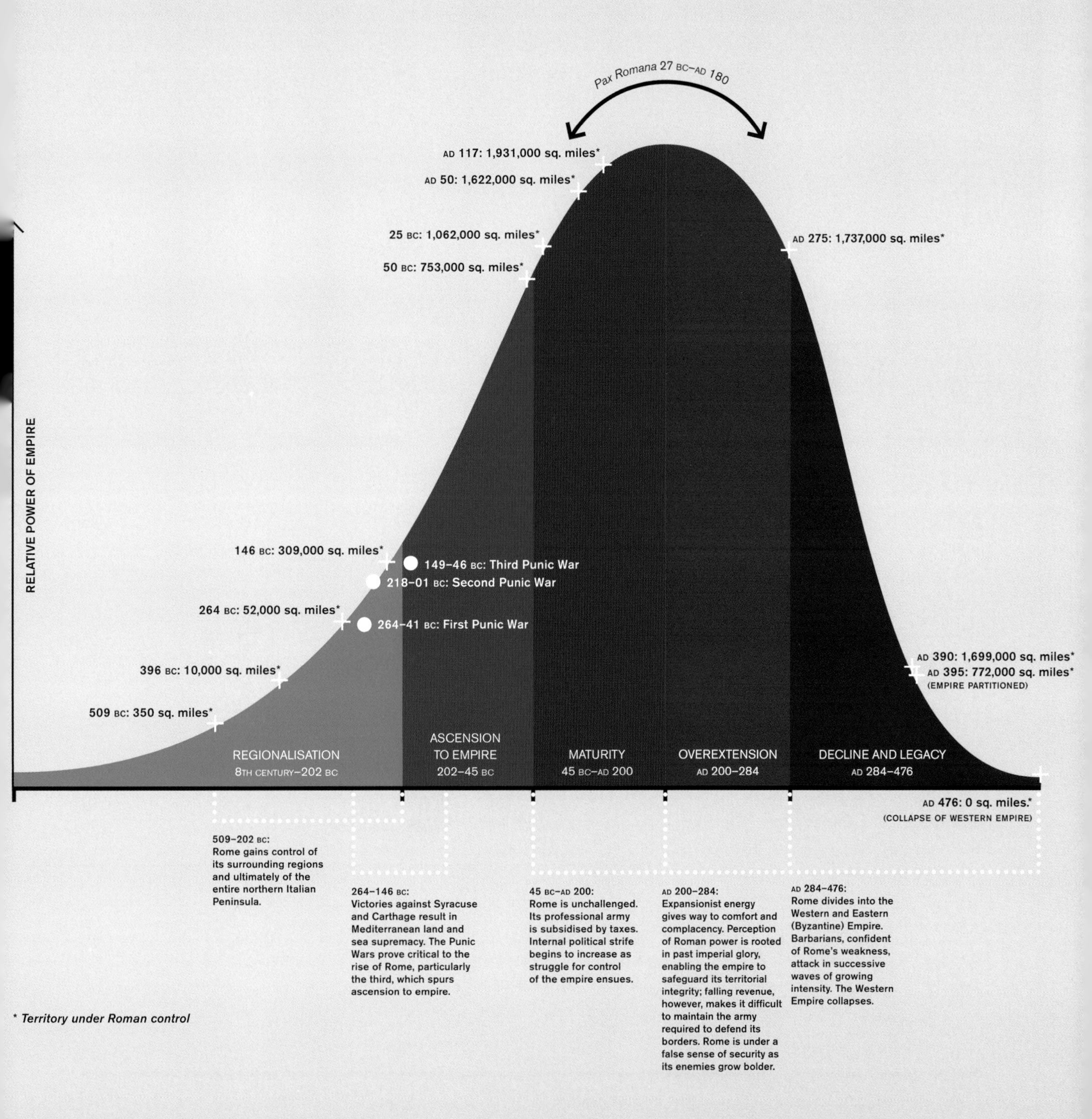

509–202 BC: Rome gains control of its surrounding regions and ultimately of the entire northern Italian Peninsula.

264–146 BC: Victories against Syracuse and Carthage result in Mediterranean land and sea supremacy. The Punic Wars prove critical to the rise of Rome, particularly the third, which spurs ascension to empire.

45 BC–AD 200: Rome is unchallenged. Its professional army is subsidised by taxes. Internal political strife begins to increase as struggle for control of the empire ensues.

AD 200–284: Expansionist energy gives way to comfort and complacency. Perception of Roman power is rooted in past imperial glory, enabling the empire to safeguard its territorial integrity; falling revenue, however, makes it difficult to maintain the army required to defend its borders. Rome is under a false sense of security as its enemies grow bolder.

AD 284–476: Rome divides into the Western and Eastern (Byzantine) Empire. Barbarians, confident of Rome's weakness, attack in successive waves of growing intensity. The Western Empire collapses.

* *Territory under Roman control*

ABOVE **Hannibal and the Carthaginian forces despoiling dead Romans following the Battle of Cannae during the Second Punic War.**

ASCENSION TO EMPIRE, 202–45 BC

The Phoenicians had founded Carthage in modern-day Tunisia in the late ninth century BC. Greek expansion into the western Mediterranean after 750 BC resulted, by 600 BC, in ongoing open conflict between Carthage and various Greek city-states – particularly over Sicily. However, by 400 BC, Carthage had carved out a small empire in North Africa, western Sicily, Sardinia, Corsica, the Balearics and parts of the Iberian Peninsula. A maritime power, Carthage enclosed the northern and southern Mediterranean shores in a triangle between Gibraltar, northern Corsica and North Africa.

The stage was now set for the regional powers of Rome and Carthage to confront each other for supremacy. Although they had previously signed treaties to exclude Punic influence from the shores of the Italian Peninsula and to limit their respective areas of commerce, it was only a matter of time before both regional powers consolidated their territory and were ready for the next stage of expansion. In the three Punic Wars of 264–146 BC, Rome and Carthage battled across the Mediterranean.

In a replay of the wars between Sparta and Athens, Rome was cast as the land power and Carthage the sea power. Initially, therefore, the Romans were at a disadvantage, but Rome overcame this weakness in 260 BC after finding a Carthaginian wreck and reverse-engineering a galley – with one significant modification. The Romans' innovation

ABOVE Rome had never fought a naval battle, so when it had to build a navy, it made sure that its legionaries had a sure way of boarding the enemy's boats. The *corvus* was Rome's secret weapon at the Battle of Mylae. A spiked grappling hook, it enabled the Roman navy to close with the Carthaginians and transfer the legionaries, who could then fight on their enemy's decks.

was a ramp called a *corvus* (Latin for 'raven'), which had a spike at one end. When lowered onto the bridge of an enemy ship, it locked the vessels in mortal combat and turned a sea battle into a land battle. It thus played to Roman strengths. Rome went on to win a series of conflicts, although the *corvus* was abandoned after Rome lost almost two whole fleets – because it either rendered ships unstable in rough seas or interfered with their navigability.

Nevertheless, Rome saw many strategic gains, and in response the Carthaginian general Hamilcar added more of Iberia to the empire, bolstering Punic power and compensating for losses. After his death, his son Hannibal famously decided to attack the Roman heartlands through the apparently impassable Alps north of Rome, and triumphed at Cannae in 216 BC: 60,000 Romans lay dead on the battlefield, with Carthage's losses amounting to one-tenth of this horrific number. (This number of casualties is equal to British losses on the first day of the Battle of the Somme in World War One, which must be viewed in the context of an army that came from a much larger population.) Any other state would have capitulated, but Rome was not just another city-state: it persevered.

With the Roman stronghold in the Mediterranean now firm, Hannibal's wait for siege engines to be sent by ship from Carthage was nevertheless in vain. Thus he could not advance on Rome, and by 214 BC the Romans under Scipio Africanus – commanding

more than 200,000 soldiers – wrested Punic Iberia into Rome's control. Hannibal's forces north of Rome witnessed many tribes in his alliance desert, sensing the ebb in Carthage's fortunes. Finally, in an ironic mirror of the Carthaginian menace outside its own capital, Rome attacked the vicinity of Carthage. At the battle of Zama in 202 BC, Scipio's forces overwhelmed those of Hannibal, and Carthage sued for peace.

The city itself survived until the aftermath of the Third Punic War (149–46 BC), when Rome set out to quell anti-Roman sentiment along with growing Carthaginian wealth and power. The Romans razed the former imperial capital to the ground, and then went on to expand throughout the Mediterranean, subduing other regional powers, acquiring immense resources and developing into the greatest empire the world had ever seen.

LEFT **An eternal emblem of the Roman Empire in full glory, the Colosseum in Rome remains one of the world's most iconic structures. Completed in AD 80 after ten years of construction, and modified up until AD 96, the arena featured public spectacles on a massive scale and had a seating capacity of some 50,000.**

MATURITY, 45 BC–AD 200

By the time Julius Caesar rose to prominence during the first century BC, Roman weapons technology had become an integral part of Rome's military strategy. Each 300-strong legion would have at its disposal thirty small *ballistae* and catapults. There is some evidence that the Romans might have even constructed a machine-gun-like *ballista* that could fire 100–150 bolts per minute at an enemy. The effect of such technology, deployed against barbarian tribes with no effective response, cannot be underestimated. It may fairly be said that Rome enjoyed a technological advantage on the battlefield along the lines of that claimed by US forces today.

The Romans also employed field fortifications to shape the battlefield, and to provide safe havens from which to launch offensive operations. Furthermore, they became masters of siegecraft. By the time of Caesar's death in 44 BC, Rome controlled the entire Mediterranean, either directly or through a series of vassal kings.

Rome had experimented with a number of different forms of rule since its founding as a kingdom. Early on, it became a republic, and under Caesar, Crassus and Pompey, the unofficial and even partly covert First Triumvirate was established in 60 BC. It resulted in the Great Roman Civil War of 49–45 BC. Caesar, the victor, became *dictator perpetuus*, but was assassinated the following year. The Roman Republic would eventually weaken further. The fully sanctioned Second Triumvirate that followed in 43 BC under Octavian, Marcus Aemilius Lepidus and Mark Antony proved as unstable as the first, and descended into further civil war. Eventually, Octavian emerged as the sole victor, having added the Egyptian lands formerly controlled by Mark Antony (along with Cleopatra) to the new Roman Empire. He became Augustus Caesar in 27 BC.

Augustus fits the description of a political hero well, having not merely restored order to Rome after a century of civil war but ushered in an era of peace, far-reaching reforms, imperial growth, and artistic and literary brilliance: all hallmarks of a golden age. Augustus's reforms extended to taxation, social mores and the elimination of official corruption. He restructured the army and extended the frontiers to boundaries that he hoped would remain in place forever. His grip on power was absolute, but at the same time he restored the republic, devolving power to the Senate and magistrates.

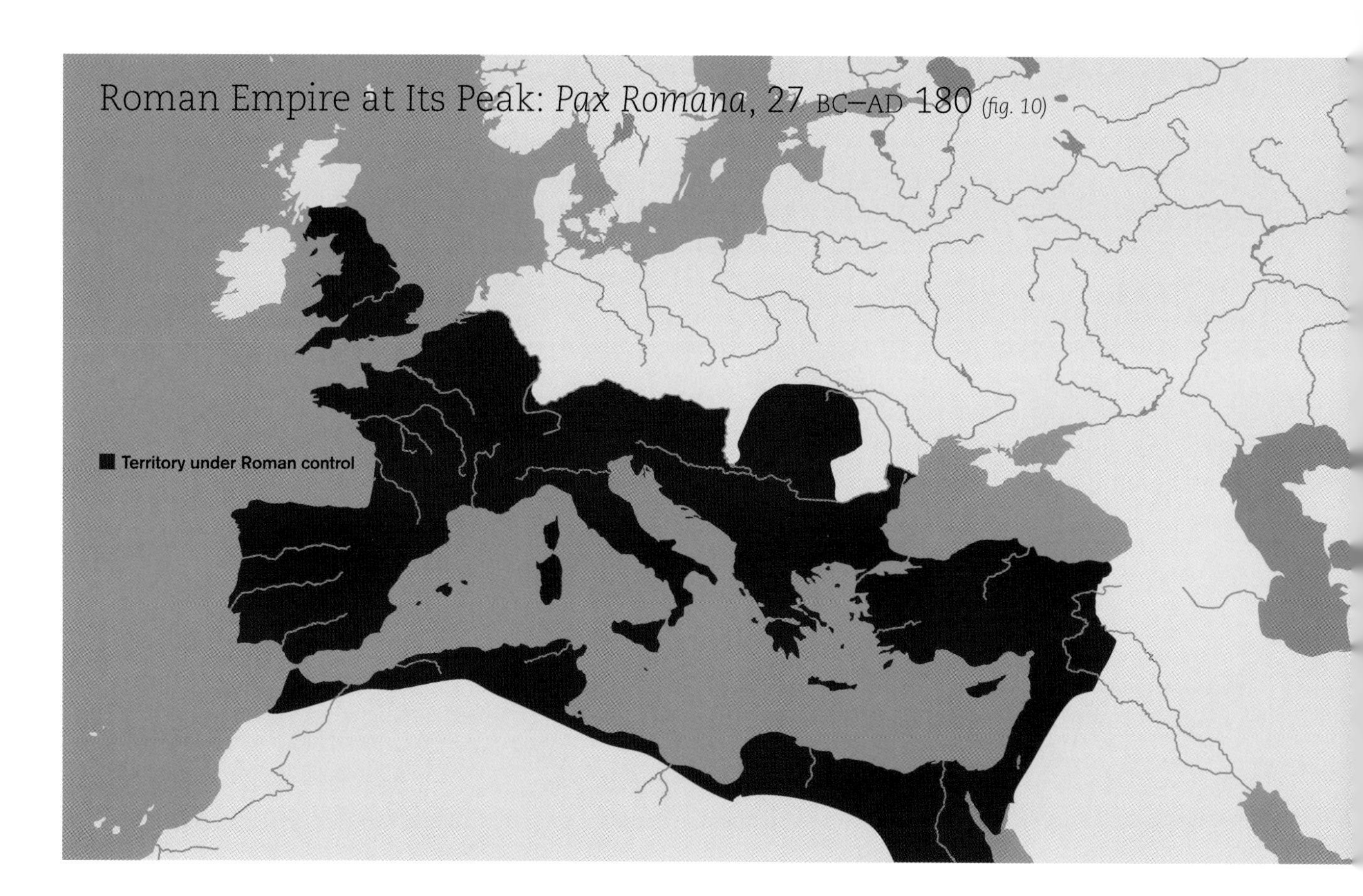

This was the start of *Pax Romana*, which would continue to see great leaders preside over the empire. It was the age of Tiberius, Claudius, the 'Five Good Emperors' (Nerva, Trajan, Hadrian, Antoninus Pius and Marcus Aurelius) and Septimius Severus. The peak of this expansion was marked by Hadrian's programme of wall-building to mark the empire's frontiers, from northern Britain to the Rhine and the Danube, and from Syria to North Africa. (The map in Figure 10 illustrates the breadth of the Roman Empire's control during this period.)

Augustus decreed that military power and economic stability together were needed to sustain the empire, a policy that would be retained by subsequent leaders. The size of the standing army was set at 300,000 men (although at times it increased to 400,000, with an unpaid reserve provided by retired veterans). Relative to the 50 million inhabitants of the Roman Empire (all of whom, slaves excepted, were granted citizenship by Caracalla in 212 BC), and given that this figure represented only 0.65 percent of the number of soldiers in the barbarian armies threatening the empire's borders, the size of the Roman army during this period can be considered surprisingly small.

Augustus's decree indicates that Rome had reached a balanced, mature stage of empire. Whereas initial regional and imperial expansion was driven by an army of conscripts rewarded with the spoils of war, its reorganisation into a professional standing army represented a high cost to the state. Yet, through conquest, Rome's wealth also grew exponentially during this time, and it could bear the cost. As the rate of new conquests levelled out, however, and the border remained constant, the cost of the army increased disproportionately – hence the need to balance the budget and limit the military's size. This dynamic is a fundamental of empires: expanding empires are always able to field larger armies than those that have matured. For many years, Rome was able to compensate for this imbalance by virtue of the superb skills and technology of its forces, the best of their time. (One modern-day equivalent is the George W. Bush-era US, in which a neo-Conservative military model as articulated by Secretary of Defense Donald Rumsfeld preferred a small, effective army that could be sent anywhere by air. It was for this reason that so few troops were used to invade Iraq in 2003, and operational risks increased manifold as a result. The consequences were dire, resulting in failure to facilitate peace in that country.)

The 250-year *Pax Romana* proved one of the most stable periods in military history, with overpowering martial strength and internal stability placing Rome in an impregnable position.

OVEREXTENSION, AD 200–284

From AD 200, Rome began to overstretch militarily and become complacent at home. It became difficult to fill the ranks of legionaries, so barbarians were admitted in greater numbers, as both ordinary soldiers and officers. Conscription was resorted to in some provinces, representing a step towards feudalism and away from the empowered status of the first Roman legionaries.

The empire felt secure, as one would expect at this stage: the borders were guarded by a ring of fortresses and settlements manned by legionaries, and supported by retired legionaries who had settled near their original posts. Initially, the Roman legions' fearsome reputation would have been sufficient to protect the imperial borders, but over time the tribes in the north became more adventurous, and began nibbling away at the vulnerable haunches of the empire. At first they had no hope of defeating the Roman army, but eventually they began to adapt and employ Roman methods in combination with their own strengths in cavalry warfare.

With the exception of the Praetorian Guard (the Augustus-created corps that protected the emperor and acted as a police force), all the legions were stationed at the frontiers; a system of effective roads allowed forces to be redeployed to meet new threats. However, as happens in the case of empires at the mature stage, all attention and motivation began to focus inward in the form of leadership challenges and away from the requirements of security and military integrity. In this way, Rome began planting the seeds of its demise, which would be manifested as decadence, decay and civil war from the third century AD onward.

DECLINE AND LEGACY, AD 284–476

Barbarians with colourful names – Goths, Vandals, Visigoths – besieged Rome for over two centuries. As the threat of their armies increased, and as they mastered the techniques of cavalry and missile warfare, discipline levels in the Roman military subtly declined, along with the effectiveness of the infantry-dominated legions. However, the Romans, ever open to military adaptation, modified the legions to counter the new, improved enemy.

The increasing reliance on cavalry was prevalent within the Roman army and the barbarian ranks alike. (Some of them now had stirrups and saddles, greatly increasing their effectiveness and striking power.) The cavalry made up 25 percent of the legion's strength, and was also supported by much more effective missile techniques used to bombard the enemy before a charge. Half the number of auxiliary foot soldiers consisted of either slingers or archers, and mounted archers were particularly devastating.

A key dilemma facing the legion was that for infantry to face a cavalry charge successfully, it needed to be packed closely; yet doing so made it vulnerable to mass missile fire. Extending the spacing between each soldier in the line, on the other hand, made it harder to repel a cavalry charge and concentrate a legion's force onto a specific point for a decisive breakthrough in enemy lines. Roman legionaries thus lost their dominance on the battlefield to cavalry-dominated armies. Rome's military edge was blunted, just as it faced increasing threats from outside its borders.

The Roman population, meanwhile, had grown to 50,000,000. It is likely that such demographic growth would have occurred outside the borders of Rome as well, driving the barbarian nations to regionalise. Eventually, at any rate, they began to look towards the weakened Roman Empire.

In 324, Constantine, emperor of the Western Empire, captured Byzantium after a series of struggles against his Eastern rival Licinius. He subsequently moved the imperial capital there, away from the threat of the northern barbarian hordes, and renamed the city Constantinople. In 476, Rome – having dominated much of the known world for 1,000 years – fell after years of relentless invasions by Germanic tribes. (Figure 11 summarises the many invasions of the Roman Empire leading up to its eventual collapse.) Its demise returned Western Europe to a period of relative regional obscurity. Two great and interconnected legacies endured, however. The Byzantine (Eastern) Empire only fell in 1453, and would even, under Justinian I, reconquer the Italian and Western Balkan Peninsulas and North Africa. The second legacy was Christianity, to which Constantine had converted personally and which he also made the imperial religion.

Invasions of the Roman Empire, Second to Fifth Centuries BC
(fig. 11)
Angles, Saxons, Jutes
Franks
Goths
Huns
Ostrogoths
Vandals
Visigoths
Western Empire
Eastern Empire
Jutes
Angles
Saxons
Vandals
Huns
Huns
Chalons 451
Western Roman Empire
Ostrogoths
Visigoths
ROME
Adrianopole 378
Ostrogoths
CONSTANTINOPLE
CARTHAGE
Vandals
Eastern Roman Empire

The British Empire

Because of Rome's longevity and vivid history, it is possible to clearly observe a shift from stage to stage. The changing stages of other empires are also clearly visible, of course. The British Empire provides another fine case study of the Five Stages of Empire. (See Figure 12 for a graphical representation of the model.)

REGIONALISATION, 950–1652

Owing to its island status, Britain remained insulated from major invasions after the fall of Rome (although it was still exposed to numerous raids over centuries from the Vikings). It retained Christianity as a religion and, indeed, would later act as a pollinator along with Ireland, re-seeding Northern Europe with the religion. From the sixth century onward, the Anglo-Saxons began to consolidate power through a number of independent kingdoms in England. Normandy and England began to cultivate ties in the tenth century, which strengthened with the return of the half-Norman Edward the Confessor to England. Upon Edward's death and the Norman Conquest led by William the Conqueror in 1066, the Normans would begin to supplant the Anglo-Saxons as England's ruling class. The histories of England and France thus became inextricably entwined, and England controlled northern France until the end of the Hundred Years' War in 1453. The population, meanwhile, continued to grow from approximately 1 million in 1086 to around 6 million by 1300.

Wales was defeated in the thirteenth century (and formally incorporated into the English kingdom approximately 300 years later). The Normans invaded Ireland in 1169, although their control would weaken in the thirteenth century, disappearing altogether by the 1400s. England reasserted its will over the island under the Tudors in the sixteenth century, although full English control over Ireland would only be gained after a series of bloody conflicts. The kingdom of Scotland coalesced in the eleventh century and remained an independent state with complex ties to its southern neighbour. The Wars of Scottish Independence during the thirteenth and fourteenth centuries, when England attempted to subjugate the Scots, ultimately resulted in Scotland retaining its sovereignty. (The two countries would merge some 400 years later, as a result of the Acts of Union in 1707.)

The English Reformation under Henry VIII initiated the creative, scientific and cultural momentum that would reach an apogee under Elizabeth I. The newly Protestant power grew in military and naval strength as well. By the end of the sixteenth century, England was ready to indirectly challenge Spain, the great superpower of the time, by covertly encouraging British privateers to attack Spanish merchant ships.

Spain, meanwhile, had been intent on eliminating the Protestant Elizabeth I, whose late (and Catholic) sister Mary I had married the Spanish monarch Philip II, briefly uniting the two countries. England supported the Dutch Revolt against the Spanish Empire, which Spain interpreted as an act of war. The Spanish antipathy towards the heretic English queen, as well as the harassment by England of Spain's ships and territories, crystallised as a plan to invade England from the Flanders coast with twenty-

The Five Stages of the British Empire *(fig. 12)*

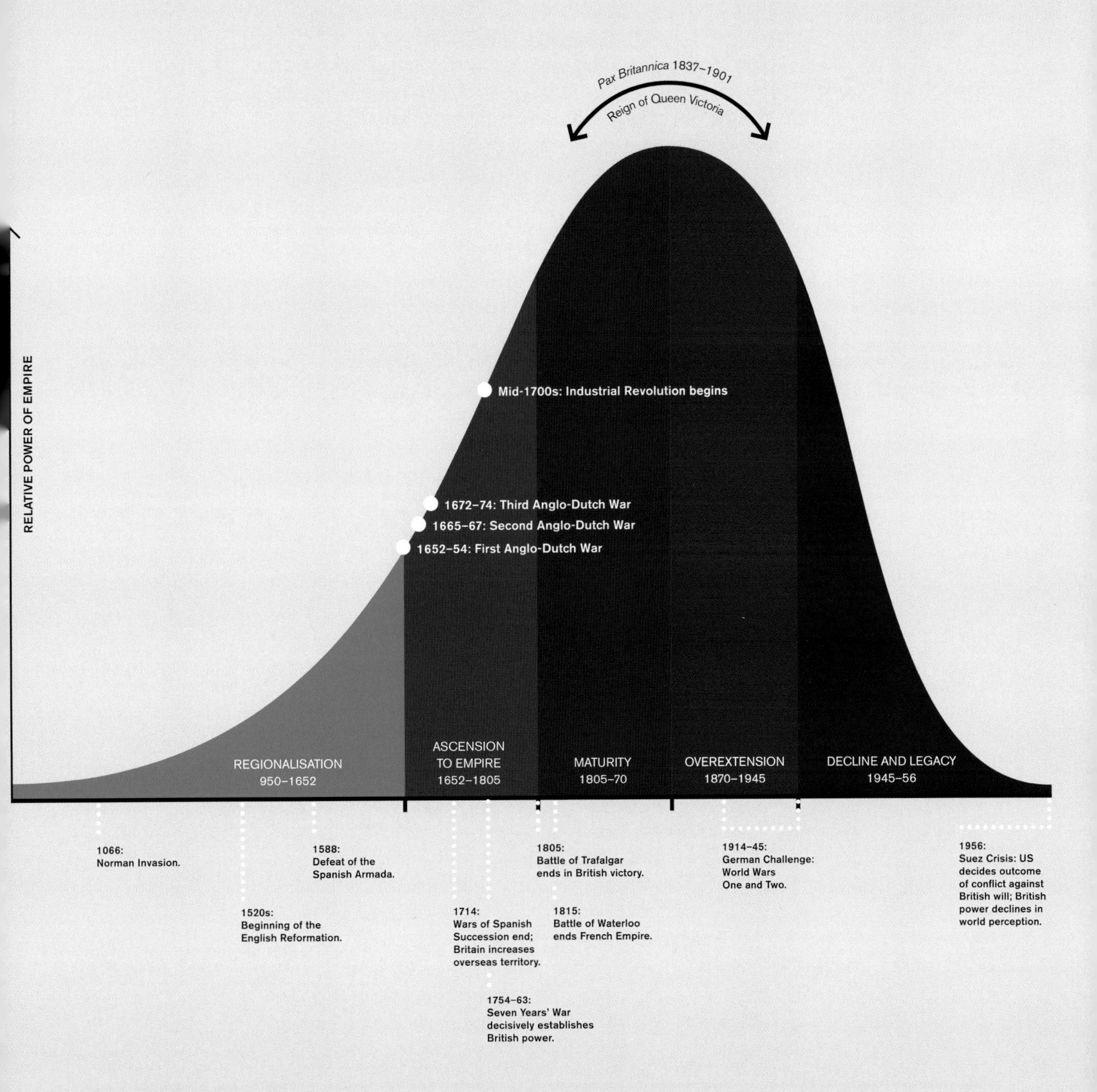

two warships and 108 other vessels refitted for battle. The ships were to rendezvous with land forces sent by the Duke of Parma, then governor of the Spanish Netherlands. England, however, broke up the formation of ships by sending in fire ships, which did not themselves do much damage but caused the Spanish ships to scatter in attempts to avoid being burned. England's navy then chased the fleet along the English coast. The Armada headed for Spain but ran into storms in the Atlantic. All told, over half the Spanish fleet (around sixty-five ships) was wrecked and an estimated 20,000 of its sailors killed.

England and Spain remained in a state of war until James I (James VI of Scotland) took his place as the British monarch and made peace with the Spanish in 1604. The English treasury had been bled severely in the interim, but at the same time Spain had abandoned the notion of defeating Britain, appreciating that the difficult waters surrounding the island nation – along with the expanding and skilful British navy – precluded any prospect of a successful invasion.

England had begun an attempt to expand beyond its borders in 1578, but did not establish any successful overseas settlements until the founding of Jamestown and the Virginia Colony in 1607. Further colonisation of present-day Canada, the US and the West Indies followed – the first paving stones of the expansionist British Empire.

Yet it was the protracted struggle between Britain's Protestants and Catholics during this period that was to ultimately establish the nation as a regional power. Civil wars raged in 1642–46 and 1648–51, in which a Protestant Parliament based on democratic values squared off against a Catholic monarchy under James I. By 1651, Oliver Cromwell, leader of the Parliamentarians, had, with his New Model Army (and navy), brought the war to a conclusion. Charles I was executed and Britain began almost immediately ascending to empire.

ASCENSION TO EMPIRE, 1652–1805

For England to ascend to empire, it had to surpass first the Spanish, then the Dutch and finally the French. This struggle for empire, which spanned the globe, began with the defeat of the Spanish Armada in 1588 and cumulated with Horatio Nelson's victory at Trafalgar in 1805 against the Spanish and French, confirming England as a global maritime empire with absolutely no challengers on the horizon – a position only equalled in history by the Roman Empire 1,600 years earlier.

Britain launched the first of three Anglo-Dutch Wars in 1652 against the Netherlands, precipitated by competition for trade routes in the Dutch East Indies. These were its first wars of expansion. The third (1672–74) concluded in final victory for Britain when, jointly with France, the nascent empire prevailed over the exhausted Dutch. These contests were to decide which of the two Protestant sea powers would dominate the oceans. It was thus ironic that, in 1689, the Dutch *stadtholder* (head of state), William of Orange, would be crowned William III of England and Ireland (William II of Scotland) at the behest of the Protestant opponents of the Catholic James II. The two nations were thus linked in alliance.

By the turn of the eighteenth century, European politics had become multi-polar, with the Protestant powers of Britain and the Netherlands balanced by Catholic

France and Spain. During the Wars of Spanish Succession (1701–14), during which Britain's principal intention was to prevent the unification of Spain and France, the great General Marlborough carried Britain and its allies to victory. Britain's presence was once more felt on the Continent, thanks to the success of its small but capable army.

By 1750, Britain dominated the seas, with France its nearest competitor and the Netherlands a distant third. While French ships might have outmatched their British rivals one-on-one, France's naval force comprised fewer vessels and lacked the British navy's experience and skill. The Seven Years' War (1754–63) saw Britain and an allied coalition defeat France and its allies, and it firmly established the British Empire. France was reduced to being essentially a continental power only. Spain, Britain's only other overseas rival, was also forced to cede colonial holdings.

British sea power continued to expand relentlessly. By the end of the eighteenth century, Britain boasted 115 well-equipped warships in its Channel and Mediterranean fleets, compared with France's seventy-six – of which only half were serviceable, a problem compounded by revolutionary purges of the French navy's officer corps after 1789. Spain had fifty-six serviceable warships and the Netherlands forty-nine small vessels.

BELOW Victorious British ships surround the burning hulk of a French ship during the Battle of Trafalgar in 1805. (James Jenkins 1819 *The Naval Achievements of Great Britain 1793–1817* Trafalgar chapter, plate II.)

ABOVE A selection of colonial troops of the British Empire in England for the Diamond Jubilee celebration of Queen Victoria, 1897.

The strategic importance of sea power in geopolitics was emphasised during both the French Revolution (1789–99) and the Napoleonic Wars of 1800–15. Few at the time appreciated this huge, necessary shift in priority, foremost among them Napoléon Bonaparte himself; the French emperor continuously attempted to develop a land power base, and paid a high price for his ignorance in his military conflicts against Britain. In addition to Britain's fleet size and nautical superiority, British admirals developed new tactics during the nineteenth century that gained the day at the Battle of the Nile (1798), at Copenhagen (1801 and 1807) and finally at Trafalgar (1805). Britain's methods were so decisive that they confirmed the country's control of the seas for the next 100 years as the world's pre-eminent empire.

MATURITY, 1805–70

At the Battle of Waterloo in 1815, British forces under the Duke of Wellington's command (and in alliance with Prussia) defeated Napoléon and destroyed 450 years of French aspirations to dominance in Europe. Freed from costly wars against France, the mature empire of Britain entered the golden age of *Pax Britannica*.

Economic and military dominance marked this stage: Britain was an unrivalled superpower. Despite the loss of its colonies in what became the US, Britain still held territory in Canada, the Caribbean, Central and South America, Africa, India, Southeast Asia and Oceania, and would go on to acquire still more territory in those regions. The Industrial Revolution was moving into its most productive phase, and it became synonymous with the rule of Queen Victoria (1837–1901).

Moreover, Britain had bent China to its will after two cynical Opium Wars, effectively gaining a monopoly on the narcotics trade with the lucrative right to sell the drug to Chinese addicts. In the process, Britain also acquired Hong Kong and other treaty ports, and opened the way for the other European powers to follow it into China.

The maturity stage of the British Empire induced sweeping social changes, foremost of which was the abolition of slavery in 1833. (Britain had outlawed the slave trade itself in 1807, but slaves could still be held within the empire thereafter.) Vigorous campaigning over decades by such figures as William Wilberforce had proved very effective, but the increasing pace of industrialisation also played a decisive role, as Britain became the first nation in the world to loosen its dependence on agriculture and slavery. Victory in the wars against France during the early nineteenth century also helped spur Britain to maturity, which in turn encouraged the deepening of social conscience characteristic of this stage of empire.

Britain's immense wealth had allowed it to develop the greatest industrial infrastructure in the world. However, by 1870, the industrial economy was beginning to lose steam, just as new challenges from Germany and the US appeared on the horizon.

OVEREXTENSION, 1870–1945

Imperial peaks are rarely evident at any other time than retrospectively. In the case of Britain's army and navy, there was no change in relative power projection until the turn of the century, when the Germans – having defeated the French to become continental Europe's premier land power – looked to challenge Britain for supremacy at sea. Britain's economy was showing some signs of decline relative to both the US and particularly Germany, which had a very active chemical industry. Furthermore, from the British Empire's peak in 1870 to the turn of the twentieth century, the world witnessed the rise of a third new geopolitical force to be reckoned with, that of Japan. Thus British dominance was shifting in the balance of power.

The arms race leading up to World War One, moreover, was very difficult for Britain, and placed it in a position of major overextension. The new powers (the US and Germany), along with the older ones (France and Russia), embarked on massive naval expansions that can be characterised as having two phases: pre-dreadnought and post-dreadnought. The battleships built between the mid-1890s and 1906 ('pre-dreadnought'), which had consumed their fair share of resources, were outmatched in every way by the appearance of the British navy's HMS *Dreadnought* in 1906, to the extent that new resources were immediately devoted by all sides to the research, design and construction of dreadnought-class warships.

This revolution in battleship technology drained coffers and opened the path to the horrors of 1914 and beyond. The dreadnought race was terminally punitive on British finances, which had spiked with the Industrial Revolution. The major developed countries were now heavily industrialised, and increasingly competitive.

RIGHT **HMS *Dreadnought* in 1909, when it was the flagship of the Royal Navy's Home Fleet. A symbol of Britain's naval power, the ship sparked a massive arms race. In 1915 it earned the distinction of being the only battleship ever to directly sink a submarine, when it rammed a German sub.**

In 1903, 25 percent of British government expenditure was devoted to the navy. At the same time, Britain faced an expensive war in South Africa, putting its total defence spending in the region of 60 percent of expenditure. (This number is, to be sure, staggering; at the same time, it pre-dated the era of government spending on social infrastructure.)

The Germans, buoyed by their successful command of chemical engineering, were in an expansionary economic phase, enabling them to challenge Britain. Germany announced its second Navy Bill in 1900, which had as its objective the development of a war fleet with no ceiling on costs. The Germans perceived the British as overstretched in China and South Africa, and felt confident that Britain would not deploy its armed forces unless pushed to the extreme – that it had, in other words, lost the will to fight and become soft.

When war broke out in 1914, Britain was required to fund both its own war effort and those of its European Allies; the burden was simply too great, forcing it to borrow from the US, its main supplier of arms. By 1916, this dependency – coupled with unsupportive US policies – had brought Britain to its knees. It was forced to liquidate its overseas holdings in Latin America and repatriate its money. Naturally, American money filled the gaps, confirming the US as the dominant power in the Americas.

In 1917, Britain was only weeks away from having to consider backing out of the war. Just then, however, Germany announced it would resume unrestricted submarine warfare despite having agreed not to do so after US threats to sever diplomatic relations in 1916 over the issue. Moreover, Germany announced that American ships could not be considered neutral as they were delivering supplies to Britain. The US declared war on Germany in April 1917, and its entry into the fray would bail out Britain. American financial aid to Britain, in fact, proved critical in ending the war and defeating Germany.

Britain's post-war debt to the US would prove a heavy burden for the next decade. Britain had borrowed the equivalent of around £850 million from the US. By 1928, 40 percent of government expenditure represented the repayment of still-outstanding war debts; by 1934, Britain had paid off $2 billion in interest but still owed the principal amount of $4.3 billion (around £866 million at 1934 exchange rates), which it finally stopped paying that year. Britain's greatest competitor had now become its economic and naval equal in every way. Indeed, competition from the US was such that many Americans thought of Britain as the future enemy, while Britain entertained no such thoughts. This is an interesting example of how an ascending empire may view the old hegemon, while the empire facing decline fails to recognise the new challenger.

DECLINE AND LEGACY, 1945–56

World War Two left Britain undefeated but broken and beholden once again to the US, without the will or the ability to hold its overseas empire together. The devolution of the empire to the Commonwealth was rapid. The final death knell was the 1956 Suez Crisis. The US, attempting to avoid a war with the USSR in the event the latter supported Egypt against invading British and French troops, threatened sanctions against Britain that would have likely crippled the British economy. Britain duly withdrew its troops.

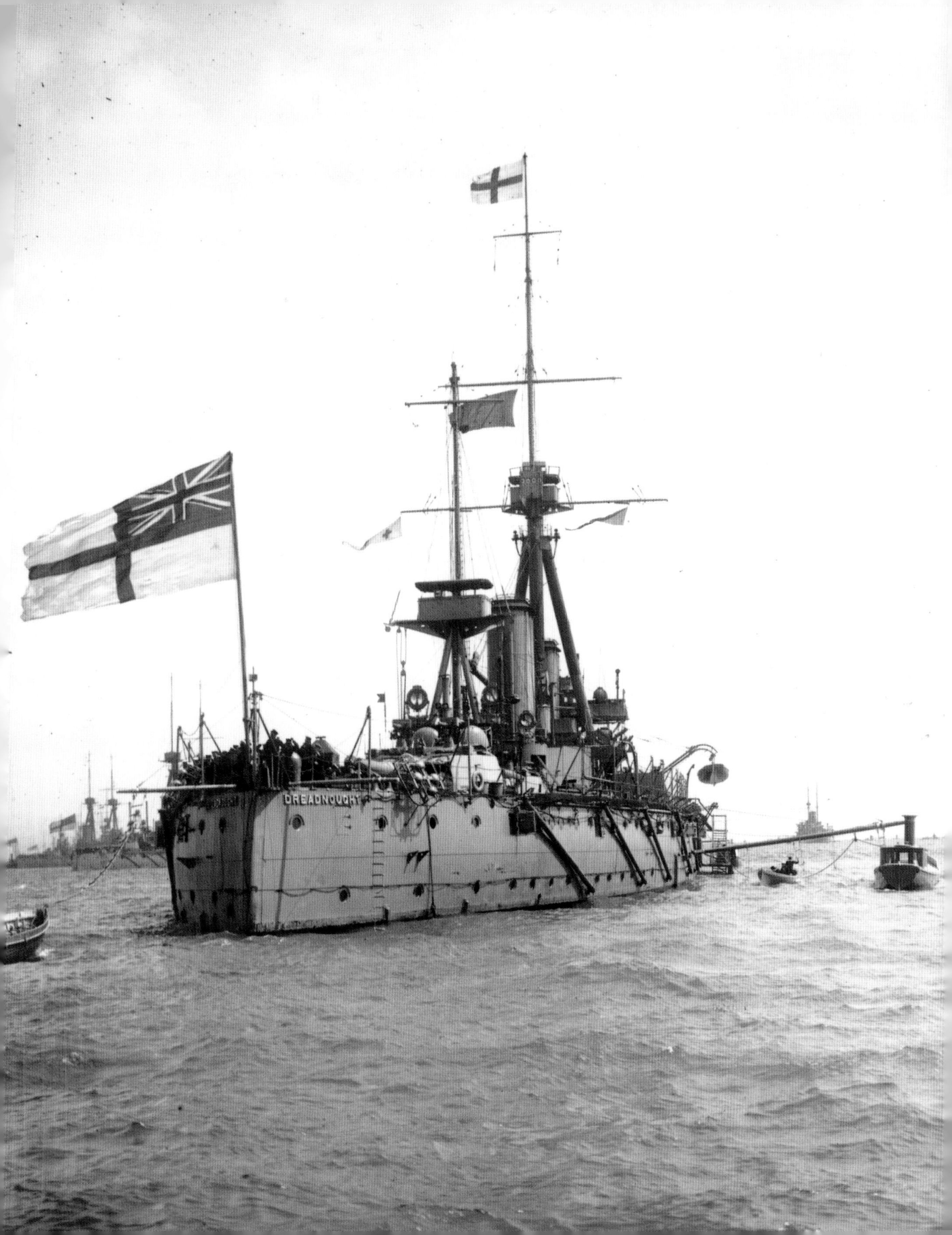
DREADNOUGHT

Its decline as an empire was thus made plain to the rest of the world. The economic ramifications of the Suez Crisis had Britain in the doldrums over the next few decades.

Suez is a sound example of the kind of event that clearly signals an overextended empire's weakness. A change in the world order is then initiated as the empire rapidly falls into decline and legacy, creating a vacuum that is soon occupied by a fresher rival.

THE COST OF EMPIRE

Empires cost enormous amounts of money to build and maintain. As such, their success is highly correlated to prosperity. History has seen evolution in the complexity not only of empires but also of the financial systems supporting them. With the Renaissance came the concept of government debt, which could finance a nation's expansion. Britain was but one empire that benefited from this novel fiscal principle. Its empire's debt-to-gross domestic product (GDP) ratio clearly follows the Five Stages of Empire model (as shown in Figure 13):

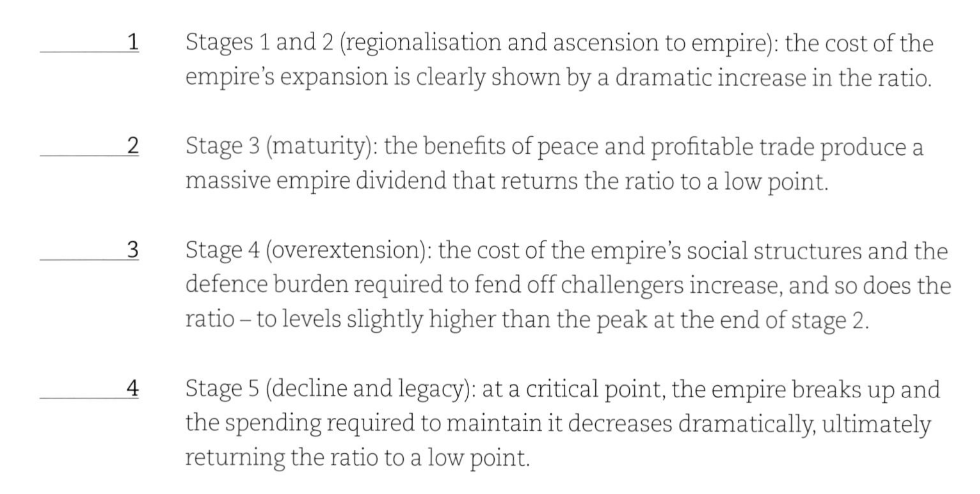

1. Stages 1 and 2 (regionalisation and ascension to empire): the cost of the empire's expansion is clearly shown by a dramatic increase in the ratio.

2. Stage 3 (maturity): the benefits of peace and profitable trade produce a massive empire dividend that returns the ratio to a low point.

3. Stage 4 (overextension): the cost of the empire's social structures and the defence burden required to fend off challengers increase, and so does the ratio – to levels slightly higher than the peak at the end of stage 2.

4. Stage 5 (decline and legacy): at a critical point, the empire breaks up and the spending required to maintain it decreases dramatically, ultimately returning the ratio to a low point.

As we have seen, Britain's investment in its navy over the course of the pre-1914 arms race proved financially daunting, and, once the war began, soaring costs forced the empire to the brink (and, fatefully, to seek US aid). It is worth recalling that, in addition to funding its war machine and then its World War One Allies, the regular cost of maintaining the British Empire was already enormous. Its size and complexity required the Royal Navy to defend trade routes that crisscrossed all regions of the globe, along with overseas naval bases and coaling stations. Never in history had a single empire controlled not only a great landmass but also the world's oceans. Its overextension thus also occurred on an unprecedented scale.

While government debt during the early stages of empire was owned by British citizens, in the last two stages it was owned predominantly by the US, which then used this hold over Britain to force it to withdraw during the Suez Crisis, effectively ending Britain's imperial age.

Power and Money
The British Empire, 1702–1991: Ratio of Government Debt to GDP *(fig. 13)*

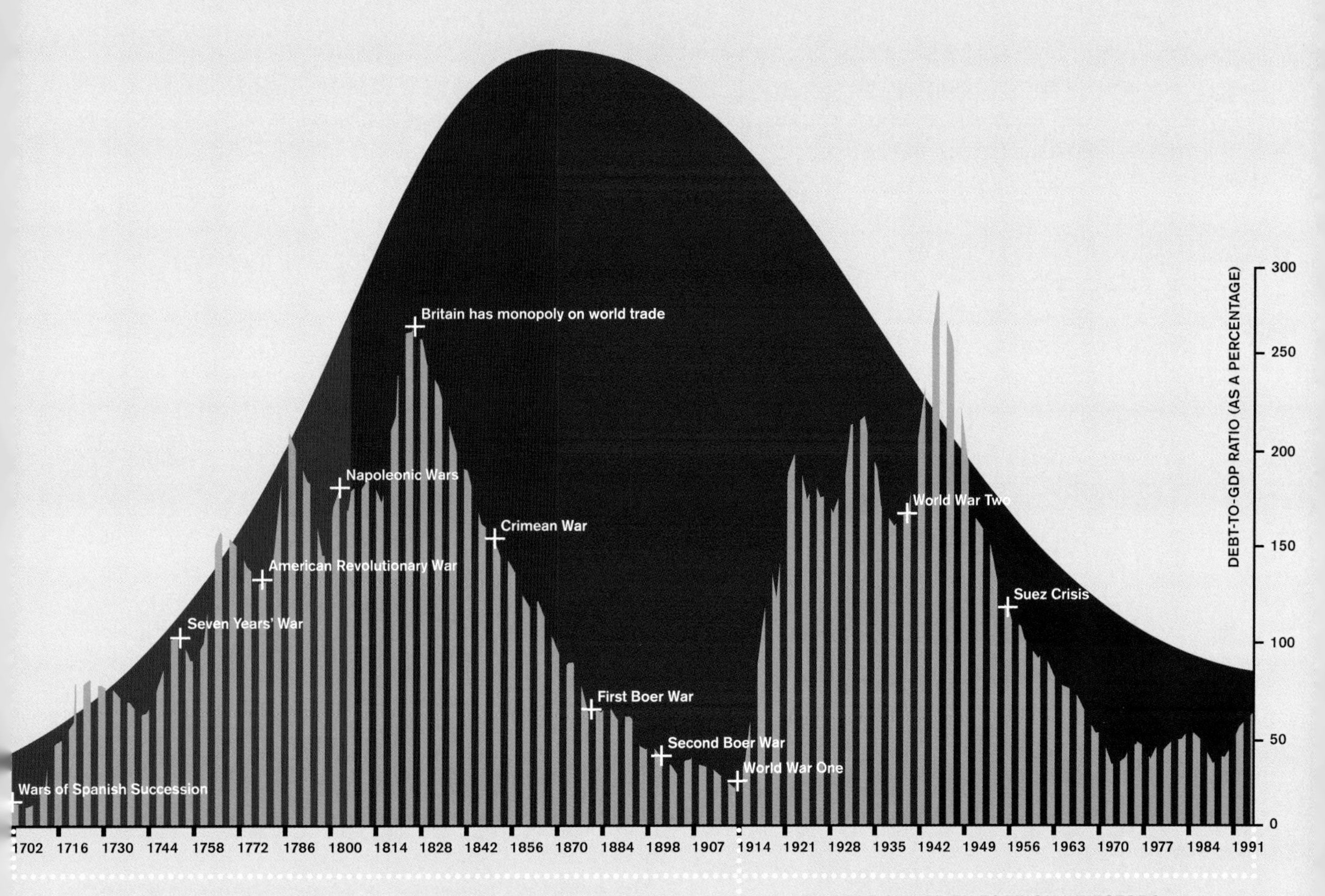

DEBT PREDOMINANTLY OWNED BY POPULATION TO GOVERNMENT

DEBT OWNED BY FOREIGN INVESTORS
(FOREIGN CONTROL OVER EMPIRE'S FINANCES AND HENCE ITS RATE OF DECLINE)

Ascension to empire

Government debt ratio increases as regional power develops into empire.

Maturity
(*Pax Britannica*)

Massive wealth accrued and debt repaid as empire acquires monopoly on trade.

Overextension

Ever-increasing amounts of funding required to meet challenges from rising powers.

Decline and legacy

As empire declines, expenditure decreases until debt is repaid.

The First Super-Empire: Rome West and East

I have discussed the concept of empires and their five-stage cycle of rise and decline, defining 'empire' as the highest level of social organisation in terms of the allegiance of a population. Considering the next step in the expansion of human societies as they grow more complex and as their populations multiply, we come to an even larger organised social system: the super-empire. Super-empires result from an advanced level of social development that transcends the limitations of empire. The fractals of a super-empire consist of empires themselves.

The Roman Empire furnishes us with the original example of the super-empire phenomenon (see Figure 14). What would become the Western Empire after the fifth century AD began, as we have seen, in the eighth century BC, with the rise of Rome to soaring heights and then its gradual decline over 1,000 years.

REGIONALISATION, 8TH CENTURY BC–AD 324

As the Western Empire deteriorated under barbarian invasions, the Eastern Roman (Byzantine) Empire arose, and thus the concept of 'Rome' would endure for another millennium. (Reference to 'the Byzantine Empire' is a modern convention; the empire's inhabitants did not use the name and called their homeland simply 'Rome'.)

ASCENSION TO EMPIRE, AD 324–565

In AD 324, Emperor Constantine founded Constantinople, also called 'New Rome', on the site of Byzantium, where modern-day Istanbul now stands. When 'old Rome' finally collapsed during the mid-fifth century, its eastern contingent continued to grow in power. Constantinople became the sole capital of the Roman Empire.

MATURITY, 565–602

By 565, the now-Christian civilisation (Constantine had converted in 312) had reached its apex, ruling the Mediterranean from Egypt to the borders of Persia. The Byzantine Empire thus outlived its parent but, while it wielded considerable power, it never grew as large as its Western predecessor. (It would enjoy greater longevity, however, lasting for over 1,100 years until its fall in 1453.) The Western and Eastern empires were distinct political entities that developed separate social and economic systems and left very different historical and cultural legacies. Yet they were two parts of a whole: the first super-empire.

At its height, the Byzantine Empire's looming overextension became apparent to Emperor Justinian I, arguably its finest leader. Justinian's army was engaged in wars across the imperial domains, with a military force comprising only 150,000 men, so he invested in a costly programme of fortifications along the weakest frontier. As a result of this enormous expenditure, his troops were often paid late, if at all, and morale suffered. Meanwhile, the long, ultimately victorious struggle by the Byzantines to reconquer the Italian Peninsula against the Goths had devastated the economy there. However, this re-absorption of much of the former Western Empire evoked Rome's erstwhile greatness and contributed to a growing sense of complacency among the empire's rulers.

The Roman Super-Empire *(fig. 14)*

Rome (Western Empire)
The Byzantine (Eastern Roman) Empire

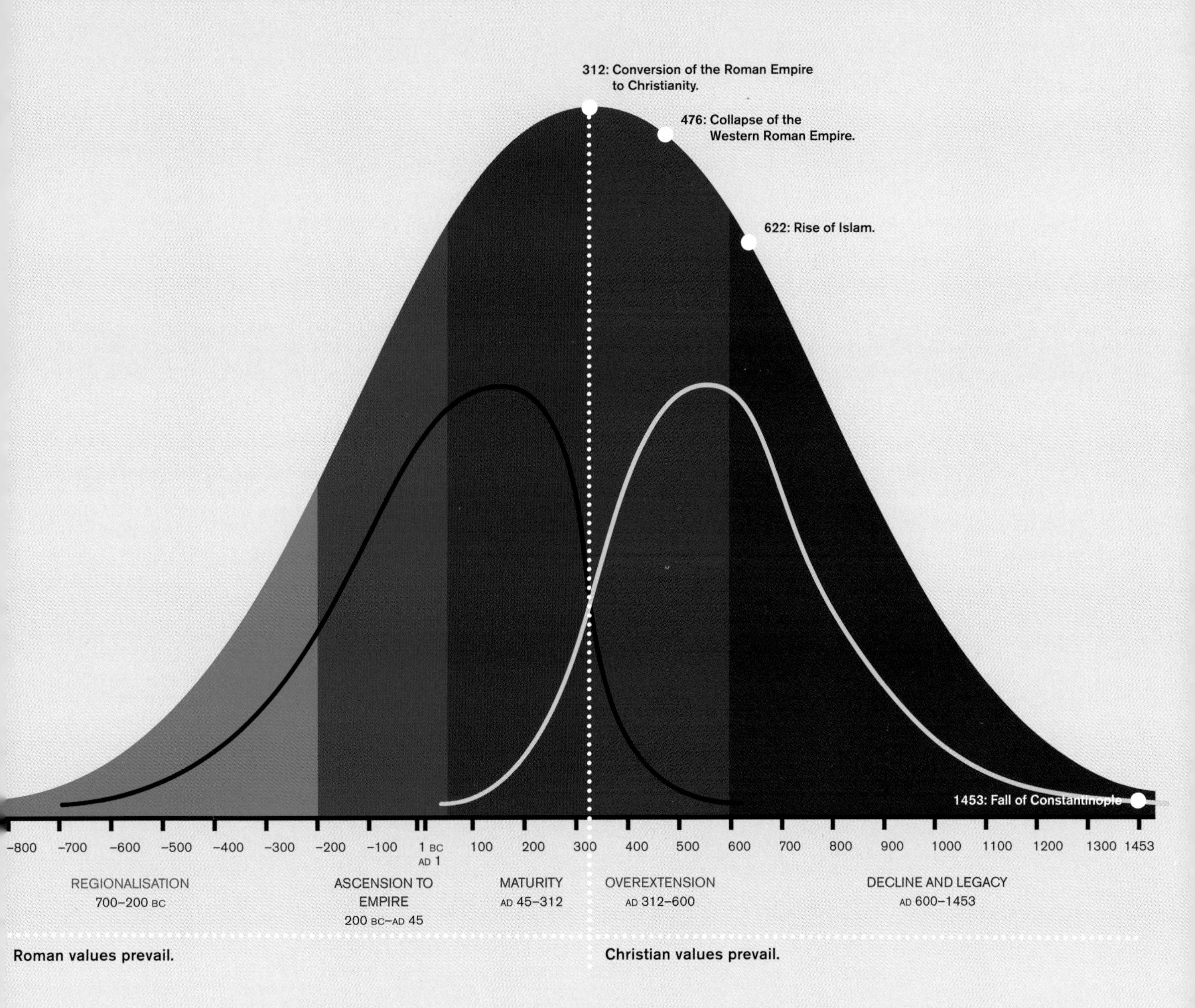

It was in this context that two political forces – outgrowths, in fact, of the factions supporting various chariot-racing teams – competed for power within the state: effectively, a peak civil war. It is possible that the Greens and the Blues (so called according to their preferred team's uniform colours) each held different beliefs regarding theological subtleties and social class. The Greens may have represented the lower classes and merchants, while the Blues, who enjoyed the support of Justinian himself, may have been associated with the landowning aristocracy. Riots driven by intense emotions aroused by the sporting spectacle were not uncommon, and they included a particularly notorious incident in 532 that resulted in 30,000 deaths. The competition between the Greens and Blues continued for decades.

OVEREXTENSION, 602–49

In 602, the army's impatience with the decrees of Emperor Maurice (Justinian's son-in-law) resulted in his overthrow and execution. The Green faction hailed his replacement, the popular but ultimately inept junior officer Phocas – whose rule would end eight years later in failure, with a contraction of the traditional boundaries of the empire. Following Maurice's death, the Persian king Khosrow II – whom the last Justinian emperor had helped regain the Sassanid throne to the benefit of the Byzantines – attacked his late benefactor's empire. Khosrow claimed the act in the name of vengeance for Maurice against Phocas, but he was, in fact, opportunistically attempting to regain territory ceded in exchange for Byzantine aid against his Persian rival years earlier. By the time Phocus was killed by his successor, Heraclius, in 610, the Persian army had seized Chalcedon, a suburb of Constantinople near the mouth of the Bosphorus. The Byzantines continued to fight back, however, and over six campaigns Heraclius carried the empire's forces from the edge of total disaster to victory, in 628 – an enormous feat of military prowess that ranks with the achievements of Alexander or Caesar.

No sooner had peace been brokered with Persia, however, than a new threat emerged. Islam had arisen in 622, and its armies now menaced the empire. Arab Islamic forces would descend on both of the weakened Persian and Byzantine Empires, which were exhausted by nearly three decades of war. Persia succumbed to an invasion in 633 by forces sent from Arabia by Muhammad's successor, Caliph Abu Bakr. Roman Syria came under attack the following year and Damascus fell to the Muslim armies. Jerusalem and Antioch were conquered in 637, and by 649 the Byzantine Empire had lost all of its eastern lands to the Rashidun caliphate, including Mesopotamia, Armenia and Palestine.

DECLINE AND LEGACY, 634–1453

The long-lived Roman military system was primarily responsible for enabling the residue of the Byzantine Empire to repel the ferocious and committed armies of early Islam, through its continued focus on discipline, organisation, armaments and tactics. Technology played a key role in the form of 'Greek fire' – an ingenious incendiary liquid (the ancient equivalent to flamethrower technology) that resisted extinguishing even on water. It was used with great effect against galleys and to defend fortifications (as during the second Arab siege of Constantinople, in 717–18).

ABOVE The sixth-century *Hagia Sophia* (Greek for 'holy wisdom') is one of the most illustrious surviving examples of Byzantine architecture. Immediately following the conquest of Constantinople, the Ottoman Sultan Mehmet II converted the church into a mosque. Restored and expanded numerous times over the centuries, it became a museum in 1935 by decree of Mustafa Kemal Atatürk, founder of the Republic of Turkey.

The tidal wave of Islam washed around the contracting borders of the Byzantine Empire until around 805, when the Abbasid caliphate entered a brief period of consolidation and internal disorder, allowing the empire to regain some of its recently lost territory. Decades of campaigning and border skirmishing followed. The majority of the more significant battles ended in Arab victory. The Byzantine Empire rallied in the late ninth century with some key victories, however, and managed to maintain dominance in the eastern part of the empire. From the late tenth century, the empire was resurgent and reconquered many of its former lands.

The Seljuk Turkish leader Alp Arslan would end this period of relative security beginning in 1067, and inflicted a disastrous and decisive defeat upon the Byzantines at the Battle of Manzikert in 1071. Emperor Romanus IV Diogenes was captured, and almost overnight the Asiatic Byzantine dominions were lost. Alexios I Komnenos attempted to revive Byzantine fortunes by appealing (via Pope Urban II) to Western Europe's Christian states for assistance in driving back the Muslim foe, and the First Crusade was launched. It would ultimately lead to the conquest of Jerusalem in 1099 and the establishment of the other Crusader states of Edessa, Antioch and Tripoli. The resultant abeyance of Seljuk expansionism allowed the Byzantine Empire to enjoy still another spell of recovery and the reversion of territory in western

Asia Minor to its control. Alexios's sons, John II and Manuel I, continued the Komnenian resurgence, bringing the empire to new heights. By the late twelfth century, it was again the strongest economic and military power in Eastern Europe and southwestern Asia.

The empire began its final decline shortly thereafter, marked at the beginning of the thirteenth century by the ugly and brutal sacking of Constantinople by the forces of the Fourth Crusade. (The Great Schism between the Roman Catholic and Greek Orthodox churches had reached a point of no return by then, and set Christian against Christian.) In an attempt to convert the empire from Orthodoxy to Catholicism, a feudal Crusader state – the Latin Empire – established itself in 1204 with Constantinople as its base. It was overthrown in 1261 by the forces of Michael VIII Palaeologus, who restored the empire's original Orthodox character.

By 1300, however, as the Ottoman Empire arose, the exhausted Byzantines were again in a weakened state. The empire would never regain supremacy, and in 1453 the Ottomans finally despatched it. Constantinople's fall involved a massive siege. It included such tactics as the deployment of the world's most powerful cannons, and the transport of warships overland on greased logs across Galata on the northern shore of the Golden Horn, thus neatly circumventing Byzantine barriers to this famous inlet of the Bosphorus. The city fell after nearly two months, along with Emperor Constantine XI Palaeologus and the last vestiges of the Roman imperial legacy, by now thoroughly Christianised.

The decline of one Christian super-empire, however, also marked the ascension of another one, more intricate and of comparable longevity.

The WCSE

The next stage in the evolution of the super-empire phenomenon has its origins in Europe at a time when the Byzantine Empire was still a major force. The structure of the entity I call the WCSE was far more complex than the Roman Super-Empire, and it endures to the present day – albeit much changed. As the second of the world's super-empires, it is a forerunner of the ASE, which will be discussed in PRESENT. Applying the Five Stages of Empire model to the WCSE provides a new view of the last millennium of European history. (See Figure 15 for a graphical representation of the Five Stages of Empire model applied to the WCSE.)

REGIONALISATION, 962–1500

At the end of the early Middle Ages, a pause in the spread of the Islamic Empire allowed the severely reduced Mediterranean Christian countries to commence their expansion back into the region. As the peoples of Western, Central and Northern Europe were converted to the faith espoused by the Catholic Church in Rome, the dominant religious theme of the burgeoning super-empire came into effect. Over the next few centuries, the key regional states would solidify and begin to play out the power games of super-empire. Otto I, the first Holy Roman Emperor to be so crowned, took power in 962.

The Five Stages of the WCSE and its Constituent Empires *(fig. 15)*

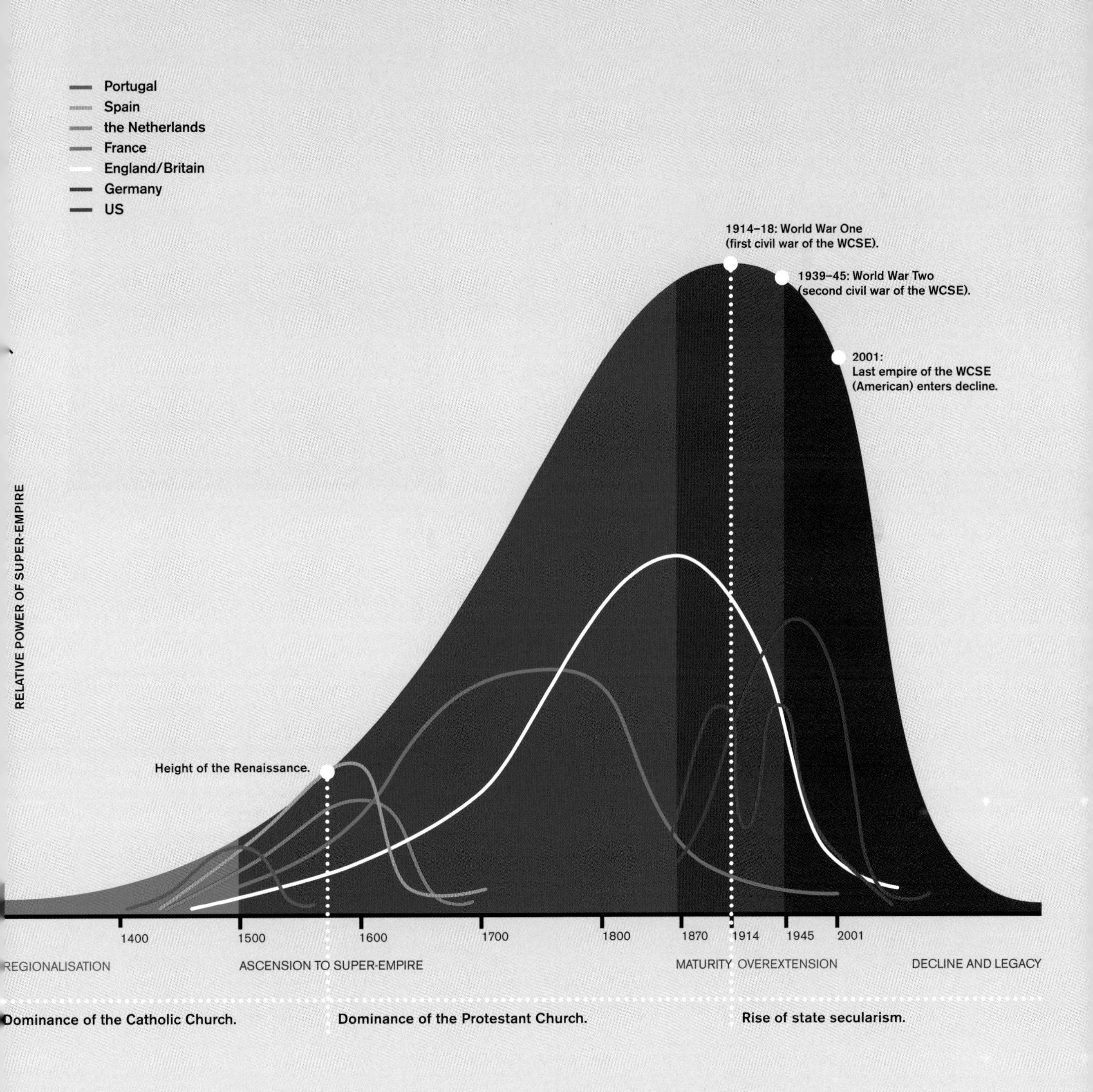

With the dominance of the Islamic Empire to the east, it was natural that European expansion would look westward across the oceans. The Portuguese and Spanish, seeking an alternative route to Asia to avoid the Ottoman Empire's restrictions on the overland spice trade, were the first European states able to project their power across an ocean. Europe became the seat of several maritime empires, and the conquest of the Americas along with migration to the New World ensued.

An empire in its expansive phase will see a concomitant flowering of cultural and scientific creativity, as was the situation in Europe during the Renaissance (which had begun in the fourteenth century, acquiring tremendous momentum by the sixteenth). The Renaissance actually comprised a series of revolutions in thought. First was the rediscovery of the sciences and traditions of ancient Greece and Rome, reinstating the significance of imperial cultural heritage. The second stage witnessed the assimilation of the culture and learning of other empires, including the knowledge base of the Islamic world and, in particular, that of Portugal and Spain. The fall of Constantinople in 1453 was one of the catalysts for this process, causing Classical Greek Christian scholars to migrate westward.

By the end of this regional phase, European spiritual and political values were dominated by the Church. The balance of power shifted from initially dominant Portugal to Spain, which derived its wealth from the precious metals it mined in the Americas.

ABOVE LEFT **This Tlaxcalan drawing, taken from the *Codex Lienzo de Tlaxcala* (*circa* 1550), depicts Spanish *conquistadores* and their Tlaxcalan allies attacking the Tarascan home city of Michoacan. The Spanish, shown upper left, ride horses and are armed with swords and lances. The Tlaxcalans, at the bottom, are armed with obsidian-bladed swords and feature among them two noble warriors clad in backbanners (a fan on the left, a bird at the centre). On the right are Tarascan archers, with a lone, lynched figure at the centre.**

ABOVE RIGHT **Portrait of Martin Luther by Lucas Cranach the Elder. Luther (1483–1546) ushered in the Protestant Reformation with the publication of his *Ninety-Five Theses*.**

ASCENSION TO SUPER-EMPIRE, 1500–1870

The major energising force behind an empire in ascension is the dynamic proliferation of its population. In the case of Europe, that process was driven by the plethora of new seed types that had been discovered during the initial phase of fifteenth-century exploration and brought back to the Continent. Coupled with new farming techniques, it fed a new population expansion.

It is important to note that this phase of expansion coincided with the rise of the breakaway Protestant faith in 1517, when the German priest and theologian Martin Luther challenged the authority of the Catholic Church and of Holy Roman Emperor Charles V. By its very definition, Protestantism signified that its adherents were not Roman Catholics: Protestants opposed the notion that the Church, a hierarchical power structure, was the mediator between God and human beings. Instead they believed that humans had direct links to the divine. This revolutionary belief cast off the traditional framework of Christianity that had centred on Rome and allowed the Church to wield power over Europe. It also birthed a wave of religious fundamentalists – Anglicans, Lutherans, Calvinists, Presbyterians – who in time would subdivide further into such denominations as Baptists, Methodists and Pentecostalists. Britain and the Netherlands were the first to embrace the Protestant faith, setting these countries on a path of direct confrontation with the dominant powers of Spain and France.

The Protestant movement generated the religious zeal that would eventually drive the colonisation of the Americas by settlers seeking to create a new society in harmony with God. Freed from the limitations of Church doctrine, Protestants experienced a creative ferment, one manifestation of which was the growth of British naval power (marked by the defeat of the Spanish Armada). At the same time, the Dutch led the charge to access the wealthy trading system of the Arabs, Indians and Asians, where half of the world's GDP flowed – and in the process developed into the next European sea power.

European naval power rose to great heights in the Age of Discovery, with Portugal, Spain, the Netherlands, France and England all taking to the sea. Over the course of 200 years of war against the Spanish, Dutch and French, Britain achieved dominance at Trafalgar in 1805 to become the world's greatest maritime empire. Notably, Britain never dominated continental Europe in a military sense because of its small population. Its small army did achieve amazing success against France in the Iberian Peninsula and then in the Crimea, but the projection of continental land power was its weakest aspect. Britain was foremost a sea power; France and later Germany would eventually vie for land power in Europe.

MATURITY, 1870–1914

As I have said, by 1870, Britain's economic growth as a result of the Industrial Revolution had reached its zenith, and depression loomed. France had suffered a massive defeat at the hands of the Prussians. The British Empire was beginning to decline, but the WCSE remained strong because of the rising power of its American and German fractals, which offset decline elsewhere within the super-empire. Thus the WCSE did not peak until 1914, when the first wave of global trade and finance (proto-globalisation) was also at its height. The new powers sensed Britain's weakness and had ambitions to assume its mantle – one, as we have seen, as a friend with ulterior motives (the US) and the other as an outright hostile enemy (Germany).

RIGHT In William Orpen's painting *The Signing of Peace in the Hall of Mirrors, Versailles, 28 June 1919*, all the key signatories of the document that ended World War One appear, including US President Woodrow Wilson, French President Georges Clemenceau and British Prime Minister David Lloyd George (middle row, fifth, sixth and seventh from left, respectively). The German delegate, Johannes Bell, seated in the centre with his back to the viewer, with his co-delegate Hermann Müller leaning over him, is shown signing the treaty.

OVEREXTENSION, 1914–45

The end of the nineteenth century and the beginning of the twentieth brought rapid changes in the balance of power in Europe, culminating in World War One. According to the Five Stages of Empire model, civil wars for control of the empire are quite common in the maturity stage. In the context of the WCSE, World War One was just such a civil war for domination on an enormous scale. The battle lines were drawn between the Central Powers (Germany, the Austro-Hungarian Empire and the Ottoman Empire) and various allies on one side and the Triple Entente (the UK, France and Russia) and various allies – including, critically, the US – on the other.

New and powerfully destructive warfare technology, alongside the militarisation of European society, marked the peak in the growth cycle of the WCSE. By the end of the war, all sides were exhausted (particularly Britain and France). Millions of their run-down citizens, moreover, succumbed to the deadly influenza pandemic of 1918. The Austro-Hungarian and Ottoman Empires were no more, and Germany would be forced by the Treaty of Versailles to pay crippling reparations that set the stage for another round of WCSE civil war, in 1939.

The illusion may have developed that the WCSE had actually expanded, with Britain's inheritance of the formerly Ottoman Middle East. In fact, the clear winner was the US, which, without a war on its doorstep, had continued to expand its economy and had now assumed a role at the forefront of world affairs. (Nevertheless, though the British presence in the Middle East would prove troublesome, Britain would later be well placed to profit from the exploitation of oil and the manipulation of various states in the region post-independence.) Although the 1920s brought an initial economic recovery for the WCSE, its influence never reached the heights of 1914. The Great Depression that began in 1929 reverberated throughout the world. Arguably, its chief effects were to make the US more insular, and Germany more receptive to the Nazis' eventual takeover.

The religious constitution of the WCSE also underwent a transformation in the twentieth century, with the atheist belief systems of Communism and Fascism emerging to transplant Christianity as the WCSE headed for decline. Adolf Hitler would prove to be the antihero for the times, rebuilding Germany with the sole intention of launching a second attempt at controlling the super-empire. But with this challenge to the old order, Germany chose an incarnation diametrically opposed to the traditional Christian values of the WCSE. The German people were polarised into a powerful and effective war nation, but by 1945 the ambition of Nazi Germany and its Axis allies – particularly the rising ASE led by Japan – had been defeated. The European component of the WCSE was in ruins. Within a decade, the global power map would completely change.

LE ROY
GOUVERNE

DECLINE AND LEGACY, 1945–2001

By 1945 the European component of the WCSE was battered physically, psychologically and financially. The collapse of the old core empires of Europe, Britain and France was accelerated. They divested themselves as rapidly as possible of their colonial assets over the next decade. The nascent American Empire, however, after a half-century of aspiration, stepped into the power vacuum as the leader of the now declining super-empire.

The Cold War that followed was, in many ways, another civil war as the Communist system (no longer considered Western or Christian) that had broken off from the WCSE fought for control of the remnants of the super-empire with the Western contingent. After nearly five decades of tumultuous struggle that set the economies of both sides head to head, the Soviet Union collapsed in 1990. The apparent victory of the Capitalist democratic model seemed to promise a new period of stability. The US (and the WCSE it had led since 1945) appeared to become the dominant global entity.

In reality, the power wielded by the US in 2001 was a shadow of that commanded by the European nations of the WCSE on the eve of war in 1914. While the US forged ahead after 1945, Europe re-entered the cycle at the regionalisation stage, taking a long, slow path with the EU towards becoming a 'United States of Europe'. (European foreign policy and power projection have verged, as a result, on non-existence.) The US, the last empire left in the WCSE, now faces immense challenges from China and the Islamic world. It does not have the luxury of passing the baton to another WCSE entity; it is, instead, faced with precipitous decline.

It may appear baffling that the threat to US dominance posed by the Chinese and Islamic blocs is so serious, that American peak power did not endure beyond a few years (1990–2001). Yet the strength of the American Empire at its height has been a distraction. It should always have been placed within the overall context of the WCSE, which has, in fact, been declining since 1914. Now that this final offshoot of the WCSE has entered its own decline, we will see a steep drop in the fortunes of the super-empire as a whole, as the Middle and Far Eastern power blocs rise.

The past millennium belonged to the WCSE. The next, however, will be claimed by these new challengers – chief among them the rising ASE, headed by China.

RIGHT During a Martyr's Day rally in Beirut in 2009, a *Shi'i* Muslim girl holds up a portrait of slain *Hizbullah* chief Abbas al-Musawi, who was killed in 1992 during an Israeli air raid on South Lebanon. *Hizbullah* remains an armed and political force to contend with in Lebanon, and is a means for Iran to project its power within the Middle East in general and against US interests specifically.

BELOW RIGHT Reservists from the People's Liberation Army of China attend a ceremony at a stadium in April 2007 in Nanjing. The country's military system combines a militia with a reserve service, volunteers and conscript soldiers.

نصر
من
الله
Love me

PRESENT

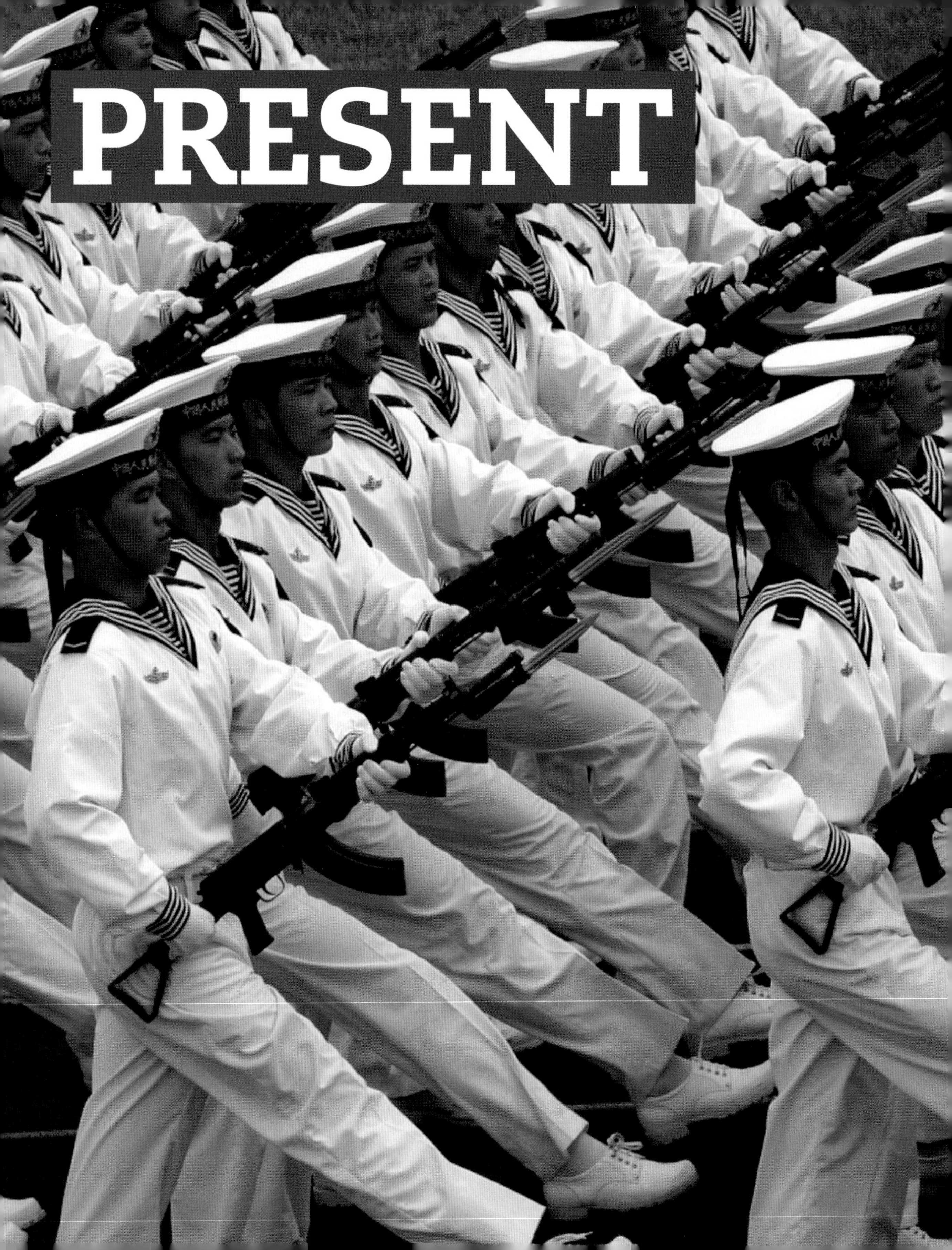

Introduction

As we have seen, each stage of the Five Stages of Empire is in expansion or contraction (see Figure 16). In PAST, the focus was on former empire cycles, but we also saw how the WCSE continues to have an impact on the world today. I shall now turn to the geopolitics of the present, looking at how the Five Stages of Empire model can enable us to establish predictive trends. The past, of course, forms the basis of today's geopolitics, but the chapters in this section will instead focus on the here and now.

In considering political and economic forces at work today, six key underlying themes are discernible. Each chapter in this section will focus on one of these themes, although they are inevitably interlinked (see Figure 17). In the last section, I shall discuss how these themes will impact on the future.

PREVIOUS **Soldiers of the North China Sea Fleet parade at a military harbour on 1 August 2010 in Qingdao, China, to celebrate the fleet's fiftieth anniversary. The fleet is one of three in the People's Liberation Army Navy.**

BELOW **Our planet looks majestic from outer space but is facing enormous and very threatening challenges.**

The Five Stages of Empire in the Modern World *(fig. 16)*

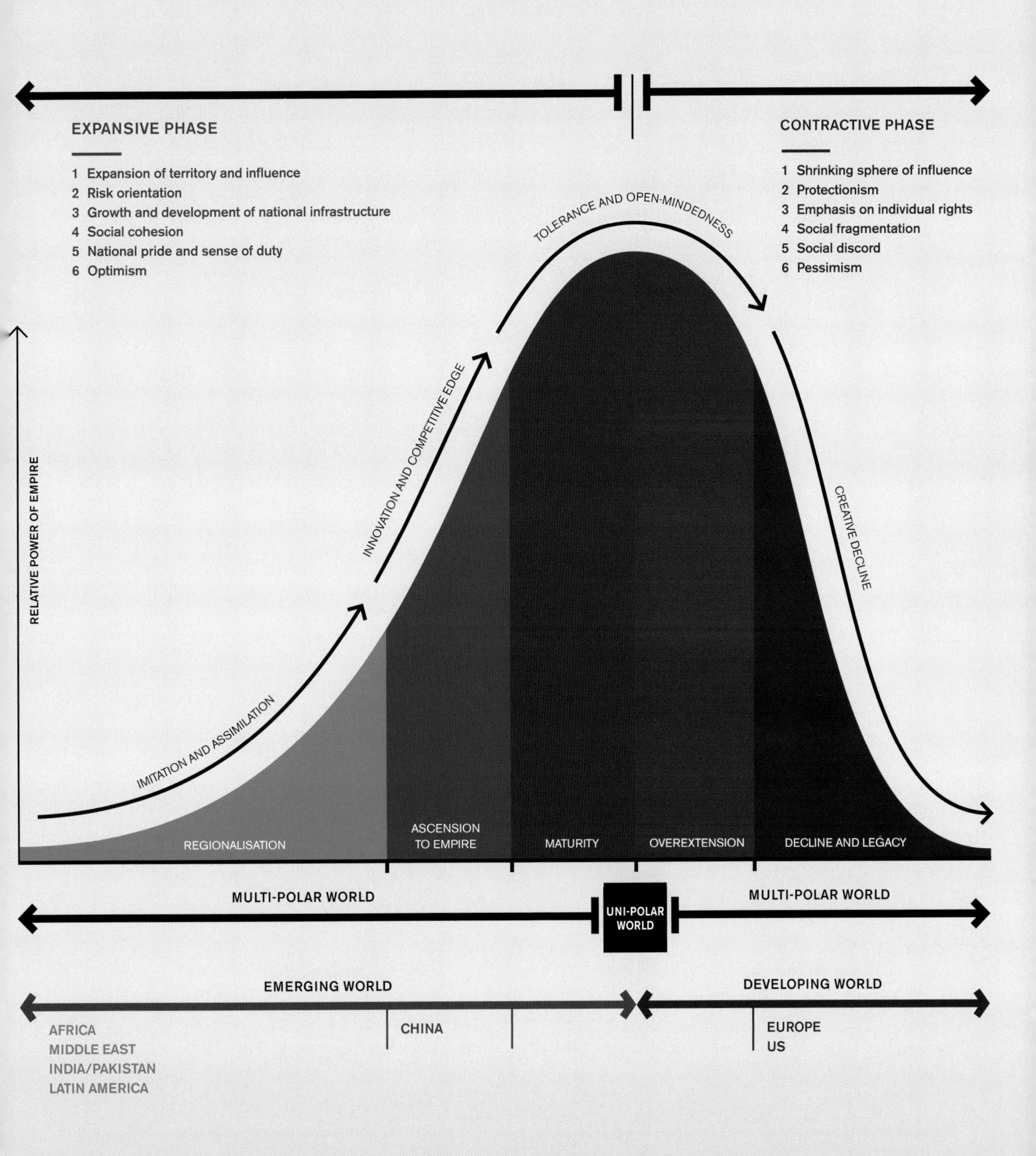

THE SIX THEMES

1 **THE MULTI-POLAR WORLD.** A marked geopolitical shift is occurring, away from the dominance of the US (as the last empire of the WCSE) to the new powers rising in the East. In this new multi-polar world, attention must be paid not only to aspiring hegemons and civilisations in regionalisation or ascension, but also to the role of smaller countries and satellites (such as Israel) in the balance of world power.

2 **COMMODITIES: THE FUEL OF MODERN EMPIRES.** As discussed, access to resources has always been essential for the growth of nations and empires. This is more than ever the case, with the Second Industrial Revolution now taking place in Asia. The demand for raw materials by both the developed and the developing worlds has precipitated competition for resources to a degree never before witnessed as the world draws from a resource pool that is historically at its most depleted.

3 **POLARISATION: THE ROAD TO WAR.** This is the process whereby a collective binds together and defines itself. It creates positive cultural values but also acts as a focus for negative emotions, such as fear and anger. These are then projected onto another collective that is perceived to embody opposite values. The increase in both national and religious polarisation is a clear precursor to and justification for conflict. Polarisation can be thought of as the measure of a global friction coefficient: as polarisation rises, so does the likelihood of conflict.

4 **GLOBAL MILITARY BALANCE.** The history of empires in any one period has been strongly tied to the balance of military power. The world today is no different. The greatest threat of destruction remains the proliferation of nuclear weapons, but we are also now witnessing a conventional arms race that is slowly building in Asia. Moreover, the threat of biological and chemical weapons, along with the possibility of continuing advances in the US missile defence shield strategy, could once again alter the balance of power.

5 **DISEASE AND EMPIRE.** In the past, disease and epidemics have not been random events but rather the product of collective social trauma at different stages of empire. The speed with which we now move around our planet enables infection to spread with potentially catastrophic consequences, which modern medicine may be unable to contain.

6 **CLIMATE CHANGE.** The changing biosphere is linked to the acceleration of human population growth and industrialisation, with profound geopolitical implications. The question is not whether climate change is a reality but the rate at which it will radically affect the world around us, and the cost to both the natural world and human civilisation.

The Interlinkage of the Six Themes *(fig. 17)*

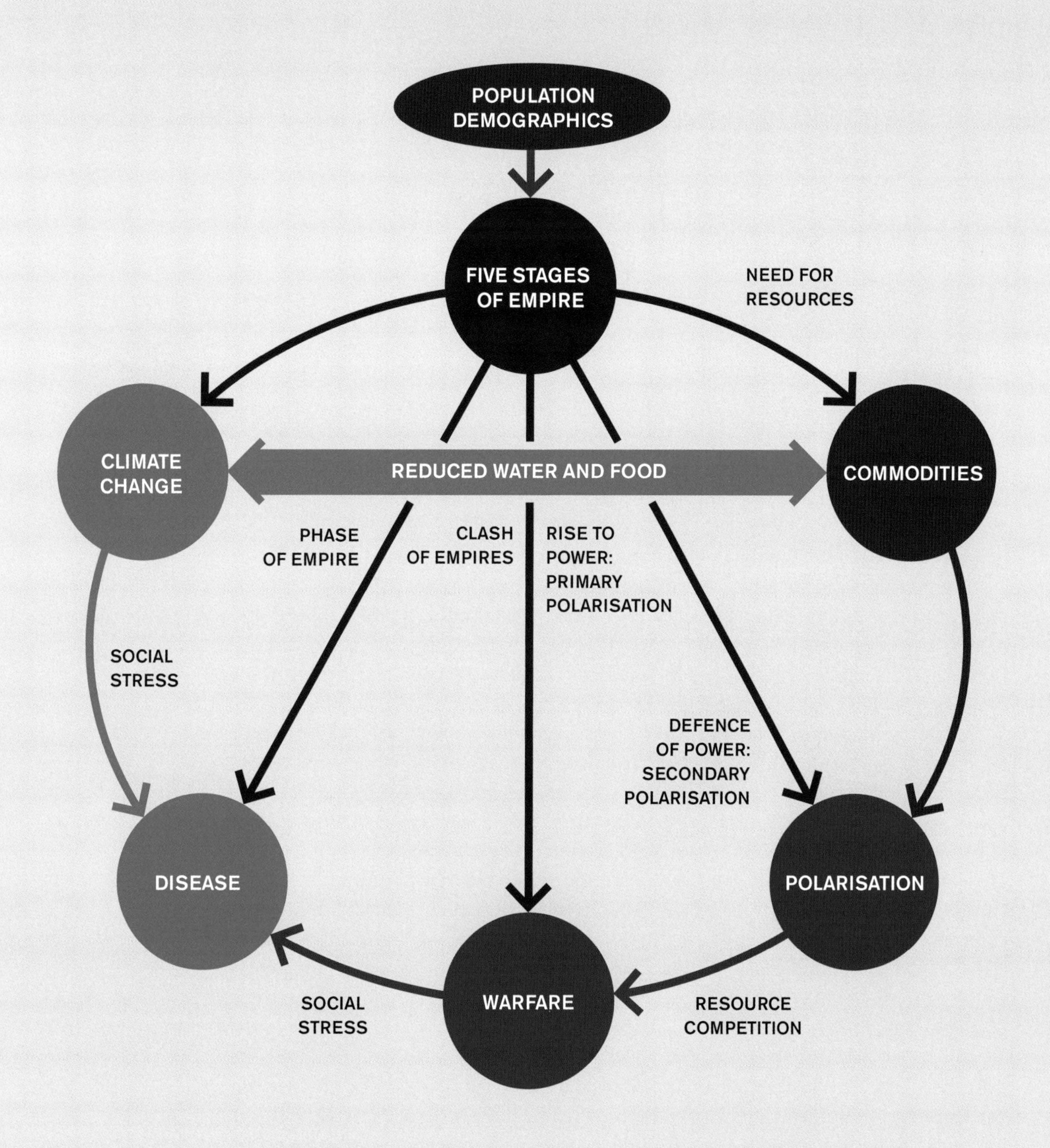

Chapter Five

The Multi-Polar World

The Emerging New World

EMPIRES LOST, EMPIRES GAINED

The world we find ourselves in today was shaped by the WCSE tipping into decline following its 1914 'civil war'. The ebbing of the WCSE is creating a vacuum that is being filled by the burgeoning powers of China, India and the Middle East, causing profound power shifts.

Views of the New World.
RIGHT **Sheik Zayed Road, which is the most widely used highway in the Arab world, connects Abu Dhabi with Dubai and the rest of the United Arab Emirates.**

BELOW RIGHT **The new Bund Area in Shanghai, China, illuminated ahead of the fifty-day countdown to the 2010 Shanghai World Expo.**

In the Middle East, a resurgence of Islamic consciousness began in the late 1920s in response to Western rule in the region. It was led by such groups as the Muslim Brotherhood under Hassan al-Banna, and formed the basis for what is now called 'Islamism'. The speed with which Britain relinquished its empire after World War Two took the world by surprise. The modern Middle East was formed as states were created and/or became independent following the break-up of the British and French Mandates and the founding of the State of Israel. As the Middle East entered the regionalisation stage of empire, particularly with the rise of the Islamic Republic in (*Shi'i*) Iran in 1979, the religious schism in Islam between the *Sunnis* and *Shi'is* began acquiring a new dimension.

In India, the centre of British imperial economic power, the drive towards independence by Hindus and Muslims that had begun after World War One was sidelined during World War Two in order to focus on the common threat posed by Japan. At the end of the latter war, the exhausted British Empire was made to honour its commitment to Indian independence – although not without first presiding over partition at a cost of over 1 million lives. Partition set the tone for the regional conflict between India and Pakistan that has since witnessed three major wars, one minor war and numerous confrontations, along with one of the closest brushes with nuclear confrontation since the 1962 Cuban Missile Crisis. Tensions remain, and although India is a far greater economic power than Pakistan, the nuclear factor along with Pakistan's increasing risk of destabilisation gives cause for great concern.

In China, the Boxer Rebellion (1898–1901) marked a turning point in the country's resistance to Western interference and influence. The protracted civil war between the Communists and Nationalists began in the late 1920s, ending with the founding of the People's Republic of China in 1949. The turbulent years under Mao Zedong gave way to a new political paradigm that was gradually implemented by Deng Xiaoping in the years after Mao's death in 1976. By 1996, China was ready to ascend to

empire. Now Communist in name only, the world's most populous country is relentlessly Capitalist, determinedly authoritarian, geopolitically ambitious and ever hungry for resources to feed its growing industrial power.

The rise of these three power blocs –the Middle East, India and China – may strike casual observers as a recent phenomenon of the last decade. Yet, as discussed, the foundations for this shift in power were laid around 100 years earlier at the end of World War One, and the momentum behind it is very powerful. When we discern events on this scale, we are either at the end of a trend that has suddenly changed direction or at the point where a trend is accelerating. As we will see, the Five Stages of Empire model points to the latter conclusion.

THE DEMOGRAPHIC FACTOR

As has been discussed, demographics are essential in situating a nation within the spectrum of expansion and contraction. The critical replacement ratio required to maintain a constant population is around 2.1 children. However, recalling that younger populations make for more dynamic and risk-orientated nations, average age is as important as population in assessing imperial development. In 1850, people aged sixty-five and older in Britain comprised 4.6 percent of the total population; in France, it was 6.5 percent, and in the Netherlands 4.7 percent. By 1900, those figures were 4.7 percent, 8.2 percent and 6.0 percent, respectively. Therefore, during this period, the great majority of the population were productive.

Today, our lifespans are extending well beyond retirement age, resulting in a 'productivity gap' whereby it becomes increasingly difficult for the productive population to support those who are no longer working. This has resulted in a great burden on Western welfare states. Europe's old-age dependency ratio between retirees and workers was 23 percent in 2005, projected to rise to 36 percent by 2030. In the US, the equivalent figures are 19 percent, growing to 32 percent by 2030. The Japanese productivity gap is even more pronounced, at 30 percent and 53 percent, respectively. Such declining demographics will inevitably slow down the economic growth of these nations.

As described earlier, gender also plays an important role: the number of men above 51 percent in a population can be viewed as extra risk capital for a developing power – available for deployment if and when it chooses to expand. In the WCSE, demographics are at best stable (as in the US) and at worst declining (as in Russia), although immigration from less developed countries may mask this process and even support economic growth to some degree. Tellingly, the US has maintained its replacement ratio at 2.1 and remains aggressive in its military power projection. The European ratio stabilised between 1995 and 2005 at around 1.4 and is currently at around 1.5. The countries of Europe are accordingly reluctant to project military power and thereby risk casualties.

Figure 18 shows the world population by region in 2008, while Figure 19 shows the change in world population by region between 1950 and 2050. In the latter, the flattest curves, at the bottom, belong to the developed world: Europe, the US and Japan. Of particular interest is Russia, with a negative curve that presents enormous problems for its current prime minister, Vladimir Putin, who has ambitions to enlarge the country's sphere of influence. Russia's former (Soviet) satellites remain of critical importance to Moscow, not only because they provide a buffer zone around the heartland but also because of their potential human resources.

In 1950, Europe comprised 22 percent of the world's population. It now represents only 11 percent and is set to decline to just below 8 percent by 2050. At the same time, Asia's population will continue to make up the major share of the world's population, at 60 percent now with only a slight decline being forecast by 2050.

BELOW A hospital nursery in Beijing, China. Despite the one-child policy that was introduced by the Chinese Government in 1978, which it claims has prevented 25 million births since its implementation in 2000, China's population is booming.

The great demographic expansions that are currently under way outside the WCSE and Japan are driving the shift of power from the developed to the developing world. By 2030, China, India, Brazil and the Middle East are likely to have added some 2 billion people to the global middle class.

China leads the way, with a massive surge in population since the 1970s that is set to peak in 2025. Thereafter, the country's older population may slow down the growth of the People's Republic and move it into the maturity phase of empire.

China's 54.5 percent male population – or 120 men for every 100 women – also presents a gender-balance crisis of huge proportions with, effectively, a deficit of 40–60 million women in Chinese society. However, there is a policy debate about this deficit

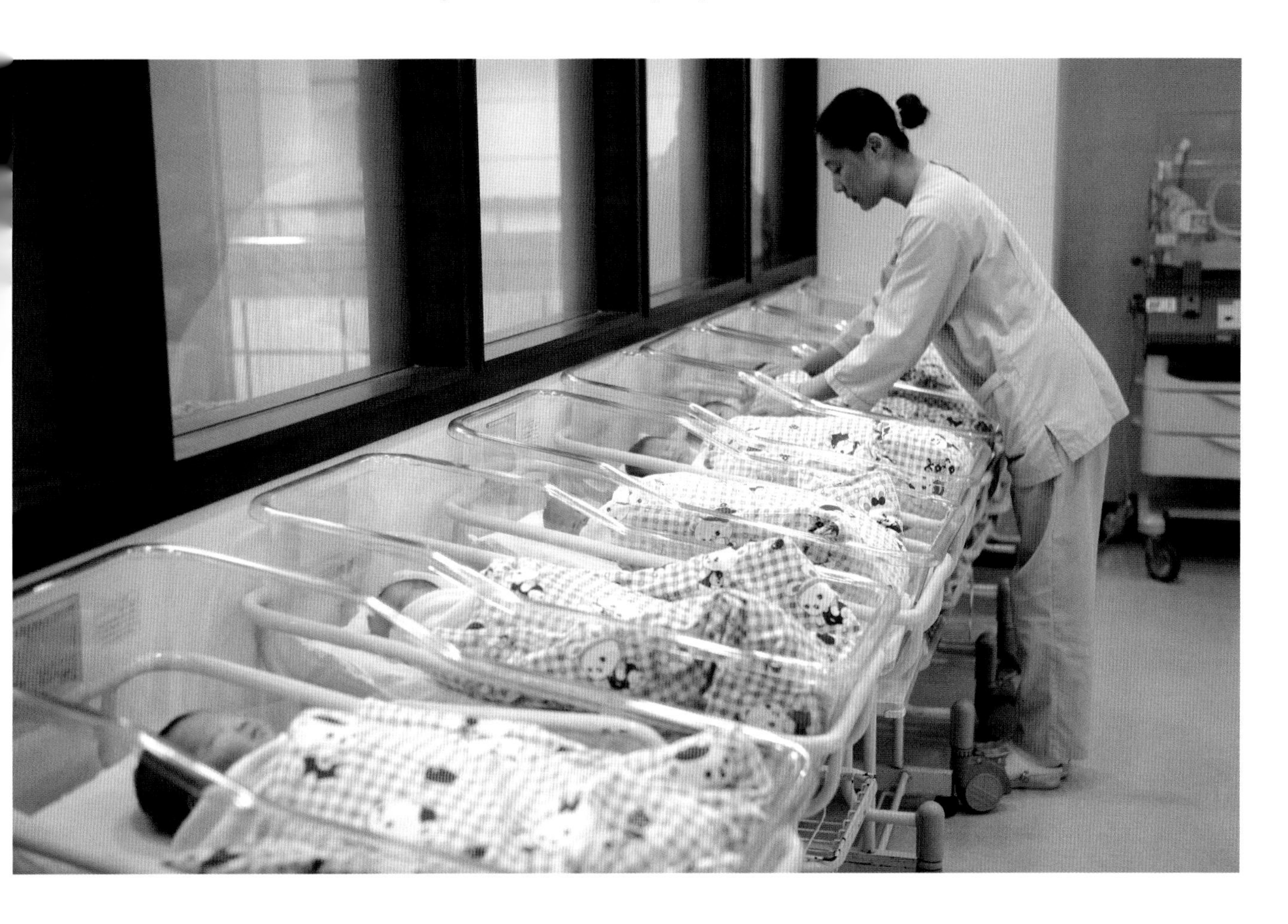

Total Population (Number of People), 2008 *(fig. 18)*

Data by UN Population Division, World Population Prospects.

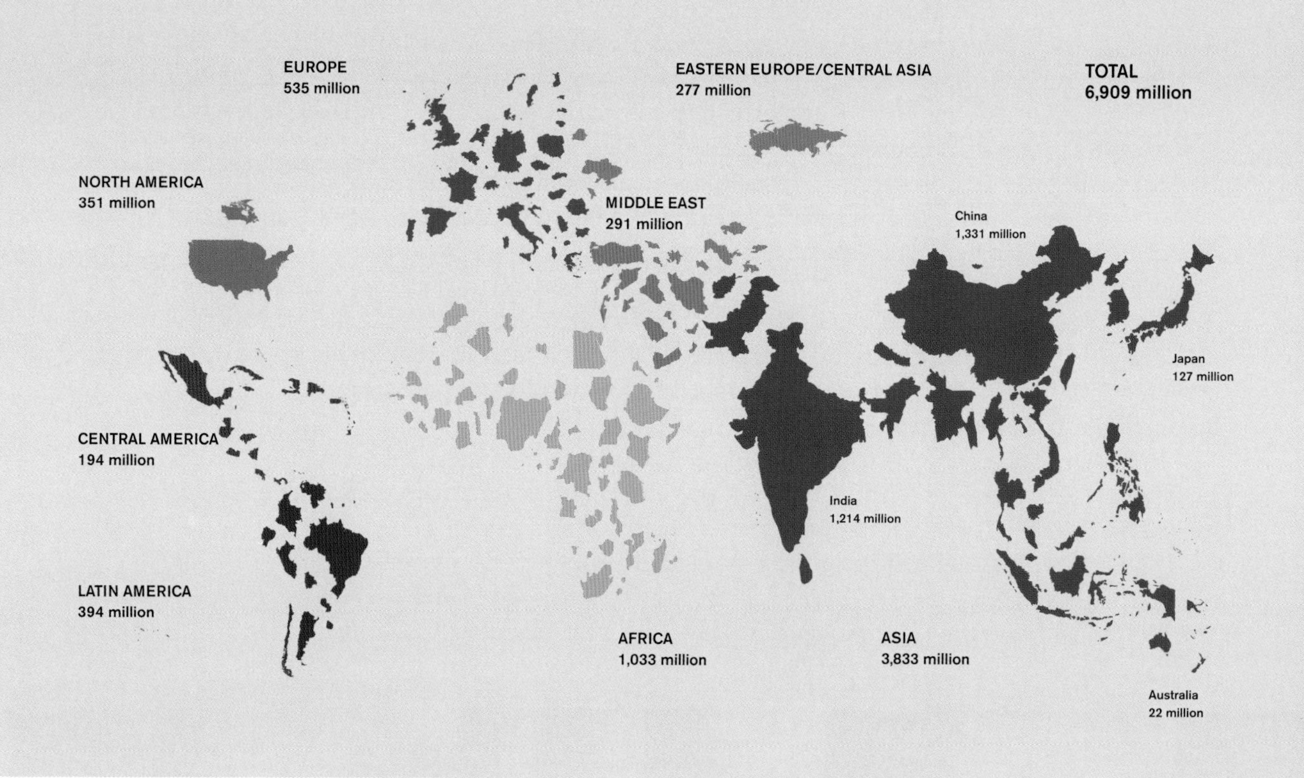

that suggests that the gap may in part be the result of unregistered births. This being the case, it is difficult to determine precisely how large the gender gap really is in China, although the imbalance is weighted on the male side. In any event, the ramifications may include a more aggressive collective mentality geared to national expansion.

India is some fifteen years behind China in terms of demographic development and relative gross national product (GNP) growth, and it will need to tackle the challenge of acquiring resources that have become scarcer on account of China's considerable head start in procuring them. A country with a strong preference for sons, India has its own gender gap. In 2009 there were, on average, 112 boys born for every 100 girls.

Most striking demographically is Africa. The continent that many in the West reflexively, presumptuously and mistakenly write off as disease-ridden, famine-blighted and decimated by tribal wars actually commands explosive demographics. Indeed, according to UN population projections, by 2050 it will expand to 19 percent of the world population. Its wealth of under-exploited resources is attractive to the Asian nations stoking the Second Industrial Revolution, chiefly China. Geopolitically, Africa has exponential growth potential and will play a key role over the next decade in terms of regional empire dynamics and within the wider context of the new multi-polar world.

Global Demographic Drivers/World Population, 1950–2050 *(fig. 19)*

Source: Population Division of the Department of Economics and Social Affairs of the UN Secratariat, World Population Prospectus. The 2004 revision and world urbanisation prospectus.

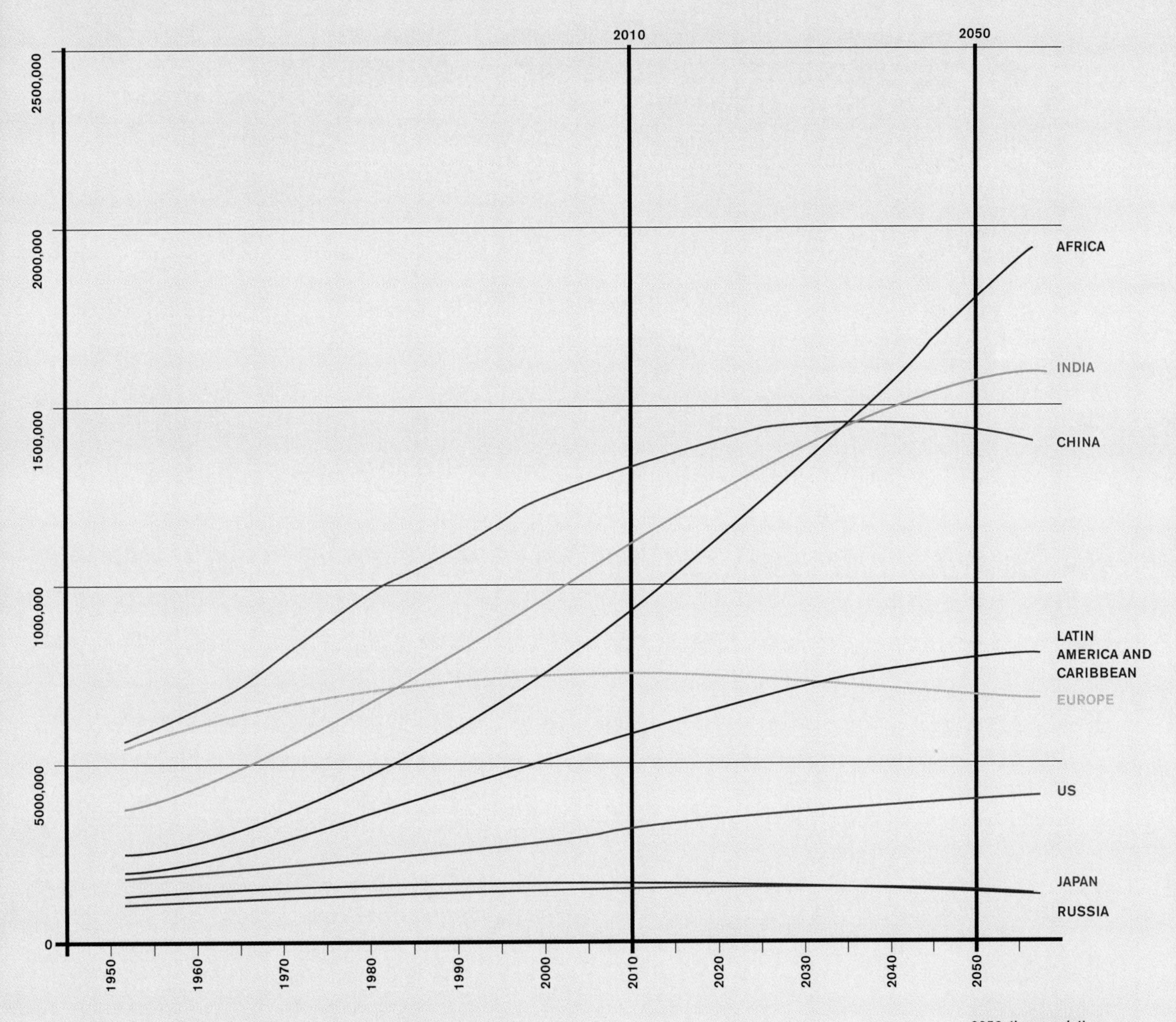

2050: the population of the world is expected to increase by 50 percent during the next forty years to more than 9 billion.

THE GEOPOLITICAL DIVISION OF THE WORLD

Geopolitics today can be mapped across seven distinct zones: Africa, the Middle East, Latin America, Asia, Europe, North America and Russia. Each of these zones is at a particular stage of empire, either in its entirety as a region or as a collection of states that are at various stages of empire themselves.

Of special note are Australia and New Zealand, which have effectively broken away, economically, from their ties with the West and should now be considered as nations in the regional stages of empire. Figure 20 shows the geographical distribution of the Five Stages of Empire.

Emerging Powers: Africa, the Middle East, Latin America and Asia

AFRICA: EARLY REGIONALISATION

The African continent consists of many nations whose affiliations with each other and the world at large are not decisive. When we speak of 'Africa', in fact, we are referring to fifty-three countries spread across 20.4 percent of the world's surface, comprising 15 percent of the world's population and encompassing widely varying ecosystems, including desert, jungle and savannah. The continent is in the earliest stages of industrial development and, as discussed, its demographics are robust.

Powerful divisions are deepening in Africa between the Islamic states of the Sahara Desert and Mediterranean, and the Sub-Saharan region where Christianity, the religion of the first European settlers, predominates. As the population grows either side of this north–south divide, exerting pressure on resources, we can expect the south to regionalise as a South Africa-centred union and the north as a Libya/Egypt-centred union. (The main dividing factor, however, would be not so much religious – because Islam continues to grow throughout the continent – as ethnic: that is, Arab/non-Arab.)

The key issue for southern Africa is the institution of legitimate, efficient government, free from the turmoil of tribal politics. The borders of modern African states are the legacy of colonialism, taking no account of traditional tribal boundaries. Upon independence, multiple tribes found themselves forced to coexist within a single state. In the ensuing decades, individual states have vied for political supremacy, but the welfare of the dominant tribe has been placed above that of the country itself. Westerners unable to comprehend this need only look to early European history, when tribes were combining and recombining to become nations and were forced to overcome similar problems. In the majority of cases they did so in the face of a common enemy, the threat of which reduced the importance of tribal differences and encouraged integration.

Seventeenth-century India was not dissimilar to Africa today, with numerous disparate domains spread across a subcontinent containing great actual and potential wealth. The British consolidated the small, independent states into a single nation as they sought to extract India's commodity wealth. This was the genesis of the self-governing India (and Pakistan and Bangladesh) of today.

China views Africa as a commodity source into which it can expand its influence so as to fuel its growth. Since as early as 1996 it has been funding investment projects in Africa (see Figure 22). China may well become the 'common enemy' that serves to unite African states, because it has moved into the continent solely to export its minerals and resources. In addition, China-funded projects have tended to import Chinese migrant labourers to work on them. This will doubtless continue to generate resentment among the local population. Furthermore, as cheap goods from China (from electronics to household items) are dumped onto African markets, local production, such as it is, risks closure. On the one hand, Africans benefit from being able to buy affordable goods; on the other, local industries are unlikely to survive the competition.

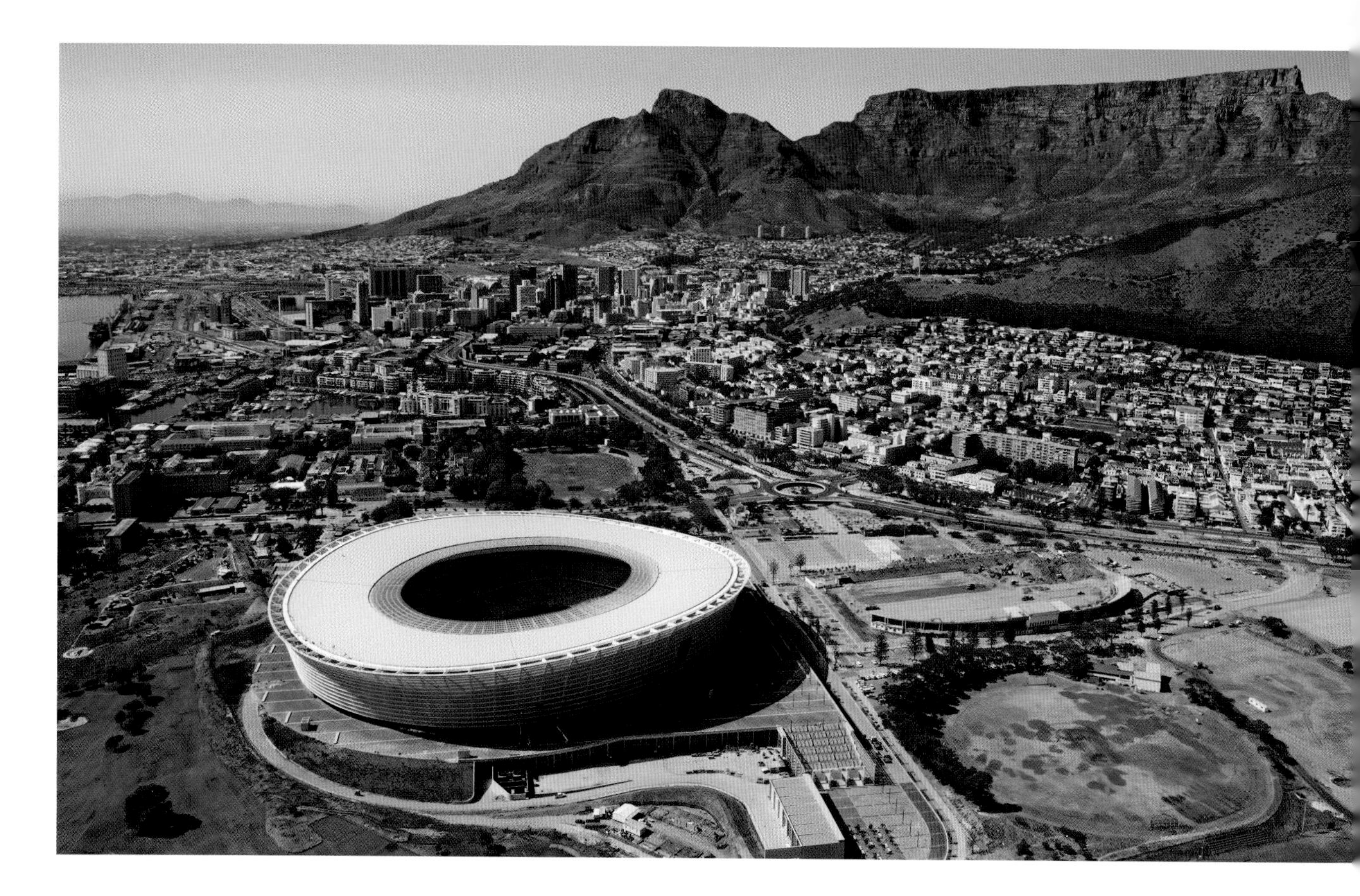

ABOVE Cape Town Stadium at Green Point with Table Mountain in the background. South Africa is emerging strongly from its apartheid years and is set to be the power focus of Sub-Saharan Africa.

Unless African states improve the quality of their governance and accumulate national wealth through self-development and from the resources at their command, they will fall prey to the lure of China's money, and in so doing entice the Chinese empire deeper into the continent.

Africa is also of considerable strategic significance as its sea lanes are vital to global trade. Although cargo and oil tankers make use of the Suez Canal for passage between East and West, the largest ships must still round the Cape of Good Hope. The cape is so crucial a passage that during the Cold War an isolated South Africa, then under apartheid rule, was nevertheless deeply woven into naval partnership with the North Atlantic Treaty Organisation (NATO). Endemic piracy in the Gulf of Aden, moreover, has also made rounding the cape an attractive alternative. Assuming South Africa remains affiliated with the West, as its influence expands, China is likely to seek to establish naval bases on the east and possibly west coasts of Africa.

THE MIDDLE EAST: LATE REGIONALISATION

The creation of the modern Middle East following World War One generated massive tensions that resulted in a string of regional wars, most of them with Israel at their centre. Sixty years after the creation of Israel, the majority of the *Sunni* Muslim powers have taken a pragmatic view towards accepting its existence (officially or unofficially), aided no doubt by the fact that Israel is equipped with nuclear weapons. The major exception is Syria, which, although enfeebled, sought to acquire nuclear weapons capacity in its own right before being thwarted in 2007. Syria is likely to re-establish its old Cold War ties with Russia, receiving arms and aid in return for providing Russia with land and naval bases. Russia would thus regain a foothold in the Middle East.

However, it is the rise of the Islamic Republic of Iran that is most critical to regional dynamics today as it seeks to establish itself as a regional hegemon. Once the countries of the Middle East achieved independence from the West, and now that Israel's existence has, to a degree, been tacitly accepted, the struggle for supremacy between the *Sunni* states and *Shi'i* Iran has taken on new significance. It is a classic struggle for regional control, within the framework of the Five Stages of Empire model.

The Gulf countries view their rivals in oil and religion with perpetual wariness. Also regarded anxiously are the considerable *Shi'i* minorities in *Sunni* states across the Middle East, such as Kuwait, Saudi Arabia and Yemen. *Shi'is* are particularly powerful in Lebanon, where they constitute a major and contentious political force led by *Hizbullah* ('Party of God'), and they form the majority in Bahrain and Iraq, where they are in ascendancy after enduring decades of oppression under Saddam Hussein's *Ba'th* Party.

Hizbullah takes its orders from Iran, and the latter's influence is also strong in Bahrain (through opposition parties) and Iraq (through *Shi'i* militias, various political parties and an evolving relationship with central government). Iran also backs the *Sunni* Palestinian resistance group Hamas, which came to political prominence in 2007 in the face of the Palestinian Authority's weakness and before increasing oppression of Israeli policies. Iran has also maintained close relations with *Sunni* Syria (which also has a significant *Shi'i* minority) since the 1979 Islamic Revolution. In turn, Syria, which has a long history of intervention in Lebanon, often acts as a conduit for Iranian supplies to *Hizbullah*.

While much is made of the end of global dependence on oil, global oil reserves are not about to end anytime soon. While Middle Eastern reserves are dwindling, there is another generation's worth yet to be tapped (estimates for Saudi Arabia stand at seventy-two years; for Iran, ninety-five). Thus the region's value to the world remains a given.

Internally, the Middle East is witnessing a massive demographic expansion driving its later stages of regionalisation. This population surge, coupled with narrow wealth distribution in both Iran and the Arab states of the Gulf, makes the region a fertile ground for both *Sunni* and *Shi'i* extremism.

One other country figures prominently in the Middle Eastern context, although it has played a subdued role in the region's affairs for the better part of a century: the Republic of Turkey. Turkey has a population of 76 million and one of the biggest GDP rates in the Islamic world. Turkish society has been resolutely secular since the early twentieth century, but it is slowly rediscovering its Islamic identity, although not without internal tensions. It still maintains strong links with Europe and the West, although its application to join the EU made in 1987 is still pending. Its concerns are currently focused on preventing the creation of a separate Kurdish state (Kurds account for as much as 18 percent of the population) and balancing Iranian power, but if US influence in the Middle East were to wane, Turkey could be expected to attempt to fill the vacuum.

LATIN AMERICA: LATE REGIONALISATION

After a turbulent and violent history, Latin America is now one of the more peaceful regions of the world, remote enough from potential Chinese proxy conflicts and insulated from Islamic terrorism. This stability invites the foreign investment that will accelerate regional growth.

RIGHT **Celebrations at Copacabana Beach in Rio de Janeiro on 2 October 2009 as it is officially announced that the city will stage the 2016 Olympic Games.**

Latin America is in ascendancy: booming demographically, with considerable commodity wealth. Brazil, Venezuela, Mexico, Chile and Argentina are all economically strong and can choose their foreign trading partners. Central and South American regionalisation dynamics will most likely continue without major external influences and, in effect, the nations in these regions will be relatively isolated from world affairs. Consistent with this process is a greater level of nationalism derived from economic growth and resource wealth. However, Brazil, the largest nation state in the region, is at some stage likely to want to join the G8, and a new chapter will begin with Latin American participation in global politics.

The awarding of the 2016 Olympic Games to Rio de Janeiro will focus the world's attention on Brazil in particular and on the region as a whole, and is thus a symbolic event of great importance. Indeed, Brazil – the fifth-largest country in the world – has consolidated its dominant role in South America. Venezuela, flush with oil wealth and a maverick leadership, may yet challenge Brazilian power, but, given the economic disparity between the two nations, the outcome in Brazil's favour is certain. There remains a risk that Russia or China might seek to threaten the soft underbelly of the US by forming close alliances with Venezuela and/or Cuba. However, the US would, in response, draw closer to Brazil, much as it has done with India to counter Chinese power in Asia.

The Brazilian economy has proved very resilient, and its relatively positive position during the credit crisis that began in 2007 is down to two key factors. First, as noted, Brazil is supported by expansive demographics, which have driven it to regional domination. (Although it is unlikely ever to ascend to empire, the country could, in time, challenge US influence in the region.) Second, the country's leaders had absorbed the lessons of the earlier economic crises that marked the last decade of the twentieth century and were better able to ride out the storm. In fact, developing nations such as Brazil are now in a position to offer lessons to the developed world on strategies to cope with the enormous challenges they face in rebalancing their debts.

In the new multi-polar world, alliances will become crucial to the economic welfare of states and empires. With the waning of the US's global influence and as Latin America grows stronger, North, Central and South America may seek the development of shared economic prosperity through the formation of a single free-trade zone.

rio2016

ASIA: ASCENSION TO SUPER-EMPIRE

For the first time since the early thirteenth century, when the Mongol Empire began a campaign of conquest extending from the Sea of Japan to the Danube, the world is feeling the burgeoning power and influence of Asia. The region is in the process of becoming the next global super-empire (see Figure 21). At present, it is comparable to the WCSE between 1500 and 1870, with China in the lead, ascending to empire with the same expansive force as sixteenth-century Britain. India, a regional power, trails behind and is in many ways analogous to France during the WCSE's ascendancy.

Comparing the nascent Asian Super-Empire (ASE) to the ascension of the WCSE makes further sense when considering the parallels of the ASE's rapid population expansion. From Kazakhstan to Japan and from Mongolia to New Zealand, Asia contains well over 60 percent of the world's population. It is, therefore, the repository of unparalleled human resources. Some of the most groundbreaking infrastructure in the world is being developed in the East, on an unprecedented scale, all of it powered by what I have called the Second Industrial Revolution.

The WCSE was a complex structure that was driven by competition between its constituent nations. Similar dynamics are in play between China, Japan and India, and their proxy allies. As in Europe, alliances might shift over time. I shall later posit that the likeliest scenario will see Japan and India forge closer ties in an attempt to limit Chinese power. For now, it is worth recalling the sobering fact that never in history has a rising economic power not chosen to grow its military infrastructure.

Japan was the first Asian nation to adopt Western influences and industrialise, a process accelerated under the Meiji Restoration, beginning in 1868. Long the leading economic power in the region, Japan's trajectory resembles that of Portugal and Spain, whose prosperity and power drove European expansion, but whose influence waned as a result of the rise of the longer-enduring British Empire. Japan's dominance is now diminishing, hand in hand with its declining demographics, in the face of China's exponential growth and ascension to empire. India is demographically strong but will struggle for access to commodity resources outside its own domain.

MODERN IMPERIAL TRAPPINGS: SEA, SPACE, NUCLEAR POWER

No Asian nation seriously invested in sea power until the West demonstrated the uses to which a powerful modern navy could be put. Japan, as an island nation, was first to awaken to this. It began to depart from its traditionally insular ways in the years after 1853 when Commodore Matthew Perry's 'Black Ships' steamed into Edo Bay and demonstrated their cannon fire. Forced to begin trading with the West, Japan was quickly brought into the developed world, and soon allied with Britain to build a world-class navy. In contrast, China and India have always maintained predominantly land-based lines of communication. The development of an effective navy is a new undertaking for these countries and inferior Russian technology has been the standard until recently. Today, however, a new Asian naval arms race is under way, led by China.

The importance of outer space has also been recognised by these three countries, which are all committed to space programmes. As with the American and

The ASE (Estimated Duration: 600 Years) *(fig. 21)*

This is China's sixth empire cycle since 1800 BC. The average duration of each cycle has been 600 years from start to finish.

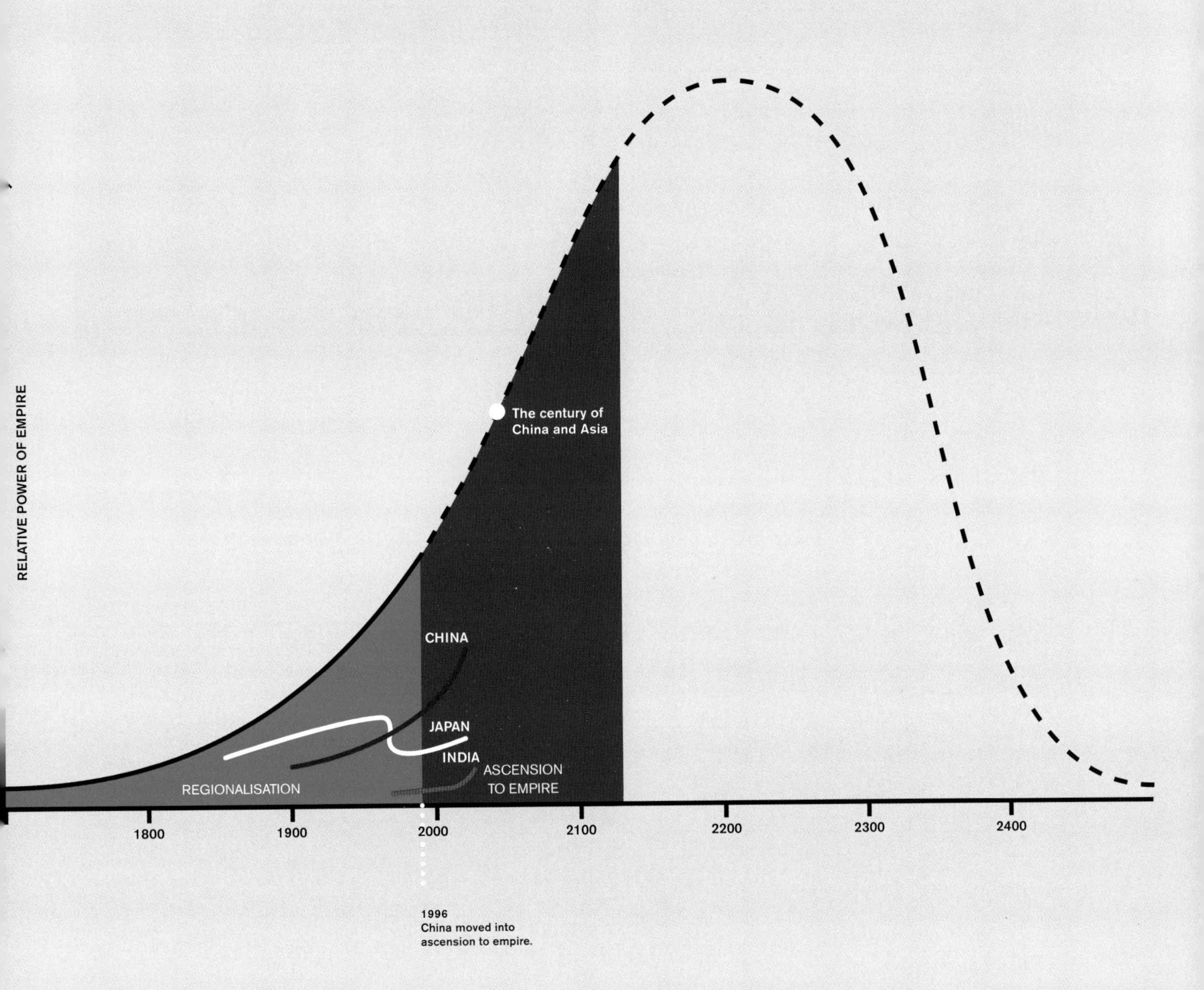

Soviet programmes before them, these appear to be about national prestige, but the underlying motivation for each is to ensure control of 'its' space above the Earth, maintaining secure space-based communications networks and observation platforms as well as missile-launch capacity. (The promise of future interplanetary exploration, which may eventually prove strategically important as the world's resources wane, also beckons.)

There are at present four Asian nuclear powers – China, India, Pakistan and North Korea – although Taiwan, Japan and South Korea could acquire nuclear capability. The region is therefore already highly militarised. Despite the direct precedents of two world wars (a key factor in the restraint that prevented the Cold War from escalating), there is considerable risk that political miscalculation might result in the first deployment of nuclear weapons since Nagasaki. Of particular concern is the potential outbreak of hostilities between India and Pakistan. Although the relationship between these neighbouring states is currently in a relatively benign phase, escalation is a far from fanciful notion and may well be prompted by a resurgence of militant Hindu nationalism in India or radical Islamism in Pakistan.

ABOVE LEFT **The Long March-2D carrier rocket carrying the remote sensing satellite 'Yaogan IV' blasts off from the launch pad at the Jiuquan Satellite Launch Center on 1 December 2008 in northwestern Gansu Province, China.**

ABOVE RIGHT **Chairman of the Indian Space Research Organisation, G. Madhavan Nair, as the country's first lunar mission, the unmanned spacecraft 'Chandrayaan-1' is successfully launched from the Satish Dhawan Space Centre in Sriharikota on the southeastern coast on 22 October 2008.**

THE MONEY QUESTION

The economic systems of India and Japan are highly compatible with those of the West. However, China's more centralised system has allowed it to acquire commodity-based assets at premium prices on long-term contract, which market-sensitive Capitalist competition does not permit. The Chinese version of Capitalism could prove a distinct advantage as the world approaches the limits of commodity supply and where the core issue will be not so much the price but whether access to supplies can be gained in the first place.

BELOW A sign advertising foreclosure tours stands outside a home for sale in Las Vegas, Nevada, US, in August 2009 during the sub-prime mortgage crisis.

Meanwhile, the US is experiencing the limits of its overextension to an unprecedented degree. While in the recent past it generated a series of credit bubbles to ensure that it could continue to grow financially during this stage, the last of these burst in 2007 with the onset of the sub-prime mortgage crisis. The ongoing financial crisis (the effects of which are being felt globally) has threatened the whole of the US banking system. American deficits have been financed for some time by the rest of the world, specifically by Japan and, to some degree, China. Asian banks may capitalise on the relative weakness of American banks and companies – US assets, troubled on account of the crisis, present a buying opportunity for better-positioned Asian banks. However, there remains the problem of the dollar, which in the long term appears very vulnerable to significant declines that would rapidly close the gap between the US and Asia.

CHINA SHIPPING LINE

LEFT **The container terminal at Stonecutters Island in Hong Kong's Victoria Harbour, in February 2010. Hong Kong is among the three busiest ports in the world.**

As ever, the ascending empire holds the purse strings of the declining empire, and at some stage it can be expected to use its power to its own advantage. While Japan will most likely continue to supply the US, China will probably seek to undermine it when a strategic advantage to do so presents itself. (Recall the financial leverage that the US employed over Britain to enforce its prerogative during the Suez Crisis. The lesson for the US should be clear.)

Although Western civilisation has dominated the globe for the past two centuries, Asian creativity, culture, innovation and values will now begin to lead the world – and sooner than commonly imagined. Ten years into the twenty-first century, it is clear that the future belongs unequivocally to Asia.

CHINA: ASCENSION TO EMPIRE

It is tempting to believe that, with a GDP of $4.3 trillion compared with the US's $14.2 trillion, China is a long way behind the WCSE's last empire. However, the gap narrows when purchasing power parity is factored in: the figure for China swells to $8 trillion, while that for the US remains roughly the same at $14.4 trillion. (These figures are for 2008.) Although China still trails at some distance, its economic challenge to the US is clear. Indeed, over the next decade, the future decline of the dollar and expected appreciation of the *yuan* will see the real GDP gap between the two states close quickly.

China commands a massive industrialised power base and is displaying energy in its development of new infrastructure analogous to that of Britain during the heyday of the Victorian era. Some Western perspectives hold that the rise of a new Chinese Empire is unlikely as internal political pressures threaten to tear the country apart. However, systems in expansion display marked resilience as overall growth improves living standards for the majority, effectively neutralising such concerns. Critics in the West would do better to fear for their own declining systems, as contraction cycles display characteristics such as gross intolerance of differences and worsening living conditions that engender scapegoating tendencies.

China, as a regional power with burgeoning demographics, needs to seek resources beyond its borders to ensure future economic growth. In 1990, with the collapse of the USSR, China also recognised that it now inhabited a world dominated by US power. The First Gulf War reinforced this belief, demonstrating to the world just how powerful the US war machine had become – particularly against inferior Soviet weaponry. Thereafter, China embarked on a strategy of 'smile diplomacy', adopting a conciliatory stance to avoid any friction with the US. However, it began moving simultaneously into territories unexploited by US foreign policy, seeking access to resources with a centralised strategy of direct investment in oil and minerals (see Figure 22). I shall discuss its particular focus on Africa in greater detail later.

In 1995, China fulfilled 23 percent of its own oil requirements, as the US had done during the early phase of its own industrial expansion. By 2007 it was importing 58 percent of its oil – a trend that will only accelerate, as Chinese car ownership attains Western levels. This figure corresponds to a 53 percent per-capita growth in oil consumption between 1996 and 2006.

On the higher-priority question of food production, given that China has 20 percent of the world's population, and only 9–10 percent of its potable water and arable land (and just 7 percent of its freshwater runoff), the country has been forced to expand its sphere of influence and seek vital strategic food resources beyond its borders.

China developed its resource acquisition strategy with a twenty-to-thirty-year horizon, and it began implementing it with typical patience, planning and foresight to a degree that is often sadly lacking in the West. The US reached its peak as an empire in 1990 and thereafter became bogged down in a steady stream of ongoing political, economic and military quagmires. The relative economic strengths of the US and China are only likely to converge further with the onset of stagflation in the West. With the US in economic decline in the wake of the current credit crisis, China has likely gained ten to twenty years on its original timeline of ascension to empire.

All told, the result is a bold and confident China that is now seeking to follow its economic success with corresponding military capability, the better to dominate Asia and to be in a position to challenge the US for control of the oceans should the need – or the opportunity – present itself. China's defence spending in 2008 was estimated at $84.9 billion, for the first time ranking as the world's second-highest military spender, still far behind the US ($607 billion) but almost doubling Japan ($46.3 billion) and nearly tripling India ($30 billion). Given a GDP growth rate averaging around 10 percent (2000–2008), and given the potential future appreciation of the *yuan*, China could conceivably narrow the defence-budget gap with the US in no more than a decade.

If such scenarios appear alarmist, consider the case of China and Sudan. Sudan was one of China's first targets for investment. It was an excellent choice, as no one in the West was remotely interested in that country's oil reserves at the time. China had 4,000 'military engineers' stationed in Sudan in 2007. Current deployment figures have not been made public by Chinese officials, but China's presence there remains considerable. It does not require a leap of imagination to foresee such 'investment' replicated by China in other strategic territories.

China will also continue to pursue an official end to its emotionally charged ongoing 'civil war' with Taiwan that began in 1927. The resolution would involve the complete reintegration of Taiwan into Mainland China.

A self-governing democratic state of 23 million people, Taiwan was founded in 1949 by Guomindang Nationalists under the leadership of Chiang Kai-shek. The People's Republic has never recognised Taiwan as anything but a breakaway territory of an otherwise united China. Since 1971, few states around the world – and no major powers – have officially recognised Taiwan, bowing to a 'One China' policy of recognising only the People's Republic (while still doing business with prosperous Taiwan). In 2005, China passed the Anti-Secession Law, granting itself the right to go to war in the event that Taiwan chooses to officially declare complete independence as a sovereign state. This is no idle threat: China's land and maritime forces are orientated around an assault on Taiwan. Indeed, the teaching of the strategic lessons of the British-Argentine Falklands War is a standard part of the curriculum in China's military academies, with careful attention paid to the possibility of amphibious landings.

ABOVE **Ma Ying-jeou, presidential candidate for the opposition Guomindang (left), and Vincent Siew, vice-presidential candidate, wave after winning the presidential election in Taipei, Taiwan, on 22 March 2008. Ma vowed to improve Taiwan's ties with China after eight years of pro-independence rule by Chen Shui-bian.**

As the US's ability to project power globally weakens (particularly in the Taiwan Strait), Taiwan will become more vulnerable to invasion by the People's Republic. Chinese strategy appears to involve building up sufficient forces in the region for a surprise attack when the opportunity presents itself. Its timing will depend on the balance in Chinese politics between party leaders with a longer and more patient timeframe and the military leadership, which is more ambitious and nationalistic. If it does take place, the successful reintegration of Taiwan would mark the ascension of the new Chinese empire.

It is also possible that, as China becomes more powerful, Taiwan may voluntarily come to political agreement with it rather than face reintegration by military intervention. Indeed, with the return of the Guomindang Party to power in 2008 after eight years as the opposition to a sovereignty-orientated administration, Taiwan has adopted a considerably more conciliatory stance towards China, declaring a 'diplomatic truce' that has warmed relations. Meanwhile, cross-strait economic ties are booming. Bilateral trade between China and Taiwan surged from $8 billion in 1991 to $102 billion in 2007, and Taiwan is one of China's most highly ranked trading partners. Moreover, remaining trade restrictions are being eased. In 2009, a hundred Taiwanese industries were opened to investment from the People's Republic, and the two countries are currently working on a cooperation agreement that will liberalise trade even further.

Direct flights between the two countries have increased since 2009 to 270 per week (up from 108), and 3,000 visitors from the mainland are now admitted daily – a tenfold rise in the quota of previous years.

To date, the Chinese navy has mainly focused on the Taiwan Strait, maintaining a growing fleet of destroyers, frigates and diesel submarines. However, it also plans to build aircraft carriers and escorts, and to develop its nuclear submarine fleet. Such naval forces would allow it to project power into the Indian and Pacific Oceans. Indeed, if European history provides any precedent, we cannot rule out Chinese ships operating in the Atlantic at some point in the future.

Commentators on China have occasionally dismissed Chinese military expansion as being geared to defensive purposes only. Yet the Chinese have made aggressive forays into Japan's territorial waters, surfaced a submarine 5 miles from a US carrier and probed the East China Sea, all within the last five years. As recently as March 2009, an unarmed US surveillance ship complained of harassment by five Chinese ships in the South China Sea. Beginning in 2008, China has sent small anti-piracy warship fleets to the Gulf of Aden to escort Chinese vessels. In 2007, it was also criticised for edging towards the militarisation of outer space when it successfully conducted an anti-satellite missile test and destroyed a defunct weather satellite in low orbit that was intended to simulate a US naval satellite. In early 2010, China used a land-based missile interceptor to destroy a missile in outer space, becoming the only country apart from the US to have employed such technology. (Observers have suggested that the timing of the test was deliberate: it came days after the US announced the sale of advanced Patriot missile defence systems to Taiwan.)

As China's industrial power has grown to the point where it encompasses a significant proportion of global manufacturing, its dependence on sea lanes of communication (SLOC) has intensified. SLOC are of critical importance to China for shipping raw materials in and finished goods out, and they have turned the People's Republic from a land power into an aspiring major sea power. To become the latter, China needs to ensure the access of its commercial ports to Gulf oil reserves and to Africa via the Indian Ocean, the narrow Strait of Malacca and the Taiwan Strait. (For example, more than 85 percent of the oil and oil products that China imports reaches the People's Republic via the Strait of Malacca.) In addition, resources transported from Latin America to China would have a wide range of route options available across the Pacific, but would become vulnerable as they entered the Philippine and East China Seas. China therefore needs to secure these regions.

To ensure this, China has developed strategic relationships with Pakistan, Myanmar, Bangladesh and Sri Lanka as its operational zone and influence have expanded. It has built and/or leased ports or naval bases at Gwadar, Sittwe, the Cocos Islands, Chittagong and Hambantota. This attempt by China to establish control over its SLOC has been coined the 'string of pearls'.

The protection of Chinese maritime interests requires a significant naval presence across the Indian and Pacific Oceans, as well as related inland seas. The US has been able to achieve these objectives; it is only a matter of time before China does so too.

ABOVE A sign reads 'Pak-China Friendship Tunnel' on a section of the Karakoram highway near Khunjerab, Pakistan. The country has signed free-trade agreements with Sri Lanka and China since 2005. It plans to sign similar deals with Iran and other countries to provide easier access for Pakistani products to these markets.

As an insurance policy against the vulnerability of its SLOC, China has also been attempting to secure land passage across great distances, thus avoiding the effect of any imposed bottleneck in the Strait of Malacca. (In 2003, Chinese President Hu Jintao suggested that 'certain major powers' could blockade the strait in the event of a conflict.) Working with Pakistan, China has been upgrading the Karakoram Highway, the highest paved road in the world, connecting Kashgar (in China's Xinjiang Uighur Autonomous Region) to Islamabad. In Myanmar, the Chinese have built a road linking the northwest to the Yunnan Province, and they have begun work on parallel oil and gas pipelines stretching 1,100 km from a deep-sea port in the Bay of Bengal in Myanmar to Kunming in Yunnan. (The pipeline taps into the Bay's rich Shwe gas fields, which China has leased for thirty years.)

This strategic expansion is but one expression of China's enormous growth as an economic powerhouse. The Chinese have learned from the West and integrated these lessons into an already sophisticated culture and history.

China in Africa *(fig. 22)*

Resource-rich Africa is the world's most geopolitically strategic location, sandwiched between the spheres of the declining WCSE and the rising ASE. As discussed, China is harvesting the African continent for its natural resources to drive its ascension to empire. This is analogous to the WCSE's use of North and South America during its own rise. The array of Chinese interests in Africa is vast, and it continues to expand across the continent. Recent and notable examples of China's involvement in Africa are listed below and opposite.

Investment in infrastructure and trade, loan guarantees, debt cancelling and the development of commercial ties all figure in China's Africa strategy, which has proved far more vigorous and far-sighted than any Western equivalents.

01/04/2006	Nigeria (1)	$2.7 bn	China National Offshore Oil Corporation (CNOOC) bought a stake in a Nigerian deepwater oilfield for $2.7 billion.
01/06/2006	Zimbabwe (2)	$1.3 bn	China Machine-Building International Corporation signed a $1.3 billion deal. China agreed to build coal mines and thermal power stations to help Zimbabwe with energy shortages. Zimbabwe will provide China with chrome.
01/06/2006	Angola (3)	$1 bn	Sinopec Group bought offshore blocks in Angola for ~$1 billion.
01/01/2007	Ethiopia (4)	$500 m	China pledged $500 million for development projects.
01/10/2007	South Africa (5)	$5.5 bn	Industrial and Commercial Bank of China agreed to invest $5.5 billion in South Africa's Standard Bank.
01/04/2008	DR Congo (6)	$9 bn	CREC signed a $9 billion deal in DRC, agreeing to build $6 billion in infrastructure for 10 million tonnes of copper and 400,000 tonnes of cobalt.
01/05/2008	Zambia (7)	$800 m	China Nonferrous Metal Company pledged an investment of $800 million in the Chambishi Multi-Facility Economic Zone.
01/06/2008	Niger (8)	$5 bn	Chinese oil and gas firm China National Petroleum Corporation (CNPC) signed a $5 billion deal with Niger's government to pump oil from the Agadem block within three years and lay a 2,000 km pipeline to export it. CNPC also said it would build a 20,000 barrels per day oil refinery, Niger's first.
01/08/2008	Ghana (9)	$562 m	China loaned Ghana $562 million for construction of the Bui dam.
01/01/2009	Liberia (10)	$2.6 bn	China Union signed a $2.6 billion contract to develop Liberia's Bong iron ore deposits, estimated at 300 million tonnes of low-grade ore.
01/07/2009	Angola (11)		CNOOC and Sinopec Group agreed to purchase a stake in an oil block offshore from Marathon Oil for $1.3 billion, but Angolan state firm Sonangol later exercised a pre-exemption right and blocked the Chinese purchase.
01/10/2009	Guinea (12)	$7 bn	A Chinese firm agreed to invest over $7 billion in infrastructure in exchange for being a strategic partner in the country's mining projects.
01/01/2010	Tanzania (13)	$180 m	China provided Tanzania with $180 million in loans as part of its $10 billion loan provision to Africa.
01/05/2010	Sudan (14)	$20 bn	Over the course of a decade, China has invested $20 billion in Sudan's infrastructure in exchange for oil.

01/05/2010	Egypt (15)	$796 m	China is to invest $796 million to develop aluminium production in Ismailia.
01/05/2010	Nigeria (1)	$23 bn	Nigeria's state-run oil firm NNPC and China State Construction Engineering Corporation signed a $23 billion deal to build three refineries and a fuel complex.
01/06/2010	Mozambique(17)	$1 bn	Wuhan Iron and Steel has committed to investing $1 billion in a Mozambique coal reserve.
01/07/2010	Kenya (18)	$360 m	To date, China has provided Kenya with around $360 million for various projects.

ABOVE Modern-day Tokyo, with Mount Fuji in the background, clearly demonstrates Japan's significant economic power.

JAPAN: OVEREXTENSION

This tough, industrious and resilient island nation in the vanguard of the ASE merits the distinction of being the first to have challenged the WCSE – twice: ending in victory against Russia and defeat against the US. Having failed in its ascension to empire, by 1945 it reformed as a Western democracy and converted its national energy into becoming an economic power in a second regional cycle that peaked in 1990–93. It then entered overextension. (As noted, Japan resembled Spain and Portugal in its initial ascension; yet its later cycle of failed regional peaks bears comparison with that of France in the eighteenth and nineteenth centuries.)

Japan is a well-ordered country with a long, proud history and a homogeneous and demographically stable population of 125 million people who, despite the decimation of their national economy by the end of World War Two, rebuilt it into an immense economic power. Japan's economy is still the second biggest in the world in nominal GDP terms after that of the US, although it drops to third behind China when purchasing power parity is taken into account.

The modern history of Japan is bewilderingly eventful. As discussed, it industrialised in little more than forty years after the arrival of the Black Ships in 1853. Later it allied with Britain and developed a powerful navy, modernised its army and began to expand its power base. After annexing Taiwan and Korea, defeating Russia and inheriting German territory in China after World War One, it went on to invade Manchuria and other areas along China's east coast, including Shanghai and Nanjing. (The brutal 'Rape of Nanjing', not easily forgotten, currently acts as a focus for Japanese-Chinese polarisation.) However, the Japanese miscalculation in executing a surprise attack on Pearl Harbor to block the US in the Pacific was enormous. Japan did not fully appreciate American industrial and military power, nor its determination and leadership, perhaps because its own leaders were blinded by insularity and a sense of cultural superiority. These are national characteristics that arguably continue into the present day, albeit in a milder form, yet just enough to affect the country's future negatively.

For Japan, defeat in World War Two was total – the first time this proud nation had experienced such a crushing blow. In the process, it returned to a pre-regional state. However, the miraculous Japanese revival that began under the Marshall Plan (a huge credit to the US) allowed it to take its place once again as one of the world's responsible democratic powers. Japanese growth continued, and by 1989, with the Nikkei Index above 40,000, it seemed a bitter irony to ordinary Americans that Japan could soon own a great deal of their country. Yet, just as the world became aware of Japan's new wealth, the country entered its 'lost decade' of deflation, and its economy was devastated. Individuals and financial institutions were ruined, and a national sense of pessimism prevailed. In 2003, then-Prime Minister Junichiro Koizumi laid the foundations for the following decade's growth, and Japan began to feel a rise in confidence and national pride once more. Moreover, it has retained its leading role in modern technology, which, coupled with the fact that Japan does not have an oil lobby, makes the country one of the most likely in the world to develop fusion power and effective fuel-cell technology.

However, Japan's population has been gradually falling since 1993 (see Figure 23). Intense industrialisation, the high cost of living and limited space have all combined to contain population growth. The imperial failure of 1945 precluded the possibility of the Japanese population expanding onto acquired land. Japan's economic performance was driven by twenty years of labour force expansion between the 1950s and 1970s. During this post-war period, the country's economy grew at a rapid rate: its labour force was highly educated and highly productive, and its unions kept wages down.

The country began to stagnate just when it needed a boost to pull it out of the economic slump that kicked in after Japan's peak in 1993, but falling demographics combined with a decade of deflation ushered in overextension. Yet Japan has still managed to provide itself with a measure of protection from the aggressive power of China that might seek to take advantage of its situation. Unlike many an empire in decline whose trading system facilitates borrowing from foreign nations, bringing about a debt crisis, Japan's borrowing, while large, is entirely domestic. In addition, it has invested in the finest modern infrastructure and as such its asset-to-debt ratio must be considered very good – particularly compared with those of Western nations. Thus Japan remains a strong US ally in countering the expansion of its neighbour, China.

Japan's post-war constitution only permits the maintenance of a self-defence force. Any regional issues since 1945 that have called for a more aggressive stance (e.g. the Korea and Vietnam conflicts) have always been undertaken by the US.

Japanese Demographics and Power, 1870–2010 *(fig. 23)*

Red line indicates projected figures.

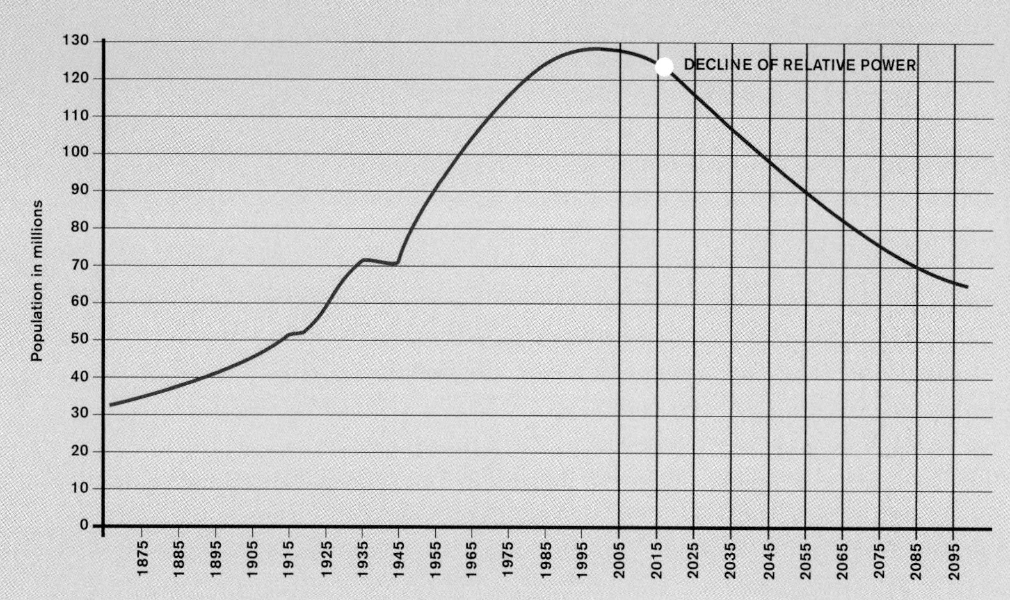

The peak of economic Japanese power came in 1989, when the Nikkei reached 40,000. The combination of the ensuing collapse and the declining demographic drivers in Japan created two economically lost decades.

However, since North Korea became a nuclear state, unafraid to launch missiles over it and coincident with a new wave of nationalism, a strong lobby in the Japanese diet favours broadening defence commitments.

Japan's defence spending currently stands at 1 percent of GDP compared with the US's 4 percent, but, as befits a world leader in technology, it commands very advanced forces, including a navy that, at present, matches the combat power of the People's Liberation Army Navy of China. Japan's increasing wish to shoulder more of its own (and regional) defence suits an overstretched US: Japan is the US's closest global ally with respect to sharing defence technology. (Japan's navy uses the US-developed Aegis Combat System, for example, which is capable of acting effectively as a missile shield to counter any potential North Korean or Chinese aggression.) The time will soon come when Japan once more develops a fully fledged navy that is able to ease some of the pressure on the US fleet in the region. This is likely to be done with the US's blessing as China continues to develop its own blue-water navy.

However, as the Asian arms race gathers momentum, it must be borne in mind that the Japanese collective memory can best be relied upon to fully appreciate the potentially disastrous consequences of military aggression gone wrong. As such, Japan can be expected to be the voice of reason amid the volatile tendencies of the other nations in the region, and it is very unlikely to be the instigator in any potential conflict. Ultimately, though, its need for resources will compel it to protect its maritime trading interests and its sovereign integrity. Also not to be underestimated is the likely positive effect on Japan of direct Chinese competition, which could act as a catalyst to move

Japan into a potentially more stable growth phase – but only if the country can structurally reorganise its moribund institutions.

BELOW Mumbai's Taj Mahal Hotel burns on 27 November 2008. Up to a hundred people were killed and about the same number wounded in coordinated attacks by gunmen armed with powerful assault rifles and grenades in India's commercial capital, with two five-star hotels being among the targets.

INDIA AND PAKISTAN: REGIONALISATION

Without competition from China, India might have had a clear run at a dominant role within the ASE. Instead, this country – so vast that it is referred to as a subcontinent – remains a regional power. The fifteen years by which it lags behind China might as well be fifteen decades, given the energy that China has devoted to cornering the world's already overstretched resources to feed its growth. However, despite the challenges that it faces, India's pursuit of commodities will force it to interact within a larger sphere of influence, giving rise to a new chapter in India's geopolitical relationships.

The Republic of India has moved ahead of its regional competitor, the Islamic Republic of Pakistan, in almost every way. Nuclear tensions between the two countries have, fortunately, been calm in the last few years, although the spectre of these two countries exchanging nuclear strikes will not be easily banished. On a rare positive note, in the aftermath of the gruesome terrorist attacks on Mumbai in 2008, India showed remarkable restraint towards a Pakistan already struggling with the rise of Islamic radicalism and the war in neighbouring Afghanistan threatening its capital and

institutions. The interrogation of the sole surviving terrorist revealed the involvement of *Lashkar-e-Taiba*, a Pakistan-based extremist organisation. This group is also implicated in the 2001 terrorist attack on India's Parliament in New Delhi, which resulted in a tense stand-off between the two countries.

France again is a useful point of comparison when considering the former WCSE and the current ASE. Japan's parallels with the European country are pronounced with regard to both of their sometimes successful, but more frequently unsuccessful, military challenges. With India, the parallels centre on the subcontinent being, like France, a regional and territorially vast power with a pressing need to expand its natural resource base while competing in second place with a neighbouring power of a similar order.

Unfortunately, India's relationships with its neighbours have for the most part not been harmonious, and China is readily able to exploit this weakness by creating strategic distractions on India's borders. (The ever-contentious, mainly Muslim northwestern region of Kashmir, for example, is disputed and administered between India, Pakistan and China.) Indeed, it was China's dark client state, North Korea, that helped Pakistan to develop nuclear warheads. Pakistan's nuclear-technology godfather, A. Q. Khan, has since been at the heart of a wave of proliferation that has encompassed Iran, Libya and Syria.

Pakistan provides a salutary lesson of why nations seek nuclear weapons. The country's 184 million-strong population is dwarfed by its neighbour India's 1.2 billion, and its GDP in 2008 was $431 billion compared with India's $3.3 trillion. Pakistan would have no chance of survival in a conventional war, but since becoming a nuclear state it stands shoulder to shoulder with India.

Yet Pakistan is internally unstable. It is allied with US interests in containing the menace of the *Taliban* in Afghanistan and within its own porous northern provinces, and its homegrown extremist groups. However, the idea of the US acting on its own prerogative inside Pakistan against the *Taliban* or *al-Qa'ida* is deeply offensive to many Pakistanis. Moreover, the ability of the country's leadership to govern following the resignation of former President Pervez Musharraf and the assassination of former Prime Minister and presidential hopeful Benazir Bhutto in 2007 is in question. Bhutto's widower, Asif Ali Zardari, is currently the country's increasingly unpopular president.

In addition, Pakistan's secret service, the notorious Inter-Services Intelligence (ISI), is a wild card in determining the country's future. It is often described as operating independently of the army, president or prime minister, and thus being unaccountable. It is also widely suspected of having ties with Islamist elements, including the *Taliban* and *al-Qa'ida*. Many people, both within Pakistan and outside it, also suspect ISI involvement in Bhutto's assassination.

The longer the US and its allies take to stabilise Afghanistan, the greater the polarising effect on Pakistan and the more vulnerable it will be to control by Islamic militants. The consequences of such a political shift in Pakistan would result in the alienation of the US. It would also be threatening for India, raising tensions in the region and therefore also the likelihood of a full-scale nuclear war. Afghanistan, therefore, is of

RIGHT Former Pakistan Prime Minister Benazir Bhutto waves to thousands of supporters at a campaign rally minutes before being assassinated in a bomb attack on 27 December 2007 in Rawalpindi, Pakistan. The opposition leader died from wounds to the neck and head after speaking at an election rally in the northern city.

BELOW RIGHT Chinese workers decorate the Chinese side of the Nathu La Pass on the border between India's Sikkim State and China's Tibet Region in July 2006, as the historic Silk Route reopens following the formal resumption of trade between the two countries.

热烈祝贺中印乃堆拉山口边贸通道开通
CONGRATULATIONS ON OPENING OF CHINA-INDIA BORDER TRADE THROUGH THE NATHULA PASS

major significance, and Chinese strategies in this area must remain under scrutiny.

Extremism, as practised by some adherents of Hinduism, is also present in India. It underlay the destruction of the fifteenth-century Ayodhya Mosque in 1992 by supporters of the Bharatiya Janata Party (BJP), for example. The BJP went on to win two general elections, but its appeal has since largely waned, and relationships between Hindus and Muslims are arguably better now than they have been in recent memory. Nevertheless, a blanket rise in Islamic nationalism in Pakistan could stir unrest again.

India's Sikh prime minister, Manmohan Singh, has done much to improve relations with Pakistan, and Sino-Indian relations under his tenure are also very positive. Singh and Hu Jintao have exchanged visits to each other's countries, and in 2006 the Nathula Pass connecting India's Sikkim state to the Tibet Autonomous Region of China was reopened, having been sealed since the 1962 war between China and India. This renewed link will increase trade between the two countries. China is already India's biggest trade partner, and Asia's two largest states are expected to attain a total trade of $60 billion. China and India even held joint naval exercises in 2007 and 2008, with a third being in the planning stages.

At the same time, India has recognised that the strategic sea lanes in the Indian Ocean place it in a unique position with regard to the flow of Chinese commerce. India currently operates the decommissioned British jump-jet carrier HMS *Hermes*, renamed the INS *Viraat*, which turned fifty in 2009. Recent refurbishment will permit the carrier to remain fit for purpose until 2015. India is also expecting delivery of a much-delayed Russian aircraft carrier, the refitted *Admiral Gorshkov*, which will be renamed the INS *Vikramaditya* and deployed in 2012. (The accompanying fighters have already arrived.) In 2009, India expressed strong interest in purchasing one of two carriers being built for the British navy, which the UK government was considering selling off as part of planned spending cuts. Furthermore, the Indian navy has declared its long-term ambitions to develop a blue-water force complete with six aircraft carriers. In 2005, India announced the construction of its own aircraft carrier, which is expected to be completed in 2012. At the time, a senior US official spoke anonymously to the press, revealing the US's intention 'to help India become a major world power in the twenty-first century'. They added, 'We understand fully the implications, including military implications, of that statement.'

India is also remaining vigilant in the Indian Ocean, wary of China's string of pearls strategy, and the establishment of bases on its periphery and the advantage that these confer on other states. For example, one of the 'pearls' – the Gwadar base – favours Pakistan, augmenting that country's coastal power and relieving the vulnerability of its main port at Karachi to the kind of blockade that occurred during its 1971 war with India.

India has modernised a former British naval base in the Maldives with surveillance craft and equipment for monitoring Chinese sea traffic. It is also partnered with Singapore, Malaysia and Indonesia to patrol the Strait of Malacca, and regularly conducts joint exercises with such countries as the US, Britain and France.

India's economy will continue to grow, and as it does we can expect to see the traditional caste system give way to a Westernised meritocracy. The scale of the country's human resource pool is enormous – it is not uncommon, for example, for employers such as the national railway company to receive replies in the tens of thousands for every opening advertised in newspapers – but the challenge for India will be in acquiring the natural resources vital for the acceleration of its industrialisation.

Declining Powers: The US, Russia and Europe

As has been discussed, the WCSE is in decline and legacy. I shall look here at the current position of the US, Russia and Europe. These are the main geopolitical forces around which are arrayed a number of satellites, such as Canada with regard to the US, and Eastern European and Central Asian states with regard to Russia and the EU. (In terms of the rising ASE, South and Southeast Asian states such as Sri Lanka, Singapore and Malaysia are satellites of India and China, respectively.) Australia and New Zealand, once linked unambiguously to the WCSE, are now increasingly oriented around the ASE.

THE US: DECLINE

The US entered decline in 2001. This may have been accelerated by the financial crisis that began in 2007, centred on the credit cycle and banking system. The country's twin deficits – its expansion of the money supply and its burdensome defence spending to maintain ongoing conflicts and developing threats – are clear hallmarks of decline. Furthermore, as there is no empire in ascension within the WCSE, the US is declining without any successor. It is, in effect, the end of the 1,000-year domination of the WCSE. That being the case, both a power vacuum and the beginning of a new world order have been initiated. The position of the US within the WCSE, in fact, bears comparison with the Byzantine Empire's relationship with the Roman Empire. Both the US and the Eastern Empire were offshoots that would have appeared powerful to their neighbours, but paled in relation to their parent empires at their height. As the US plays such a key role in today's geopolitics, it is worth tracing its power trajectory.

REGIONALISATION, 1775–1898

Warfare and religion marked the earliest days of the British colonies that would become the US; its Puritan founders sought to worship God in peace and freedom, but life on the new continent hardened them considerably. Relations between the colonists and various indigenous tribes in the seventeenth century began peacefully, but the settlers would side with certain tribes in their wars against others. The French and Indian War of 1754–63 (fought in alliance with Native Americans on both sides) was followed by the American Revolutionary War of 1775–83, the Mexican-American War of 1846–48, the Civil War of 1861–65 and others against resisting Native Americans into the early twentieth century.

These conflicts were part of the regionalisation stage of US history. In effect, the US was forged in the crucible of war, although throughout this tumult the power of the democratic ideal inherited from the US's British roots held firm.

The bloody Civil War would ultimately crystallise US industrial development and political values. In line with the Five Stages of Empire model, this was a war of

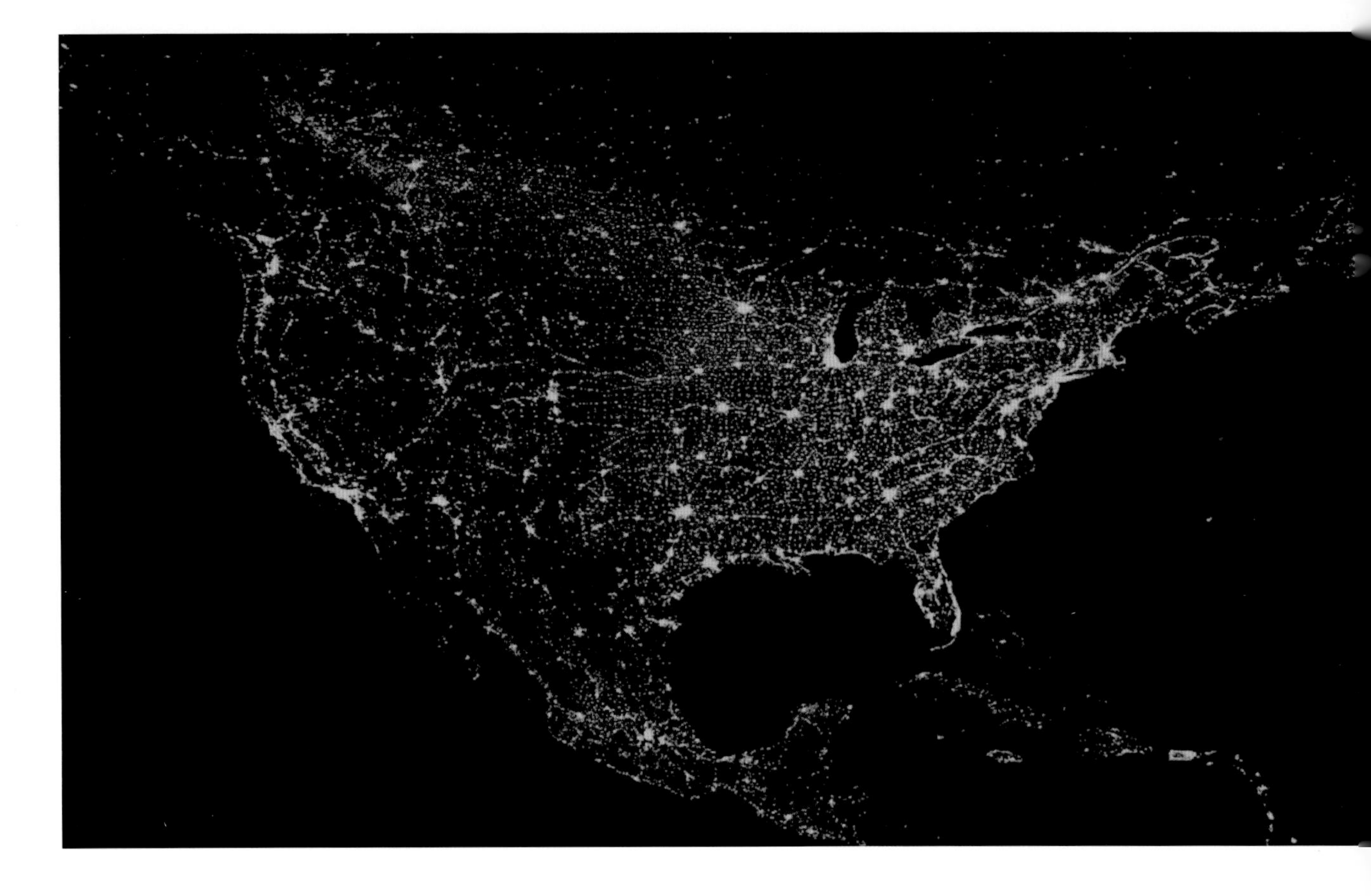

ABOVE The US at night viewed from a NASA satellite, providing clear evidence of the country's prodigious energy consumption.

regionalisation, the final stage in the consolidation of the US as a power and the beginning of its ascension to empire, The key to the conflict lay in the demographics and economics of each side. The twenty-three states of the Union had a population of 21 million, including just over 429,000 slaves in the 'border states' of Delaware, Kentucky, Maryland and Missouri, where slavery was permitted. The eleven seceding Confederate States of America had a total population of 9 million, of which 3.5 million were slaves.

The North was industrialised to a far greater extent than the agrarian South and hence was able both to sustain a long war and to pursue further economic development without the need of slavery. To sustain the war effort, the succeeding states would need to maintain their overseas trade, exporting tobacco, cotton and sugar, and importing arms – a difficult proposition in the face of the North's naval blockade. President Abraham Lincoln's masterstroke, the Emancipation Proclamation of 1862, declared the freedom of all slaves in Union-occupied Confederate territory. Britain, an anti-slavery power for some three decades before the US, declared its support for the North (as did France). In 1865, slavery would be abolished in the now-reunified US.

ASCENSION TO EMPIRE, 1898–1945

Post-war recovery was rapid, driven by the northern states' war-energised industrial power. The US was further unified by its railroad infrastructure and booming economic growth. Food could be transported quickly from field to consumer, and raw materials and finished products conveyed more efficiently. Costs were lowered and production areas linked to the country's markets.

By 1865, the US's 'military–industrial complex' (a term coined nearly 100 years later as a warning to the nation by President Dwight D. Eisenhower in his final address) had come into being. Also beginning in the nineteenth century, enormous influxes of European immigration expanded the US's demographics. The aspirations of these new Americans injected fresh energy into the country, enabling the absorption of the West through settlement.

ABOVE Civil War 'contrabands' (fugitive slaves) who were emancipated on reaching the North, sitting outside a house in Freedman's Village in Arlington, Virginia, in the mid-1860s. Up to 1,100 former slaves at a time were housed in the government-established village in the thirty years during which it served as a temporary shelter for both runaway and liberated slaves.

Colonial expansion continued in 1898 with the Spanish-American War: the beginning of the US's ascension to empire. Newly acquired territory from that conflict included Panama, and the great canal that the US would see built there compensated for the lack of coaling stations required to round the Cape of Good Hope. With Panama, the US now had an easy means of moving warships between the Atlantic and Pacific Oceans. it also moved into Cuba, as well as Hawaii, Midway and Wake Islands, Guam and Samoa, establishing coaling stations *en route* to the Philippines (ceded by Spain during the war). This strategic chain of coaling stations – the US's own string of pearls – made the Pacific an American ocean, and allowed it direct access to Japan and China. In a substantial recognition of the new American Empire, Britain accepted that Latin America came under the US's sphere of influence.

US infrastructure and industrial power developed at a dizzying pace between 1825 and 1910. A population explosion, which was largely driven by immigration from Europe, made possible by a plentiful food supply and increasing mechanisation, turned a regional power into a superpower. The world's leading industrial nation had abundant natural resources, an educated population, political stability and foreign investment.

We have already discussed how the US profited from World War One, as it came to the aid of the older WCSE empires against Germany. The depression of the 1930s was a difficult time for the Americans, although, in retrospect, it can be viewed as a period of consolidation following major expansion. However, it was the US's involvement in World War Two that completed its ascension to empire.

BELOW A tug pulls a ship through the newly constructed Panama Canal. Begun in 1904 and completed in 1914, the 77 km conduit for shipping joins the Atlantic and Pacific Oceans. Annual traffic has risen from around 1,000 ships in the early years of its use to nearly 15,000 in 2008.

MATURITY, 1945–90

From the end of World War Two (the second civil war of the WCSE), the US assumed the dominant role in the WCSE formerly occupied by Britain, and it subsequently battled the USSR in the Cold War (1945–91) for control.

A key factor in US global supremacy after 1945 was the rise in power of the dollar. The world's first regulated international monetary policy, pegged to the gold standard, was introduced at the 1944 Bretton Woods conference in the US and became operational in 1945. Ultimately, however, the system proved untenable; gold was now a floating asset, and the US dollar became the global standard for trade and commerce. The US thus assumed a place at the heart of the global economy for the following decades. This role, along with its ever-increasing military power, defined the American Empire.

During the Cold War era, notwithstanding the risk of a nuclear exchange between the two superpowers, the US and the USSR competed in a series of proxy wars on their peripheries. This monumental struggle began to exert an economic toll on both empires. The USSR was already well on its way to economic ruin, but the US's current-account deficit began to mount from 1984 as well. By the time the Soviet Union imploded in 1991, the US had already crossed the threshold from maturity into overextension.

BELOW **Customers in an unidentified store in California gather in the electronics department to watch US President John F. Kennedy as he delivers a televised address to the nation on the subject of the Cuban Missile Crisis on 22 October 1962.**

OVEREXTENSION, 1990–2001

Ironically, just as the American Empire should have been entering a golden age following the demise of the only other superpower on Earth able to compete with it, its ability to sustain itself began to wane. Individual presidents and their administrations are credited or criticised for determining the direction of US power, and, of course, their influence is inescapable. However, understood within the context of the Five Stages of Empire model, these figures were elected according to the demands of the empire cycle; individual presidents, who loom so large in the American discourse on domestic politics, are really mere actors facilitating a given stage of the cycle.

RIGHT **United Airlines Flight 175 from Boston crashes into the South Tower of the World Trade Center and explodes at 9.03am on 11 September 2001 in New York City. The crash of two planes hijacked by terrorists loyal to *al-Qa'ida* leader Osama bin Laden, and the subsequent collapse of the twin towers, resulted in the deaths of some 2,800 people.**

Unlike many empires at the peak of their power curves, the US maintained a military complex of extraordinary size, taking little advantage of the peace dividend that resulted from the end of the Cold War. Hence it amassed an enormous defence budget that failed to take into account future liabilities. Its policies were short-sighted, failing to weigh present spending against mounting future debts. Expenditure on defence was to play a large part in the decline of the US.

Simultaneously, the historical heart of US power – the country's manufacturing base – began to erode as globalisation enabled companies to take advantage of the far cheaper production costs available in the developing world. To compensate for this investment drainage, the US financial system was permitted to mutate into an entity that could carry the debt burden resulting from the country's overspending. This logic set ticking a time-bomb that would explode in the next decade. Indeed, by 2007, the ratio of financial-sector profits to non-financial sector was a staggering 50 percent as the country shifted from a manufacturing powerhouse to a service-based economy – just as Britain had done during its own decline.

DECLINE, 2001–PRESENT

The tipping point for the decline of the US may be dated to 11 September 2001, when nineteen terrorists attacked both the World Trade Center in New York and the Pentagon near Washington, DC. The perception of the American Empire as invulnerable consequently came crashing down, along with the twin towers of the World Trade Center. (The failure to prevent these events, now commonly known as '9/11' – which, it has since been determined, was possible – has been widely attributed to a failure of US intelligence. Competition between the country's various agencies prevented the sharing of critical information, enabling the terrorist plot to proceed without interference. Such intra-government competition and antagonism can be viewed as lesser manifestations of the same dynamic that presages civil war.)

The attacks played into the hands of American neo-Conservatives, who espoused a policy of projection of power overseas to create a world order in accord with American values. They had come to power with George W. Bush, whose administration would be defined by neo-Conservative politics as well as by a Republican Party base with strong religious overtones. Rhetoric from the Bush White House focused on removing despotic regimes hostile to liberal democracy, followed by non-approved nations in the early stages

of nuclear capability. The simplistic objective of the administration was to influence, inform and reshape the rest of the world. The events of 9/11 provided it with a cause – the so-called 'war on terrorism' – and a focus – the invasions in 2001 and 2003, respectively, of Afghanistan and Iraq. (In fact, neo-Conservative plans to force regime change in Iraq had been drawn up long before.)

LEFT **An Iraqi woman gestures as she walks past a crouching US army soldier during a patrol in the al-Dora neighbourhood in Baghdad in July 2008. The cultural and political chasm between the 'liberators' and the local population has never been bridged.**

The administration was confident that the US's overwhelming military power would see the mission succeed, but it had not appreciated that not every 'liberated' population would readily abandon their cultural values for those of the US. Such can be the blind arrogance of an empire in decline. Following World War Two, the great general George C. Marshall had understood that the US needed to invest in the defeated populations of Japan and Germany, and spur their reconstruction as quickly and decisively as possible in order to win hearts and minds. However, the lesson was lost on the Bush administration. Famously, Bush's Secretary of Defense, Donald Rumsfeld would declare, 'The United States does not do nation building.'

Since 2001, the US has been mired in the kind of wars in which its military is least optimised to fight. With the high value that the country places on human life, every lost American soldier batters the long-term commitment of the nation to finishing the job it started. (As of early 2010, the war in Afghanistan had seen over 940 US casualties, and the Iraq conflict over 4,350.) The US is not waging high-tech warfare in the Middle East, but counter-insurgency operations that require boots on the ground – precisely what Rumsfeld's vision did not encompass. These two wars may well come to have the same repercussions for the US as the Boer War did for the British Empire, as a long, drawn-out counter-insurgency signals to a competitor (Germany then, China now) that the world's most powerful empire has become weak enough to be superseded.

With the invasions of Afghanistan in 2001 and Iraq in 2003, the US moved to a war footing. The Federal Reserve was then given the power to maintain and stabilise prices to ensure the US's continued economic dominance. In this, it was instrumental in creating and inflating the credit bubble that was designed to counter the deflationary effects of stock-market decline between 2001 and 2003. For four years, the Federal Reserve led the charge in pumping liquidity into the system to maintain the illusion that this overextended empire was as strong as ever. In June 2007, the bubble burst with the 'sub-prime mortgage crisis' – the beginning of the ongoing global financial crisis. A massive wave of mortgage foreclosures and delinquencies revealed the folly of the whole process of under-regulated, artificial credit. The American banking system was rendered practically insolvent, along with the banking systems of many European countries, necessitating government bailouts. Compounding the crisis was the continued intensification of the new commodity cycle that had begun in 2001, fuelled by Asia's resource hunger. As the world's most advanced commodity-dependent consumer society, the American Empire will continue to suffer economic aftershocks as the planet reaches the limit of its economically accessible resource base. (See Figure 24 for a summary of the US's evolving debt burden.)

The US Debt Burden as a Function of Its Empire Cycle. *(fig. 24)*

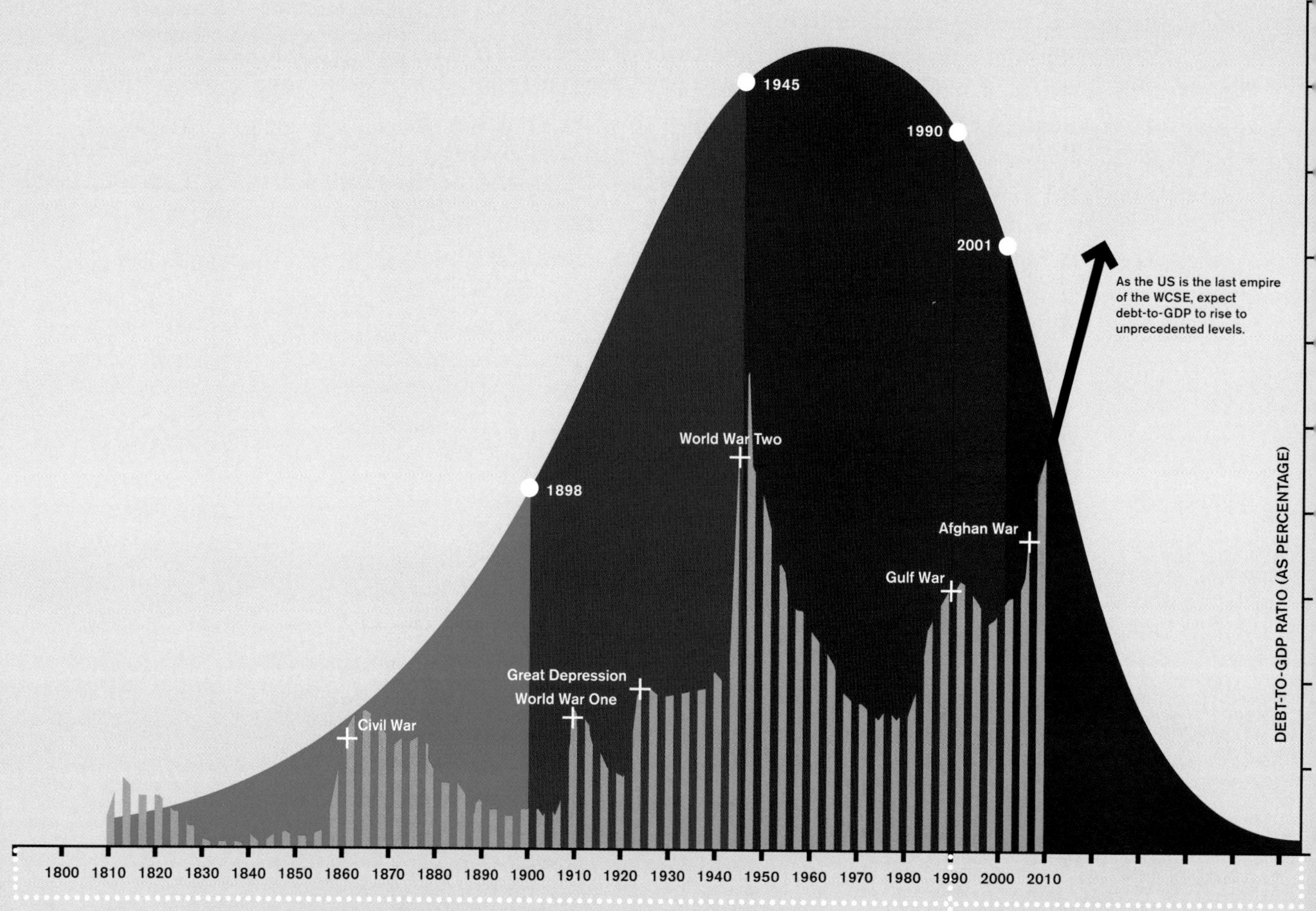

Regionalisation

Debt was accumulated during the wars of independence that established the nation, and then again during the regional civil war.

Ascension to empire

Government debt ratio increases as regional power develops into empire.

Maturity

(*Pax Americana*)

Massive wealth accrued and debt repaid as empire acquires monopoly on trade.

Decline and legacy

As the American Empire faces decline, it has no other cousin in the WCSE to whom it can pass its power, and facing the challenge from the ASE its debt-to-GDP levels will rise steeply, until a sudden catastrophic collapse when the reality of the loss of power hits home.

Overextension

Ever-increasing amounts of funding required to meet challenges from rising powers.

Projected US Population Growth (Immigration/Fertility Constant) *(fig. 25)*

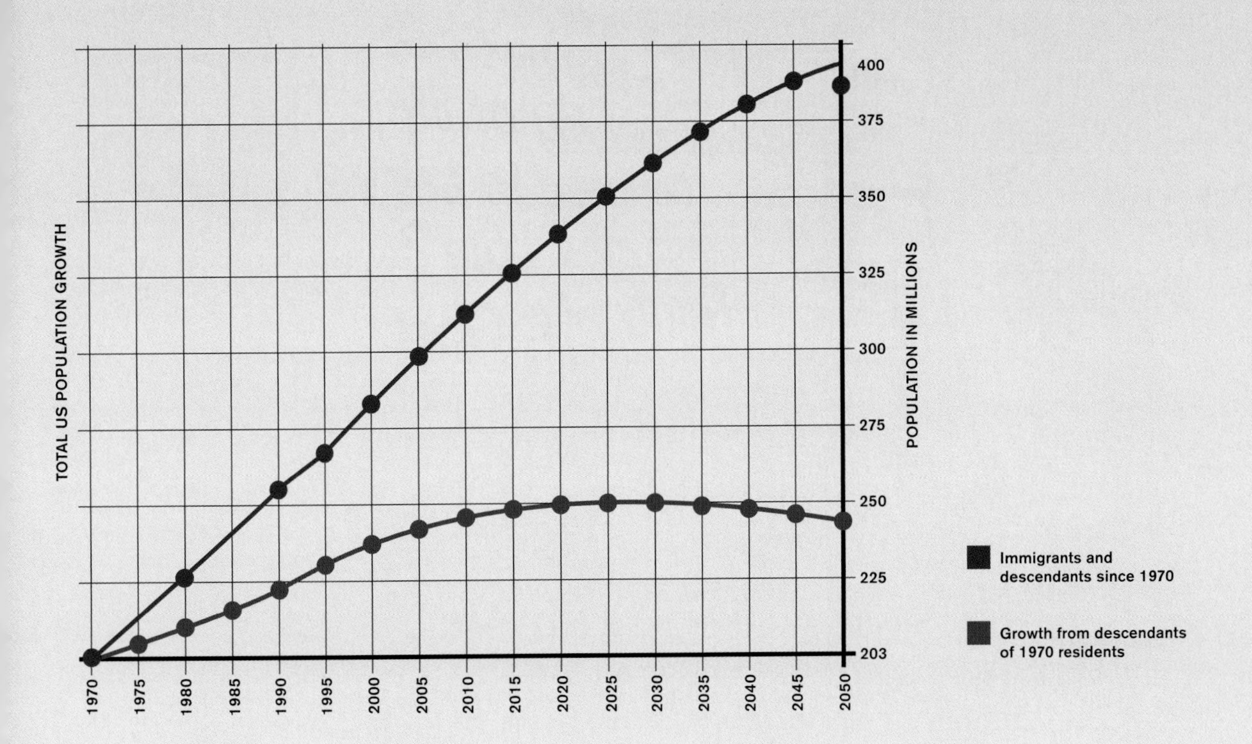

Although the US is not experiencing negative population growth, it is ranked only 131 out of 230 countries by the UN (2005–10). Although its population growth rate of almost 1 percent is comparatively high for a developed country (see Figure 25), its demographic constitution has changed rapidly. The 50 million Americans of Hispanic descent constitute just over 15 percent of the US population, behind Americans of European descent and ahead of African and Asian Americans. The US Census Bureau reported in 2008 that the non-Hispanic white population is expected to cease being the majority ethnicity by as early as 2042, when it is likely to comprise 46 percent of the population (down from 65 percent today). The Hispanic population, meanwhile, is expected to double, to 30 percent, with small rises in the African and Asian populations. Thus a great social change is under way, with the original power base of the US – the so-called White Anglo-Saxon Protestants – becoming ever more diluted (see Figure 25).

The values bequeathed by this class – the perception of what it means to be American – have been integrated into the present-day collective value system. However, this perception will be determined in the future by the increasingly Hispanic population, and their values will in turn become those of the US's future generations.

Demographics aside, how will the American Empire respond to the pressures facing it from so many other quarters – economic crisis, military failure, loss of prestige and confidence abroad, and challenges (both actual and potential) from the East?

ABOVE A visitor checks out a display at the Museum of Creation and Earth History in August 2005 in Santee, California. The museum contains exhibits that depict the story of Creationism and refute the theory of evolution.

Most likely, it will return to the core values that informed it at the end of its regionalisation phase. These included strong religious beliefs. Deep-seated Christian beliefs could spur a rise in Christian fundamentalism, which is already a burgeoning component of the American religious scene. (Christian fundamentalist voting blocs play an important role in US politics, exerting an influence that has extended into foreign policy, not least under the George W. Bush administration).

The former American core value of alliance building will also regain importance as a priority for US policymakers, who will invoke it in order to contain China's power and compensate for the waning influence historically enjoyed in various regions of the world. Moreover, the US, considerably more aggressive than its European counterparts, is and will remain far more prepared to go to war to defend its interests.

The Obama administration has made strides towards rebuilding the US's image abroad and forming alliances. It has also made some headway in terms of domestic policy. Yet its full potential to imprint itself on this new phase of US history remains to be seen. It has taken small steps, here succeeding and here failing, but has not made enough bold, decisive moves that might announce a new era (other than the very fact of Obama's success as a presidential candidate, of course).

However, the magnitude of the problems inherited by the current administration, and of the American Empire as the last of the WCSE powers to enter decline, is such that merely holding the nation's ground may be considered accomplishment enough. Soaring national debt will only worsen matters. The first debt peak was owned by the American public, while the current mountain of debt is owned by foreign powers – and particularly by China, which will inevitably seek to use that leverage to subvert US power when it feels that the time is right.

Arguably, the US's greatest problem will be adapting to a lesser status in the world, and adjusting its diplomatic approach accordingly. One of its chief weaknesses has been a poor understanding of other regions and cultures. This foreshortened perspective will persist and hamper US policies.

Since 2001, the US has focused on the threat from militant Islamism, but it has failed to anticipate – and to negate through pre-emptive political action – the rising threats posed by its old foe Russia. It has also failed to appreciate the full implications of Asia's rise, which has the power to irrevocably alter the Western way of life for centuries to come. In a mere decade, the world has transformed itself. It is a new, complex and often treacherous place, replete with a multitude of simultaneous threats, through which a financially weakened and militarily challenged US must wade.

RUSSIA: LEGACY

The vast expanse of Russian territory stretches from Europe to East Asia. The country shares land borders with, among others, Norway, Kazakhstan and China, as well as maritime borders with Japan and the US. Its connection with the WCSE dates back to the Byzantine Empire. Indeed, Russia can be conceptually grouped within the WCSE, and the case for its inclusion is made stronger by the fact that it is in legacy and reformation, just like its European counterparts. Yet, it is incorrect to refer to the country as 'Western', for its geography, culture and history have bequeathed it considerable Eastern influence.

Russia's most recent empire cycle (see Figure 26) can be plotted from its break from the WCSE in 1917 as a result of the Russian Revolution, its emergence from World War Two as a superpower, its peak of empire during the Cold War, its overextension coincident with its misguided invasion of Afghanistan in 1980, the assumption of power by Mikhail Gorbachev in 1985 and, finally, the waning of Soviet national energy until the collapse of the USSR in 1991.

REGIONALISATION 1700–1939

The Russian Empire had begun as the thirteenth-century Grand Duchy of Moscow, evolving into the sixteenth-century Tsardom of Russia and peaking as a key regional power in the eighteenth century. Its decline in the late nineteenth and early twentieth century completed one cycle of empire and marked the commencement of a new one.

By 1914, although controlling a great landmass and being ruled by an emperor, Russia was, in reality, only a regional power, unable to project influence far from its massive borders. Although it had attempted colonial expansion in line with other nations of the WCSE, being predominantly pre-industrial, it could not compete with its

The Five Stages of the Soviet Empire *(fig. 26)*

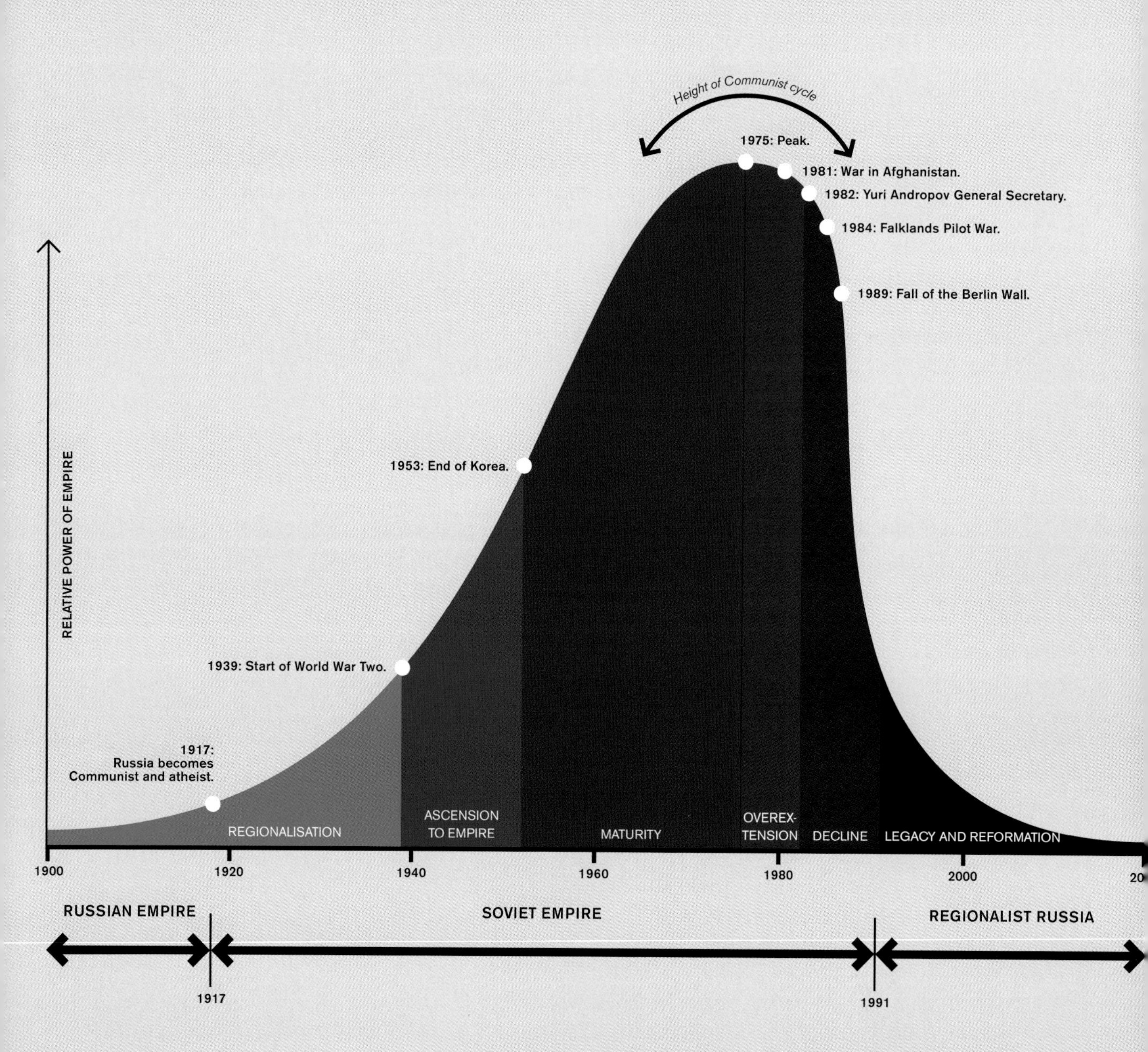

European cousins. In compensation, it had enormous manpower reserves used to staff its army. Indeed, since the nineteenth century, the country's military strength has been disproportionately robust compared with the state of its economy.

When the 'civil war' for control of the WCSE erupted in 1914, the Russian Empire sided with the old powers against the German upstart. However, World War One was also the final catalyst for Russia's own civil war of regionalisation – the revolution that overturned the old order. The Communist, atheist power that emerged set the new USSR ideologically apart from the WCSE and in a second phase of regionalisation. On a larger scale, this was the start of the fracture of the WCSE that would result in the power struggle of the Cold War.

ASCENSION TO EMPIRE, 1939–45

By the 1930s, under Stalin, the USSR was set to expand. It had seeded Communism into China and had attempted to do the same in Germany in the chaos following its great depression, but was defeated by a competing atheist ideology – Fascism. This battle for control of the atheist populace spread by proxy into the Spanish Civil War, and by 1939 it was unleashed on the world. The struggle between these atheist powers was titanic, ruthless and without moral limitations. In the process, Russia created the largest and most effective army ever known, which enabled it to crush Germany and establish the new Eastern Bloc Empire.

MATURITY, 1945–75

The USSR, which emerged in 1945 as a new imperial powerhouse, continued in maturity during the first half of the Cold War, peaking in around 1975.

OVEREXTENSION, 1975–82

By the late 1970s, the USSR became overextended as its military expenditure became greater than its revenues and the commodity cycle went into a twenty-six-year decline. However, this subtle process was not immediately obvious to either the leaders of the USSR or the rest of the world.

DECLINE AND COLLAPSE, 1982–91

When former KGB head Yuri Andropov came to power as General Secretary of the Communist Party in 1982, he recognised that the country was in need of drastic restructuring. This necessitated domestic and international openness, as well as liberalisation of the state's powerful security apparatus and failing economy. Andropov died shortly after assuming power, as did his successor, Konstantin Chernenko. It was left to Mikhail Gorbachev to pursue Andropov's vision upon becoming General Secretary in 1985. *Glasnost* ('openness') and *Perestroika* ('restructuring') became world-famous terms as Gorbachev strove to preserve the integrity of the party and the USSR by opening up to the West, in an attempt to salvage the economy and reform the state. The fall of the Berlin Wall demonstrated that these philosophies were not compatible with Communist ideology and the USSR dramatically fractured.

Critical to the peaceful decline of the USSR was the successful outcome of the pilot war in the Falklands, in which a key member of NATO demonstrated a determination and resilience that forced the USSR to re-evaluate its assessment that capitalism was weak militarily. In effect, it acted as a deterrent to the USSR using military force in the face of economic collapse. In any event, this would have negated the strategy of openness. *Glasnost* had outpaced the ability of the Communist Party to survive it, and the rationale for defending the Eastern Bloc and protecting the Soviet economy from the West had disappeared. When the full scale of the economic catastrophe facing the USSR became evident to the leadership, the Soviet state could not be restored quickly enough, nor could it find new legitimacy. It ceased to exist on Christmas Day 1991.

RIGHT St Basil's Cathedral, Red Square, Moscow. This dazzling building, which was constructed on the order of Ivan IV to commemorate the capture of Kazan and Astrakhan, marks the geometric centre of the city.

The collapse of the Soviet Union stunned the world, but in terms of the Five Stages of Empire model it was to be expected. The USSR was clearly overstretched following the peak commodity cycle in the 1970s, and it was financially, politically and militarily weakened by the 1980s. Its futile attempt to match US military spending drained Soviet resources as the commodity cycle declined and the miscalculated war in Afghanistan depleted the country's treasury, as well as damaging its collective psyche (more than 4,000 Soviet soldiers died between the time troops entered Afghanistan in 1979 and withdrew in 1989).

LEGACY AND REFORMATION, 1991–PRESENT

Following the collapse of the Berlin Wall, the appeal of Communism evaporated, and Russia attempted to adopt the Capitalist doctrine that it had once so vehemently opposed. This embrace of Capitalism in the post-Soviet era mainly profited a new class drawn from the ranks of the former KGB. They were well trained and well placed to secure state assets for themselves amid the lawlessness and wave of wholesale privatisation that succeeded Communism. Organised crime thus escalated and the oligarchs exploited the absence of a regulated system. Ordinary Russians became financially insecure and impoverished as the looting and the acquisition of property and major industrial concerns formerly controlled by the state reached fever pitch. Western interests also hurried to take advantage of the chaos, seizing the opportunity to conduct business on terms weighted in their favour. The US believed that it was witnessing the end of Russia as a contender for any kind of geopolitical power, and envisioned democracy taking root in a country that had never experienced it. Yet Russia failed to gain economic momentum as the commodity cycle entered the century's final decade. National pride was at an all-time low, and pessimism and resentment were high.

The new commodity cycle at the start of the twenty-first century breathed new economic life into Russia. Simultaneously, Vladimir Putin's rise to power as president (2000–2008) and subsequently as prime minister provided strong leadership as the country sought to reform itself.

Putin exemplifies the political hero who surfaces during a low point in a cycle, and, indeed, he has restored confidence in Russia despite the increasing authoritarianism

ABOVE Shoppers queue up to buy alcohol in Moscow. Alcoholism is endemic in post-Communist Russia. In June 2009, the Public Chamber of Russia reported that Russians each consume about 18 litres of spirits a year – more than double the 8 litres that World Health Organisation experts consider to be dangerous.

of his rule. (It is commonly believed that Putin retains ultimate control despite occupying the second-most powerful position in government, and that his assumption of the premiership was simply a means of circumventing constitutionally mandated term limits.) Having existed in a limited and experimental form for barely a decade, democracy is a façade in post-Communist Russia. As demonstrated by the constituent empires of the WCSE, it takes centuries rather than decades to implement and refine a democratic system of government.

Oil prices began a sharp upward trend as Putin took power and Russia, rich in oil and gas, received enormous revenues from its reserves. These have been largely channelled into the pockets of the leadership, but the country has also benefited. Wealthy Russia now seeks to regain a measure of its former glory, and Putin – one of the most effective world leaders of the present day – will seek to take advantage of the weaknesses in the foreign policy of the overstretched US.

Although it is likely that the Russian armed forces will receive a boost in funding and morale, their military technology lags far behind that of the US. It will be many years before the Russian army will be able to threaten the US directly. More worrying is its immense stockpile of nuclear warheads, which developments in US missile-shield technology might negate, thereby removing the necessary deterrent of mutual assured destruction (MAD) as the ultimate arbiter of warfare. To counter this threat, for now at least, Russia will most likely seek to deploy its armed forces on the periphery of the American sphere of influence.

Before the military can be extended, the economy must continue to prosper. However, Russia has relied on its oil reserves too much, dampening future prospects through resource nationalism and failing to invest sufficiently in its future. Moreover, as discussed, a nation cannot hope to ascend to empire without the necessary demographic vigour. Russia's vast commodity resources have endowed the country with a great deal of wealth in a short time, but commodity wealth alone will not suffice. Post-Soviet declining demographics place it at a significant disadvantage with respect to its aspirations to resurgence to empire.

Russia's demographic crisis is without parallel in the modern world. Its population has been shrinking on average at a rate of 0.2 percent per year since 1990, and increased mortality rates have accompanied decreased birth rates. The crumbling of the Soviet health system as well as post-Soviet social stress, unequal wealth distribution and impoverishment have taken a serious toll, exacerbated by excessive alcohol consumption (including, it has often been found, of alcohol-based liquids not meant for ingestion, such as cologne). Some 50 percent of male deaths in Russia, in the 15 to 54-year-old age range, are alcohol-attributed (from poisoning, accidents, violence and related diseases); the figure for women is 33 percent. Between 1990 and 2001, alcohol consumption in Russia has directly or indirectly resulted in over 7 million deaths.

Russian life expectancy declined from sixty-five in 1988 to fifty-nine in 2005 for men and from seventy-five to seventy-two for women. Comparable figures for Europe in 2005 were seventy-seven and eighty-three, respectively. It is not possible to overstate the demographic crisis in Russia. According to the UN, the population of the Russian Federation in 2010 is 140 million – a figure projected to decline to 132 million by 2025. Like the US, Russia may have to encourage immigration or seek the assimilation of the former Soviet states on its periphery so as to hold its ground.

EUROPE: LEGACY INTO NEW EARLY REGIONALISATION

Europe is in a state of legacy and reformation following 1,000 years of the WCSE, which peaked, as we have seen, in 1914. Europe's tailspin of decline following World War Two has led to an impressive recovery, as the continent has reformed itself into a larger entity that is able to interact more competitively on the world stage. The gradual transition from the six-state European Economic Community of 1958 to the 1993 creation of the European Union (EU), which currently has twenty-seven members, was part of an ongoing regionalising process. Regionalisation takes time, but the EU is still a formidable trading bloc with a collective GDP of $15 trillion compared with the US's $14 trillion.

Europe's benign regionalisation since 1945 has taken place under the protection of the American, British and French nuclear umbrellas, without which it would undoubtedly have been absorbed into another power, such as Russia. The integration of a multiplicity of independent states with strikingly different national attributes into a common union, each able to overcome the fear of subsuming their identity and

prerogatives to the larger entity, is a remarkable achievement. Perhaps the lessons of two world wars have impressed upon the nations of Europe how much more preferable it is to choose one's partners than to be invaded or engage in protracted brutality.

LEFT **George C. Marshall testifies at a Senate Foreign Affairs Committee Hearing on the Marshall Plan in January 1948.**

During the Cold War, Western Europe and the US found a common purpose and bonded through NATO in their defence against the perceived Soviet threat. That threat evaporated with the fall of the USSR. Europe's current incarnation might be broadly compared to a 'greater Switzerland', with insular politics and no ability to project power even near its own borders (e.g. the fractious Balkan states). The militaries of Europe, with the exception of Britain and France, are but token armies.

However, in time the shifts of power currently taking place across the world may well galvanise the timid European attitude towards militarisation, particularly given a diminished US that is unable to commit the same degree of military resources to the defence of Europe as it once did.

In this context, the expansion of NATO into the old client states of the USSR may appear attractive, but there are two serious drawbacks to this process. First, the expansion of the perimeter can lead quickly to overextension. Second, in the event of an attack on one of these NATO outposts, the inability of the entire force to respond decisively would undermine the credibility of NATO as a whole in one sweep. Similarly, the participation of NATO in the US-led war in Afghanistan has been ill thought through. In Afghanistan, NATO is operating outside its original defensive mandate, and its involvement has the active support of only a handful of its member states. As a result, the forces deployed there lack the resources to perform effectively and conclusively, and therefore risk failure. Were that to prove the case, NATO would have gambled away its greatest asset: its reputation.

It must also be borne in mind that Europe comprises many nations, each undergoing a different phase in its own cycle. Germany emerged first from the post-war trough, thanks to a combination of the Marshall Plan and German industriousness, and the 'German miracle' of the 1980s raised European hopes again. Ironically, the reunification of Germany halted German expansion, as Bonn/Berlin sought to rebuild East German infrastructure to its own standards. Germany has, for the past decade and a half, experienced internal economic integration on a monumental and costly scale. But in time it is sure to regain its role as the economic powerhouse at the heart of Europe.

After austerity and social and economic turmoil, the UK was next to recover from World War Two. Under Margaret Thatcher's leadership, the crumbling post-war economy was rejuvenated. France has recovered to a lesser degree, but expansive growth is restrained by a system that emphasises socialist policy at the expense of productivity. However, under Nicolas Sarkozy, France's president since 2007, it is yet possible that the country will undergo major changes in its work and social structures.

The relationship between Europe and the US noticeably deteriorated after the demise of the Soviet Union. This new phase of the relationship, now nearly twenty years old, in many ways resembles the pre-1914 lack of transatlantic trust. The election of Barack Obama has improved relations (the neo-Conservative direction of the Bush administration proved inimical to most of the EU states), but as the US enters decline, further dissonance may be expected with Europe. A common enemy in the form of Russia and/or China might yet draw the two sides closer once again.

Europe's historical legacy of vicious and destructive power struggles encourages a general wariness among its constituent states of any one nation becoming dominant. The structure of the EU, therefore, requires consensus among its diverse members. However, given regionalisation without a military element, one nation or group of nations will rise to leadership of Europe: dominance is inevitable even within a structure requiring unanimity. Until France looses the constraints on its economy, only two contenders present themselves for this role: Germany and the UK.

One substantial problem facing the consumer nations of Europe is the question of access to natural resources. With the exception of Britain and Norway, Europe is a net importer of resources and lacks a coordinated strategic resource policy, making it susceptible to economic blackmail by Russia, its largest supplier. Furthermore, the spectre of stagflation hangs over European economic growth: anti-inflation is the mantra of the European Central Bank (ECB), which has not understood that the imported inflation of Asia is very different from the inflation of the 1980s that existed in the West. (To wit: the West today is forced to pay import prices that are set by Asia.) The ECB's purportedly preventive actions could well halt what little growth there is in its tracks, accentuating the two-tier growth that already exists across Europe, with countries outside the Continent's prosperous northwestern hub (facing increasing vulnerability). This threat has already played itself out in the financial crisis faced during 2010 in Greece.

Geopolitically, Europe will not play a major role within the next decade, but will act as a neutral power as it continues to refine the structure of the EU. The implementation, in 2009, of the terms laid out in the 2007 Treaty of Lisbon – including the appointment of a President of the European Council – is part of this process, aiming to safeguard neutrality and driving towards greater centralisation. If a threat on the scale of the Cold War does present itself, the EU's development may well be accelerated, with decades of groundwork underlying its creation perhaps then paying dividends.

RIGHT **EU heads of state pose for an official photograph at the start of a European Council summit on 17 June 2010 in Brussels.**

BELOW RIGHT **Police struggle to contain rioters near the parliament building in the centre of Athens on 5 May 2010. A nationwide general strike gripped Greece in the first major test of the Socialist government's resolve to push through unprecedented austerity cuts, which were needed in order to avert fiscal meltdown.**

European Council - Brussels
Conseil européen - Bruxelles
17.06.2010

ΑΠΟΨΕ ΠΕΘΑΙΝΕΙ
Ο ΦΑΣΙΣΜΟΣ

Chapter Six

Commodities: The Fuel of Modern Empires

The Commodity Cycle

The fundamental premise of the Five Stages of Empire model is that throughout history, human beings have conducted their affairs according to specific patterns, repeating them without conscious awareness. The patterns operate over the long term, but, consistent with fractal dynamics, shorter-term patterns can also be observed.

The brilliant Soviet economist Nikolai Kondratiev followed a similar logic, positing that human behaviour and response constitutes a critical factor in determining cycles of prosperity. (He exerted considerable influence over the Soviet economy until 1927. Arrested in 1930, he was executed eight years later during Stalin's Great Purge.) In the mid-1920s, Kondratiev studied Capitalist market economies and developed his Wave Theory, which proposed that alternating cycles of rising or falling commodity prices (plotted as sinusoidal K waves) follow a set, predictable pattern in economics. K waves, according to the theory, averaged approximately fifty-three years each.

In practical terms, a K wave represents around twenty-five years of mounting commodity prices, peaking before entering a deflationary or downward cycle for another twenty-five years. It follows that a consumer society dependent on raw materials at low input prices would suffer during the first part of the cycle, as the cost of commodities appreciated. An economy supported by the extraction and export of raw materials, on the other hand, would benefit. As discussed, access to commodities is critical in determining geopolitical power; in the ascending part of the K wave cycle, therefore, competition for resources pushes up commodity prices, increasing geopolitical friction. This dynamic results in significant conflict at the end of a cycle – for example, the Napoleonic Wars, the US Civil War, World War One and the Vietnam War.

Empires that are in the process of building their power on commodity production are positively affected by peaks in the Kondratiev cycle, while industrialised empires that heavily consume commodities are affected positively by troughs. The corrective cycle therefore significantly influences where a given empire can be situated within the Five Stages of Empire model.

The Cold War makes for a good case study of a K wave in terms of the geopolitical ramifications of the theory. The US, which by 1945 had outgrown its indigenous supply of resources and was dependent on large-scale importing, squared off against the commodity-rich Soviet Union, which derived its wealth from commodity prices. Soviet power relative to the US peaked at the time of the Vietnam War – the Cold War's main

US Producer Price Index, All Commodities *(fig. 27)*

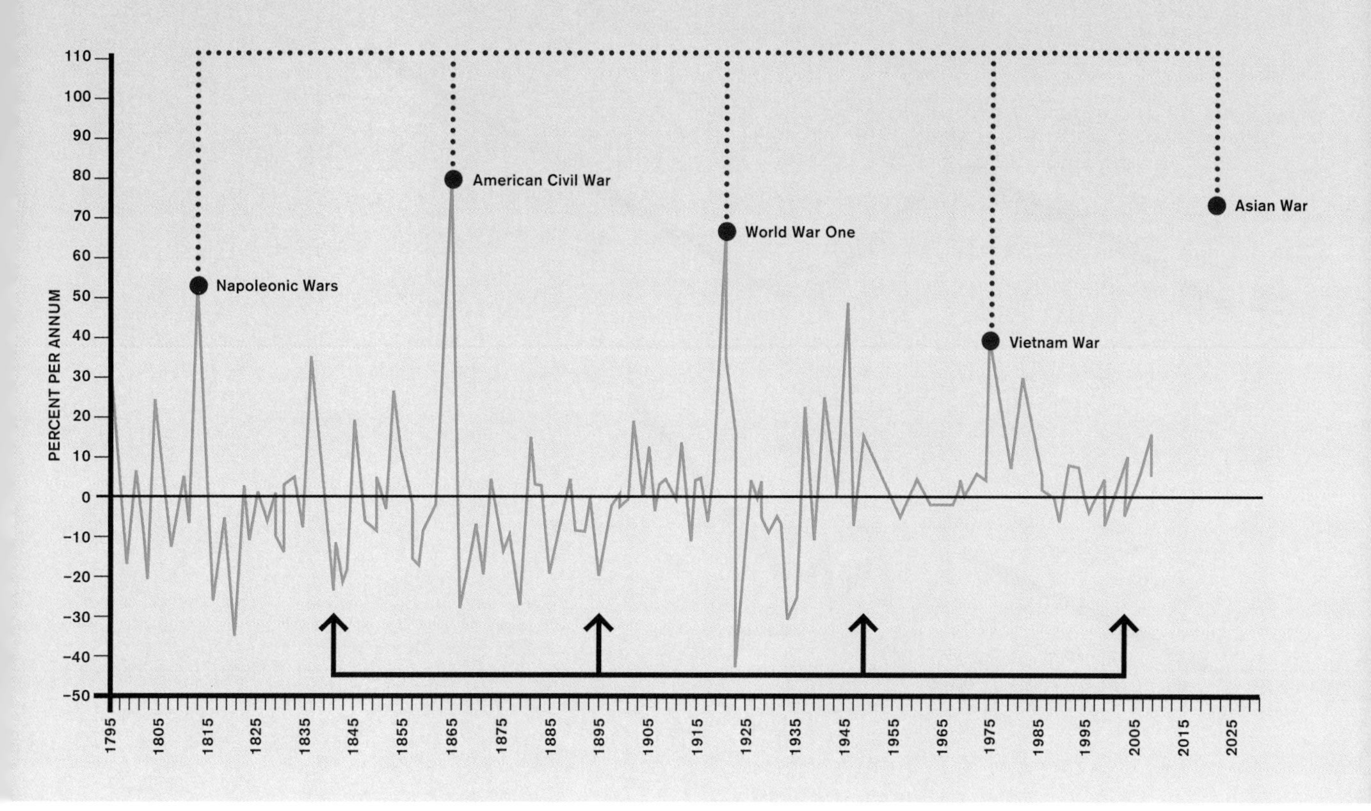

proxy conflict. The USSR then slid into a twenty-five-year decline, attaining its lowest point in 2000. This slump accelerated from 1980 onwards, following the Soviet invasion of Afghanistan. The US formed a partnership with anti-Communist Saudi Arabia to suppress oil prices, with the intention of bankrupting the USSR, a strategy that could only be achieved during an underlying bearish trend. Since 2000, by contrast, there has been a reversal in the relative power aspirations of Russia and the US. This can be expected to continue over the next two decades.

As a commodity cycle builds towards its peak, limited supply and increased demand cause prices to rise. Consumer nations compete at such a level of intensity that conflict is engendered (see Figure 27). Prices then soar even more vertiginously, creating a clear commodity spike. Alarmingly, the most recent cycle began only in 2000, yet by mid-2008 it was already exhibiting qualities (namely, the level of competition) that would normally be present at the end of a K wave. Swollen and rising levels of population and industrial demand have taken us to the limits of resource availability. Oil is, of course, the most glaringly representative commodity in this context, and food is likely to be next. This commodity cycle is unlike any other economic phenomenon ever observed; there are simply not enough resources to accommodate the global population. The situation is one of the utmost seriousness.

The Three Phases of the K Wave *(fig. 28)*

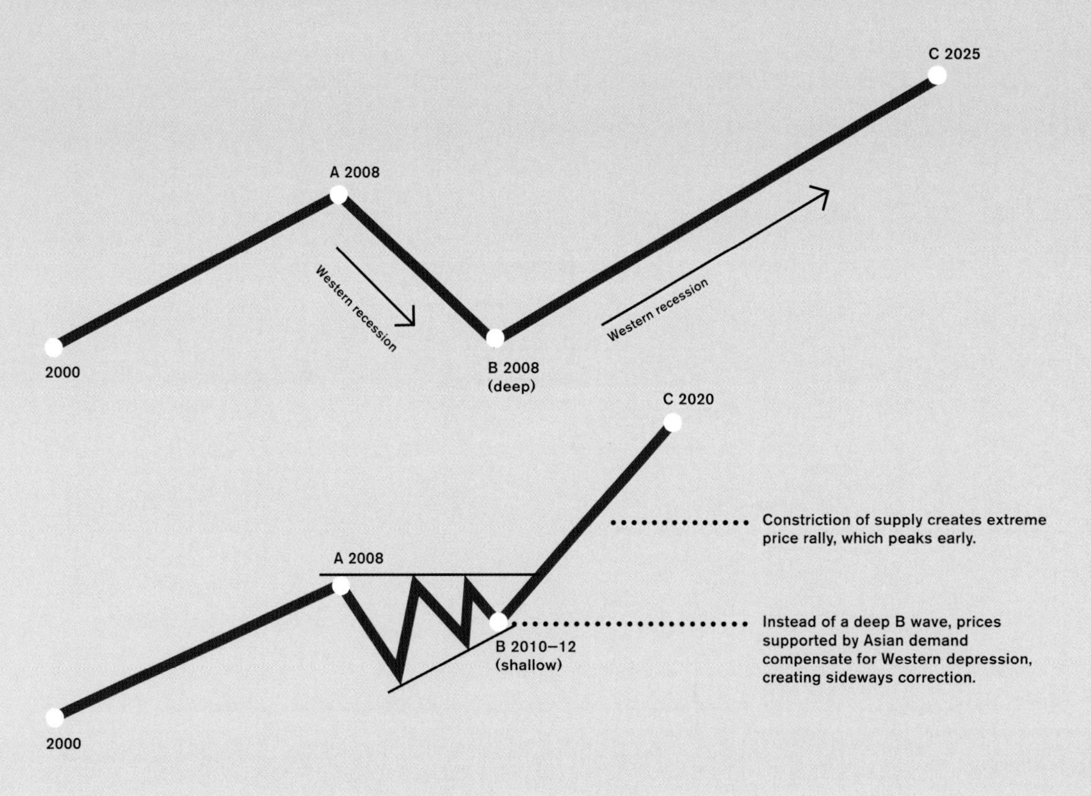

THE THREE PHASES OF THE CURRENT K WAVE

Both the upward and downward cycles of a K wave break down into three phases: the A wave, the B wave and the C wave (see Figure 28). Each phase can be observed in the current inflation phase that began in 2001 and can be expected to peak in 2025.

THE A WAVE

The current A wave occurred between 2001 and 2008. When accelerated Asian demand for resources during China's new ascendant phase was added to the stable Western demand cycle, the commodities sector became re-energised after some twenty-five years of decline in productivity. The inflationary cycle first began in the energy sector, with oil, and then spread to basic metals as new construction in China pushed up prices. In the final stage of the rally, it moved into the soft commodities sector.

The most recent rally (concerning just the A wave) pushed global commodity markets to new highs, well beyond the 1975 peak. An A wave can be considered as the first warning sign that there are too many people chasing the same limited resources.

THE B WAVE

The B wave of the present cycle commenced in 2008 after oil reached an all-time high of $150 (within a year of the 2007 'credit crunch' financial crisis). The resultant demand for oil has had a global effect but is centred on the US economy. Usually, such a deflationary phase would be expected to end a decade later (i.e. in 2018), but this may not be realised.

Over the next two decades, as the B wave takes effect, the US will undergo an economic depression, recession will linger in Europe, and China will continue to consolidate its gains and advance relative to other world powers. Pressure on commodity prices will perpetuate high input prices, and in the West, stagflation will set in: economically, the worst of all worlds. In this environment, China will hold its own, sustaining moderate growth. The West, on the other hand, will contract in growth and will experience the attendant social consequences of decline. Commodity-exporting nations (particularly those with oil resources) will continue to prosper as long as their economies are well managed and their budgets balanced. Strategic lessons can be learned from A and B waves: countries must invest in resource acquisition to prevent the consequences of a C wave. However, with the West in decline and on the brink of impoverishment, only China and perhaps the Gulf states will adopt this approach.

THE C WAVE

The C wave of the current inflationary cycle can be expected to commence in 2012, when mechanisms of supply and demand will be more intense than at any other point in recorded history. Global competition for resources will become feverish and there will accordingly be a high risk of conflict, and even the risk of another world war. China, Japan and India may well clash, and China may seize the opportunity to assert itself as the new dominant global power.

The China Effect

Since the late 1990s, China has embarked on an ambitious and successful resource acquisition policy. It has taken advantage of the fact that the West has not formulated any cohesive policy to secure present and future commodity needs and has failed to fully recognise the resource riches of Africa. China's strategy involves obtaining long-term commodity contracts from African nations in exchange for political favours and copious infrastructure investment. The West, on the other hand, has operated according to spot prices and uncertain future access.

Most Western countries have been rocked to their foundations by the financial crisis of 2007, whereas China has been shaken only lightly in relative terms. The People's Republic will continue to use this pause in the commodity cycle to attempt to corner all the resources it can, at discount prices. This buying programme, combined with the expectation that oil prices will remain high for structural reasons, stands to generate one of the shortest and shallowest B waves in K wave history. In light of intense competition for resources, the K wave peak that might otherwise have been expected to occur around 2025 must be reconsidered: the cycle is highly likely to be compressed, bringing its peak to within the next decade.

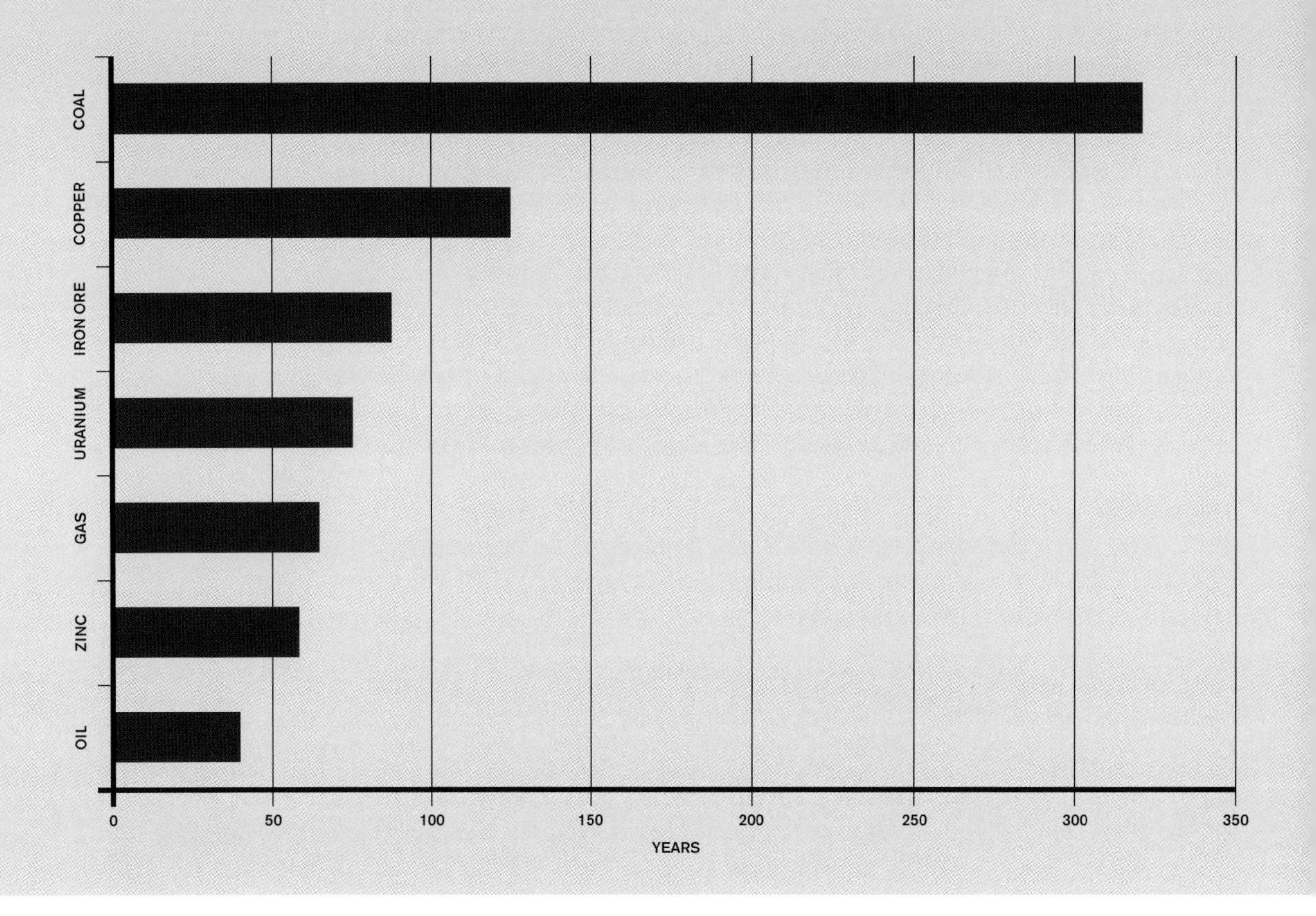

Commodity Supply and Demand

In 2001, the commodity sector had only a few specialised players, such as import and export dealers. Governments were not much concerned with the question of resource acquisition until they realised that resources were becoming scarcer (see Figure 29). Since the latest commodity cycle began in 2001, it has restructured the geopolitical landscape into the consumer-world 'have-nots' and the commodity-exporting 'haves'. The crucial question of resource acquisition is now receiving urgent attention.

The first of the commodity sectors to be affected was oil, which surged from an all-time low annual average of $11.91 per barrel in 1998 ($15.77, adjusted for inflation) to $27.40 ($34.29) in 2000. Increased demand from Asia spurred a continuous rise, to an all-time high of $91.48 ($91.35) in 2008. By the 2000 peak, oil-price activity was linked to the start of a new Kondratiev cycle, to be followed by rises in the price of base metals and of other natural resources and, ultimately, agricultural commodities.

The first empirical evidence that the commodity cycle was moving from oil to base metals was not only the appreciation of prices but also the increasing prosperity of commodity-exporting countries. Visiting South Africa regularly in the early 2000s, I began to notice on each successive journey from the airport to Cape Town that the city's once-empty harbour was increasingly filled with ships, as trade between South Africa and the world picked up. At the time, the country's economic management was less than impeccably efficient and the rand was bearish, yet South Africa was clearly prospering. I realised that any of the country's problematic management issues were more than compensated for by the fact that it was a commodity exporter, surfing a rising tide. Other examples have since multiplied across the globe, such as the oil-producing countries of the Middle East as well as Russia, Venezuela and Canada, together with a whole host of smaller countries.

The vulnerability of the West to the commodity cycle is greater than that of the East, on account of relative growth rates. Yet the entire world will be forced to engage in increased competition for resources, and the likelihood of friction between nations with contested, resource-rich boundaries will be even greater than at present (e.g. the competition between Japan and China for oil beneath the South China Sea). Conflict zones will likely extend to vital commodity trade routes.

As the commodity cycle moves into its second and third phases over the next decade, resource nationalism – the assertion of state control over indigenous natural resources – will become more common. (State-controlled companies are the norm, in fact, for both current production and proven reserves: more than 52 percent and 88 percent in 2007, respectively.)

Any direct action to challenge resource nationalism or gain access to another country's supplies must assess the energy required for such conflict. There must be enough of the commodity in question to make the challenge worthwhile.

The twenty-first century has seen a huge increase in commodity prices of a magnitude that has exceeded similar surges at any point during the previous century. The cost of oil, gas, metals and food have all mounted, whereas previous rises in the commodity sector were concentrated in only one or two categories. A 2009 World Bank study found that prices (adjusted for inflation) have risen 130 percent from their lowest point in 1999. Previous rises never exceeded 60 percent, according to the same study.

The history of empires has demonstrated repeatedly that access to resources is the primary cause of war, cloaked though it may be by ideologies or religious schisms. As we move towards the end of the first quarter of the new century, we may well witness the accumulation of commodity wealth resulting in expansive foreign policy, and watch the hope of global peace slide ever further out of reach.

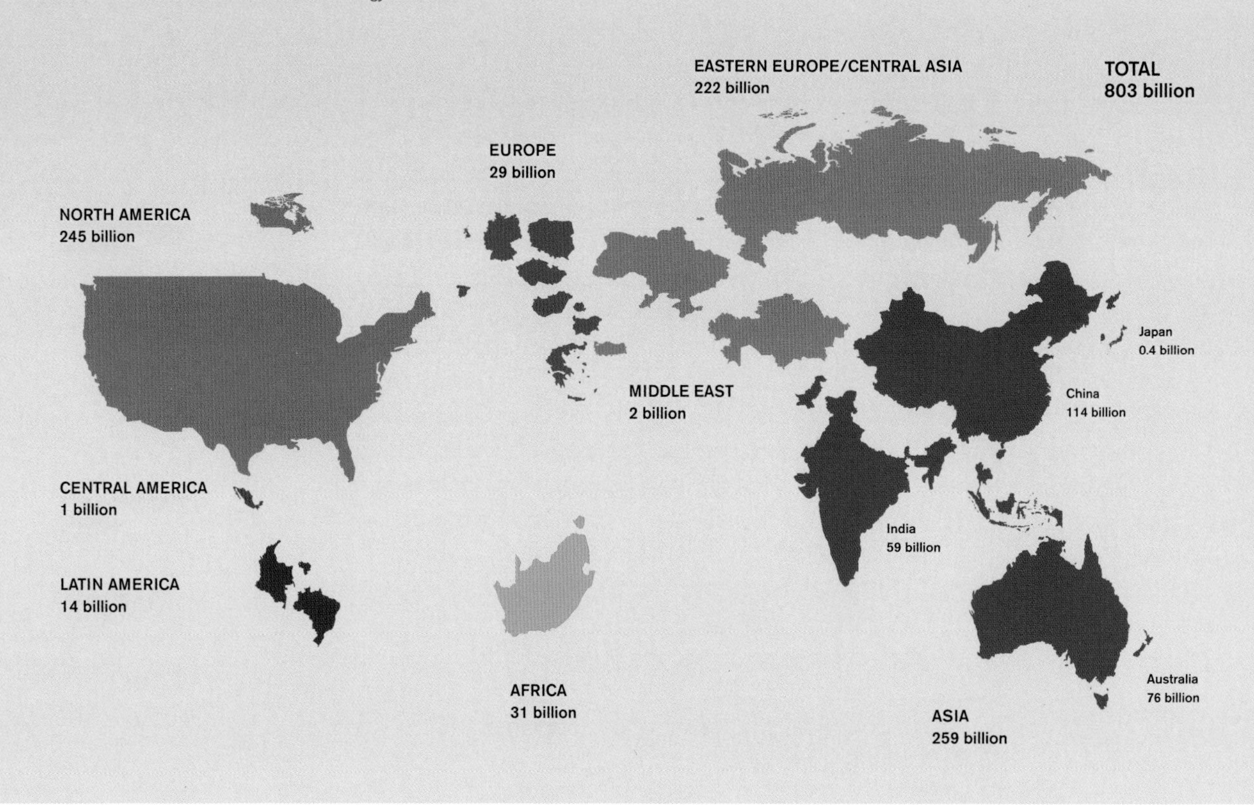

COAL

Coal is halfway along the metamorphosis from vegetable matter to oil. It is found closer to the surface of the Earth than oil reservoirs, where wood has been carbonised through a combination of high temperatures, high pressure and the absence of air. Coal is one of the younger and thus more readily available sources of energy.

The First Industrial Revolution in Britain was powered by coal, which could be found in abundance across the country. Power stations around the world still use coal today, particularly in the developing world. China, faced with the need to provide cheap power for its vast and growing infrastructure, burns coal at a greater rate than any other country, and it builds two new coal power plants each week, on average. The country produced 2.72 billion tonnes of coal in 2008, an increase of 7.65 percent on 2007, and by 2020, Chinese demand for coal is estimated to exceed 3.5 billion tonnes. (Since 2005, China has shut down its least efficient coal-fuelled power plants, accounting for 45.07 gigawatts – a figure greater than Australia's total electric capacity.)

By comparison, China has only six nuclear power plants under construction, to be added to the eleven currently in operation. By 2020 it plans to have constructed three

RIGHT A Chinese worker shovels coal at a mine plant in Hefei, in China's eastern Anhui Province, in December 2009. China is touting its carbon-limiting plan as a 'major contribution' to the struggle against climate change, but the country's already massive greenhouse-gas emissions will rise for years to come.

times as many nuclear plants as the world total, at around ten per year.

All told, China and the US are the two most ravenous consumers of coal. The former burned 38.6 percent of the world's coal in 2006, and the latter consumed 18.4 percent; and so great is China's demand for coal that it has now begun to import it. (Figure 30 shows the coal reserves remaining in 2009.)

Coal is affordable, which explains its status as the energy source of choice for emerging economies. Developed nations have turned to other, more sophisticated and expensive energy (in addition to oil, of course, which remains the primary energy source). This disparity is a key factor in the climate-change crisis, which I shall discuss at length in Chapter Ten.

BOTTOM RIGHT **The side gusher on Spindletop Hill in Beaumont, Texas, in January 1901. This was the site of the first Texas oil gusher and marked the beginning of the modern Texas oil industry.**

OIL

Russian engineer F. N. Semyenov has the distinction of being the first person to drill for oil in the modern era, using an early percussion-drilling instrument in 1844, in what is now Azerbaijan. In 1854, in Bóbrka, Poland, pharmacist Jan Józef Ignacy Łukasiewicz drilled wells 65–100 ft deep. In 1858, in southwestern Ontario, Canada, Charles and Henry Tripp founded the continent's first oil company. The following year, Colonel Edwin Drake built a well 69 ft deep, in Titusville, Pennsylvania. The 'Drake Well' sparked a search for petroleum worldwide, and Pennsylvania became the most productive oil region on the globe, remaining so for the next forty or so years.

The modern oil industry was born in 1901 on an oilfield known as Spindletop, in Beaumont on the Gulf Coast of Texas. The first 'gusher', greenish-black in colour, rose to a height of 50 m, double the size of the drilling derrick. It yielded an initial rate of nearly 100,000 barrels per day, more than all the other wells in the US at the time combined, and it awakened the world to the possibilities of oil as an industrial fuel supply.

Coal-run ships and trains converted to oil (the Santa Fe Railroad alone grew from a single oil-driven locomotive in 1901 to 227 by 1905). Petroleum would later fuel the first mass-produced aeroplanes and automobiles. The next fifty years of American growth were powered by its own oil supplies. However, by the early 1960s, the US economy had grown to such an extent that it was required to seek additional sources, and so the country became a net oil importer.

The British Empire had assumed control of most of the Middle East at the end of World War One, discovering that it had inherited the richest and most readily available oil reserves in the world. To develop them, it called upon US oil technology, forming the basis for the enduring American influence in the region. In 1956, the US intervened in the Suez Crisis as a show of its new status as a superpower. One of its motives was to supplant the UK in the Middle East with the access to oil that this implied. This was the genesis of modern oil geopolitics.

The first sources of oil were located according to the presence of 'surface seeps', where oil had been forced to the surface. As new technologies were developed, such as seismic surveying and improved drilling, oil was located and extracted at deeper levels. Initially it was only found on land, but seaborne technology allowed access to oil sites in the shallow waters of continental shelves such as the North Sea and the Gulf of Mexico.

Oil Reserves (Barrels), 2009 *(fig. 31)*

Data source: BP Statistical Review of World Energy 2010.

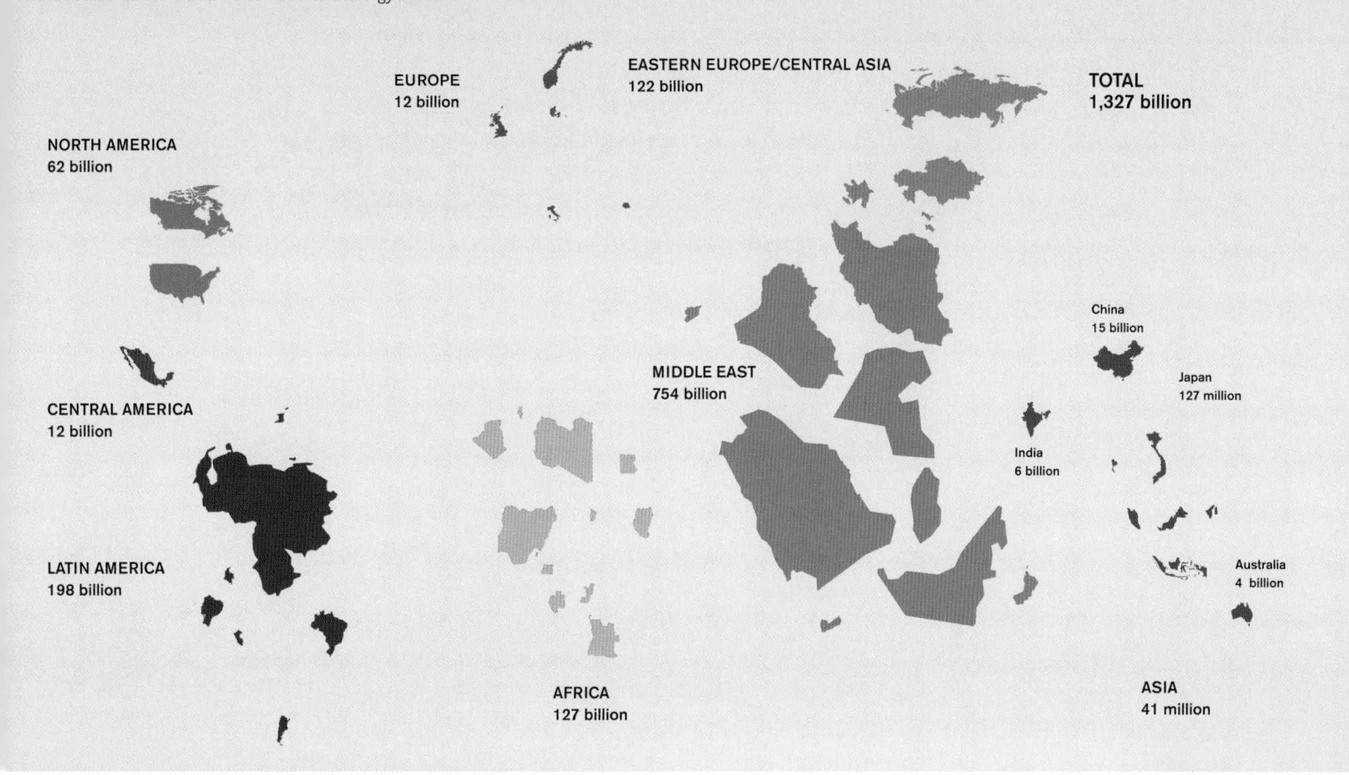

Today, the majority of oilfields on land have been found (with the exception, perhaps, of those that might exist in Africa) and are in the process of development or production.

Oil does not represent the same economic value to every producer. First, there is the consideration of 'API gravity' (so called after the American Petroleum Institute), which is a measure of how heavy or light the oil is. Light oil commands the highest prices, because heavy oil has a complex molecular composition that reduces its value. Second is the depth at which the oil is found, which affects the cost of drilling. Land drilling is much cheaper than offshore production. (In both cases, the remoteness of the location affects the ease of extraction and the distance to the open market.)

The issue of 'peak oil' has been fiercely debated in recent years. Simply put, the term refers to the point at which maximum extraction has occurred before oil production enters decline and demand exceeds supply. Current projections suggest that we are currently in the phase of peak oil, and that production of, and competition for, this resource will only intensify during the next decade. One way to phrase the question is to ask, 'Is this the end of cheap oil?' The answer is, emphatically, 'Yes.' First, there is the issue of demand, as new economies join the industrialised world and seek to grow. Then, on the supply side, a powerful case has been made that the largest and most easily accessible land reservoirs have all been found, and that they are now in the latter stages of their lifespans. Saudi Arabia, for example, is home to the largest oilfields in the world, but all of them were discovered prior to 1970 and are now water-pressurised, signifying that they are near the end of their productive lives. (Water pressurisation is only used in the later stages of well production: the oil, which has stopped flowing on its own, must be forced out by means of water injection into the bottom of the reservoir.) Moreover, the distribution of oilfields is such that there are few large ones, a reasonable number of medium-sized ones and many smaller ones, thus the economics of oil production do not benefit from scale. (Figure 31 indicates the oil reserves remaining as of 2009.)

As supply has become constricted, oil prices have increased, making it economically viable to drill in more marginal reserves, such as those deep under the ocean and in the Arctic Circle (see Figure 32). Extracting oil from such environments is hugely expensive and often requires long lead times (up to fifteen years) before production can flow. Substitutes for crude oil, such as oil sands, oil shale and biofuels, can be cultivated to a degree, but converting these into fuel is energy intensive and expensive. For these sources, oil prices would need to remain above $40–50 per barrel for them to stay economically viable. (Oil sands, also known as 'tar sands', are very dense mixtures of sand or clay, water and bitumen; oil shale is rock from which kerogen, a substance that can be converted into synthetic crude oil, can be extracted; biofuels are derived from plant oils and alcohols, like ethanol.)

However, against the backdrop of the rapid growth of Asia's economies, the reality is that it is near-impossible to find new, productive reserves that can adequately narrow the demand–supply gap. The inescapable conclusion is that the price of oil has permanently moved to new levels.

The Geopolitics of Oil

One statistic in particular underlines the current commodity drag on the US. After decades (1946–73) of virtually free energy at $20 per barrel, by 2008 the US was importing 12.9 million barrels of oil a day at an average price of $91.4 per barrel for that year (adjusted for inflation). That equates to $1.2 billion a day, or $438 billion for that year.

This enormous wealth has been transferred most spectacularly to the oil-producing nations of the Middle East, vastly enriching their economies, particularly in recent years. This dynamic, which is set to continue, will act as a catalyst for fundamental change for nations on either side of the equation. Other resource-rich countries are also profiting enormously in these oil-thirsty, oil-costly times. The US and China rank twelfth and thirteenth in the world, respectively, in terms of possessing proven oil reserves. The eleven other nations that precede them constitute a mixed bag: developed or developing, rich in multiple commodities or mainly oil-dependent, with high standards of living or low. However, their resource wealth renders all of them key players in the game of geopolitics, whether or not they tend to resource nationalism or are fertile ground for potential conflicts. They are discussed below, in descending order.

SAUDI ARABIA

Of the fourteen largest oilfields in the world discovered before 1970, most are in the Kingdom of Saudi Arabia, the foremost oil producer in the Middle East. Although Saudi oilfields fulfil 20 percent of global oil needs, they are likely to be nearing the end of their productiveness. The kingdom has most likely already seen its production rates peak, and has maintained leverage on the world stage through overstatement of its reserves.

Saudi Arabia is a *Sunni* state with a narrow wealth base, aligned with the West and with the US in particular. Yet its stability cannot be taken for granted. There is widespread inequality in the country despite its vast wealth, and the rule of the royal family (with its several thousand princes) is absolute. In time, resentment among the poor may become widespread. In addition, the country's sizable *Shi'i* minority – numbering roughly 2 million, or 10–15 percent of the population – may be inclined to revolt. (Despite modest measures to improve inter-sectarian relations, which were taken in the early 1990s under then-King Fahd and then-Crown Prince 'Abdullah, the *Shi'is* of Saudi Arabia have long experienced discrimination although have a tradition of political activism.) The question of *Shi'i* agitation will be revisited in Chapter Seven.

Saudi Arabia first experienced geopolitical significance in 1973. It led the Arab oil embargo in response to the US supply of Israel during the Arab-Israeli war of that year, although Venezuela stepped in to fill the breach, causing the embargo to backfire. All the same, the 1973 oil crisis would act as a precedent for the joint Saudi-US strategy of suppressing oil prices during the 1980s to neutralise Soviet economic power. This economic warfare proved successful, and restored relations between the two countries. (The 1990–91 Gulf War would see the kingdom grant the US permission to establish military bases on Saudi soil to enable the US to repel Iraq from Kuwait. The two countries have continued to maintain foreign policy alignment despite strains during the Bill

The Resource Substitution Profile of Oil, 2010 *(fig. 32)*

The stage at which extraction cost makes it viable to exploit various oil reserves.

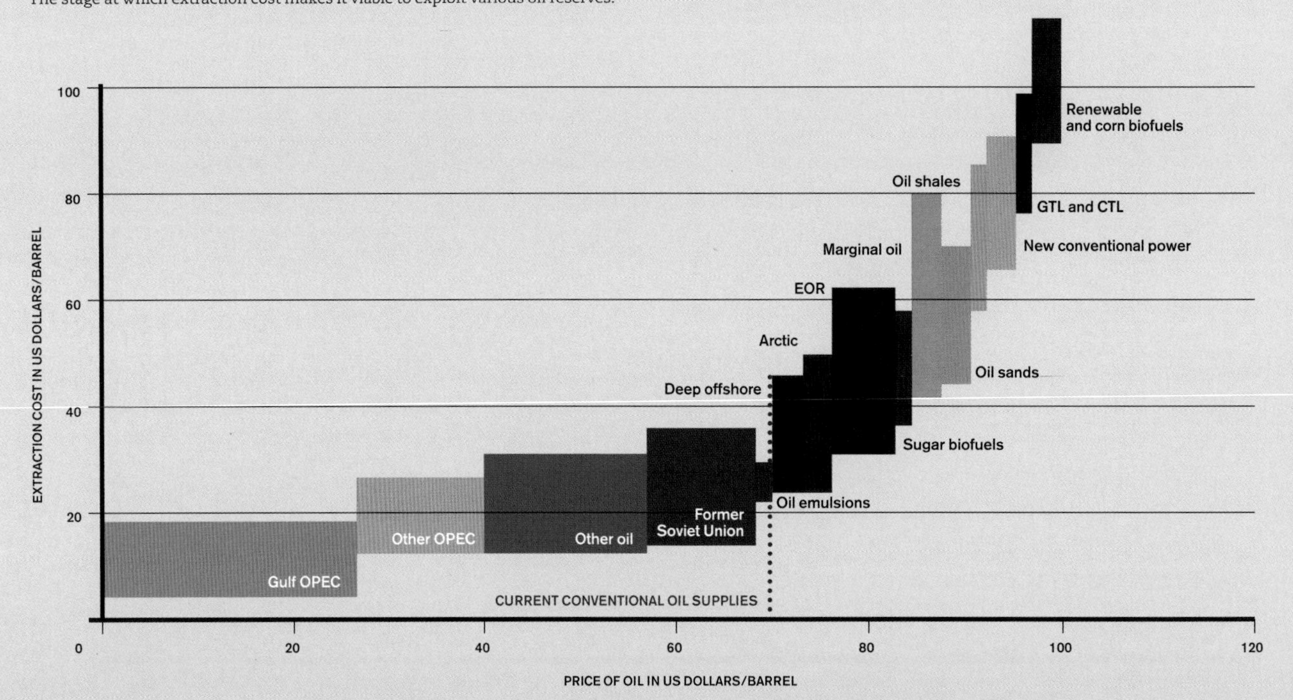

Clinton and George W. Bush administrations.) Both sides may now be tempted to replicate the strategy that served them so well against the USSR: low oil prices would aid the stumbling US economy and disempower Russia and Iran, common enemies to the partnership.

LEFT A 'donkey pump' at work in an oilfield near the Saudi Arabian border, Kuwait, in January 2003.

However, there are two balancing considerations. First, whereas in the 1980s Saudi oil reserves appeared inexhaustible, today the kingdom's interests lie in seeking to extract maximum value from its diminishing resource base. Indeed, Saudi Arabia has taken care to ensure that its national budget is based on a very conservative $45 per barrel. The second consideration is that in the 1980s, the Kondratiev cycle was in a disinflationary phase, whereas today the opposite is the case. In addition, the key strategic consideration for the Saudis is their containment of Iran's influence across the region. To this end, they have attempted to garner favour with political and militant groups in Iraq and Lebanon that oppose Iranian expansion.

CANADA

As oil prices surged in recent years, it became economically viable for Canada to extract oil from the tar sands in the western province of Alberta. (As noted above, this process is costly, requiring huge energy input. Estimates have determined profitability at just above $50 per barrel.) Yet, in 2007, Canada was deemed to have the second-largest oil reserves in the world, over 95 percent of which are contained in the Alberta sands. The remainder is extracted conventionally in other provinces, such as Saskatchewan and offshore Newfoundland.

Canada, the only developed Western state among the top ten countries with proven reserves, has not engaged in resource nationalism to the extent of, say, Russia or Venezuela. It has no national oil company – national and foreign oil firms drill or mine in Canada and pay royalties. Yet, controversially in the industry, the Alberta provincial government has modified royalty rates four times in less than two years – raising them to extract more profit, and then lowering them in 2009 on account of falling revenue from petroleum corporations cutting investment on account of the global financial crisis. (In 2009, Alberta's $1.3 billion-dollar deficit was the province's first in fifteen years.)

Socially progressive, Canada has taken a harder turn domestically in the past few years as its oil wealth has mounted. Oil patronage is rife, particularly in Alberta, giving rise to accusations of corruption. Studies on the toxic effects of the environmentally damaging tar sands industry on humans, wildlife and various ecosystems have been ignored or minimised. Nor does the province sequester petro-profits in a special government pension fund, as in Norway. (Such an initiative, the Alberta Heritage Fund, was, in fact, conceived in 1976 but has ended up subsidising special government projects instead, and in 2009 it recorded a loss of nearly $3 billion.)

These shifts in political temperament, subtle and unsubtle by various degrees, have implications for Canada's relationship with the US, its largest trading partner generally and certainly in terms of energy exports. As the tar sands industry moves to full production over the next decade, Canada may acquire a more aggressive stance on the world stage than previously seen in its history, particularly in relation to its neighbour.

The appreciation of the Canadian dollar against the US dollar – virtually on a par in 2010 – tells its own story with regard to the transference of wealth.

LEFT The tar sands mines at Fort McMurray, Alberta, Canada. These are very dense and contain a form of petroleum that could equate to about two-thirds of the total global resource. Until recently it was not financially viable to extract the oil, but new technology and rising oil prices have now made it so.

IRAN

The Islamic Republic has the third-largest proven oil reserves in the world but has been constrained by three decades of sanctions, isolation from the West, under-investment and mismanagement from converting these reserves into free-flowing oil on the world's markets. (Iran also imports as much as 40 percent of its petrol, on account of a woeful lack of domestic refineries.) A 2007 study by the US National Academy of Sciences posited a reduction to zero of Iranian oil exports by 2015.

In 2009, China imported 15 percent of its oil from Iran (544,000 barrels per day). The Islamic Republic's second-largest oil export market after Japan, China has pledged billions of dollars' worth of investment in the country's energy sector with an eye to securing these imports in the long term, thereby once again taking advantage of a vacuum in Western foreign (and resource) policy. China's veto on the UN Security Council is also highly attractive to Iran, which continually faces sanctions over its suspected development of nuclear weapons. In 2009, Beijing in fact opposed the Obama administration's calls for tougher sanctions against Iran, although in 2010, despite a measure of diplomatic engagement with Iran on the part of Brazil and Turkey, China endorsed a subsequent US-led sanctions package.

The US has attempted to persuade China to reduce its dependence on Iranian oil imports (as it has Japan, with some success). By way of encouragement, in 2009 the United Arab Emirates agreed to boost its exports to China to 150,000–200,000 barrels per day, and Saudi Arabia has indicated its willingness to follow suit, although in so doing both countries would need to circumvent OPEC (Organization of the Petroleum Exporting Countries) quotas. However, despite these measures, it is likely that Iran's importance as an exporter of crude oil will grow in time, assuming that it is not set back by conflict with the US or its allies.

IRAQ

Iraq is in possession of the fourth-largest proven oil reserves in the world (115 billion barrels) and may have many more unexploited oilfields. However, two decades of sanctions and wars have inhibited the country's exploration and production capabilities. Lack of investment remains a critical problem – Iraq does not even rank among the top ten global oil-producing states at present. It has been estimated that Iraq's reconstruction could cost as much as $100 billion, with the country's energy sector accounting for a third of that. The World Bank estimate for Iraq's oil industry requirements per year (simply to sustain current production levels) comes in at $1 billion or more in additional revenues. By 2009, the US had allocated just over $2 billion to the country's oil and gas industries to begin this badly needed modernisation. The Iraqi government has stated as its objective the export of 7 billion barrels per day by 2016. In 2009, total Iraqi production stood at 2.4 million barrels per day.

ABOVE An Iraqi Kurdish soldier guards the Tawke oilfield in May 2009. Near the town of Zakho, Dohuk Province, the site is about 250 miles north of Baghdad. The Kurdish Autonomous Region of Iraq began oil exports in June 2009 from two major oilfields, producing some 90,000 barrels per day.

However, there are some signs of revival. Over two days in late 2009, the government of Iraq held a major oil auction – one of the biggest in the history of the industry. It offered twenty-year service contracts for ten massive oilfields in the *Shi'i*-dominated south of the country, where approximately half of all Iraqi oil is located. (The Majnoon oilfield alone, one of the auction's star prizes, is among the world's largest untapped fields with estimated reserves of 12.6 million barrels.) Despite the country's current problems, the government retained major advantages, including the retention of rights to the oil while allowing the forty-four successful bidders to extract it for a low per-barrel fee of $2 above a minimum set target. (The most lucrative contracts were awarded to Russian and Chinese corporations, with French, Malaysian, Japanese and Dutch firms also scoring major successes. The US was, perhaps surprisingly, not a strong bidder: it managed to win only a share of two relatively minor deals.)

The Kurdistan Regional Government (KRG) in northern Iraq (which came into existence in 1992 after the First Gulf War) has prospered since the Second Gulf War in 2003 and the subsequent US occupation. (The Kurds have remained at some remove from the *Sunni–Shi'i* sectarian violence that has become the norm in Iraq's central and southern Arab regions.) The KRG is part of the new Iraqi federation whereby constituent regions have the right to self-rule while Baghdad assumes responsibility for foreign policy. Nevertheless, the KRG has signed production, exploration and development contracts with foreign oil companies. (Kurdish production could potentially reach 250,000 barrels per day). However, in 2009, an increasingly confident Iraqi central government ruled such deals illegal. The dispute over the sharing of oil has led to a standoff between the KRG and Baghdad, the central government refusing to pay firms contracted by the Kurds. As the US pulls its troops out of Iraq, it is likely that the KRG will continue to defy the central government, and armed conflict is not inconceivable. If Baghdad cannot assert its authority over Iraqi Kurdistan, the KRG may conduct further exploration in order to prove its oil reserves and claim oil profits for itself. Should the situation culminate in a move by the KRG to attempt separation from the Iraqi federation altogether, the alarm would be raised in neighbouring Turkey and Iran, each with their own sizable indigenous Kurdish population (including separatist elements), and the scene would be set for a wider conflict.

Iraq has a majority *Shi'i* population, long oppressed under the *Sunni*-dominated regime of Saddam Hussein. There are pervasive fears among *Sunnis* in Iraq and other Arab states that Iran has infiltrated Iraq with the complicity of Iraqi *Shi'is*, and that as a result, Iraq will end up a client state of Iran. (When, in late 2009, Iran sent troops to occupy an inactive oil well of no major importance in southern Iraq, it was met with no resistance. Local *Shi'i* leaders did express outrage and willingness to defend their territory against Iranian incursion, but other, more powerful *Shi'i* organisations in Iraq have minimised the incident.)

Iran has sent a continuous stream of money, weapons and agents across the border to Iraq, which has been porous since the US invasion. It supports militant *Shi'i* groups inside Iraq, and many of the *Shi'i* parties that dominate Iraq's Parliament have strong links with the Islamic Republic. However, the loyalties of Iraqi *Shi'is* should not be assumed to automatically attach to Iran. Iraqi *Shi'is* are divided militarily as well as politically. Local interests prevail, and there is no predominant wish to permit Iranian control over *Shi'i* affairs, whether internally or in terms of relations with foreign interests. It may be that the *Shi'is* of Iraq have used the Iranian connection to balance US control after the invasion of 2003 and the subsequent occupation, and that once US troops are drawn down, Iranian influence will not necessarily consolidate unopposed.

Tensions with Iran could further escalate as Iraq's oil industry gradually returns to strength and reasserts itself on the global stage. The two countries will then be competitors in terms of both attracting foreign investment and exerting *Shi'i* influence within the region.

THE ARAB GULF STATES

Kuwait and the United Arab Emirates are the major oil producers of the Gulf after Saudi Arabia, and their proven oil reserves are the fifth- and sixth-largest in the world, respectively. Qatar, as well as Bahrain and Oman, are smaller producers. All of the countries of the Arab Gulf are strongly aligned with the West, and none of them would be expected to incline towards power projection, preferring to invest in their own knowledge economies and other diversified industries, such as tourism. Kuwait and

Bahrain, however, have significant *Shi'i* populations and might factor into any Iranian strategy for Gulf expansionism. Compared with Saudi Arabia, the other Arab Gulf countries have a far greater distributive economic system, and standards of living are high for their citizens. (Conditions for labourers from South Asia, the Middle East and other regions, however, can vary from extremely poor to acceptable, and are frequently a source of controversy.)

BELOW **Venezuelan President Hugo Chávez and his Ecuadorean counterpart, Rafael Correa, inaugurate the Ayacucho 5 block in the Orinoco oil belt, southeast of Caracas, on 29 August 2008. Correa visited Venezuela to discuss oil and energy issues.**

VENEZUELA

President of the largest oil-exporting country in the Western Hemisphere since 1999, Hugo Chávez has attempted to harness the resource wealth of his country in order to develop a broad populist base and strengthen his regional challenge to the US. To date he has failed. Moreover, ironically, Venezuela is heavily dependent on the US for oil revenue, exporting on average 1 million barrels of crude oil and petroleum products per day – fully 9 percent of US oil imports.

The two countries have had troubled relations since the Fidel Castro-idolising Chávez became president. He moved immediately to establish greater independence for Venezuela, aligning the country more closely with the South American bloc. He severed historic military ties with the US, favoured aggressive OPEC policies that raised the price of oil, and engaged in copious and personal anti-US rhetoric, becoming a major critic of

US foreign policy. The US, in turn, was quick to recognise Venezuela's would-be successor government in the aftermath of a very short-lived coup against Chávez (which followed his attempts to bring the state oil company fully to heel). However, when Chávez was restored to power less than two days later, the US condemned the incident – although the Venezuelan president has maintained that the coup was plotted with US sanction.

Resource nationalism has steadily gathered momentum since Chávez. In 2007, he nationalised all Venezuelan oil production and distribution, forcing foreign oil corporations to cede majority control to the state. Two US firms, ExxonMobil (the world's largest) and ConocoPhillips, pulled out of the country altogether and have since been locked in a dispute with Venezuela over compensation. In 2009, in light of deteriorating infrastructure, falling revenue and national grid shortages, Venezuela moderated its stance, offering significant reductions in taxes to firms as an incentive to drill in the Orinoco Basin. Russia and China then signed deals with the country, worth $20 billion and $16 billion, respectively. (China, now Venezuela's second-largest trading partner after the US, has moved towards developing a close relationship with the country. It has invested in numerous projects, such as railway and automobile factory construction, and has also launched a Venezuelan telecommunications satellite.) In late 2009, China secured additional access to heavy-crude oilfields in the Orinoco and agreed to construct a refinery with Venezuela. It signed a deal in early 2010 for rights in one of the country's highest-rated drilling projects. Venezuela puts its total oil exports to China at 400,000 barrels per day, and it has pledged to increase this to 1 million.

RUSSIA

Russia, which possesses the world's most ample natural gas reserves, also has its eighth-largest proven oil reserves. However, this ranking only represents the consolidation of oilfields discovered under the Soviets, many of which are in the mature phase of production; the country's reserves are, in fact, in decline. Russia's energy sector has greatly enriched the nation's treasury in the last decade, but Moscow has failed to encourage new oil exploration programmes. Power has been centralised into two government giants, Gazprom and Rosneft, which lack the drive and innovation to grow the country's reserve base. In addition, owing to the natural distribution of oilfields, there are very few large reservoirs to be tapped – they make up only 30 percent of Russian reserves. Many smaller fields make up the remaining 70 percent.

Russia erred badly in its refusal to open up development of its smaller oilfields to foreign corporations. Doing so would have boosted production. Gazprom and Rosneft will consequently need to maintain Russia's oil reserves, but will experience difficulty locating reservoirs big enough to have a meaningful impact on their balance sheets. Most of the big oilfields on land have been found. Russia must therefore also master offshore drilling technology to explore its environmentally harsh northern waters, and there is at least a fifteen-year lead time before the oil can flow from sub-sea reservoirs. Accordingly, and particularly in light of the country's declining and stagnant oil production output (in 2008 and 2009, respectively), Russia has softened its inflexible stance in order to attract more foreign investment and exploration. Foreign firms will,

in all likelihood, continue to be barred from acquiring majority stakes, but export taxes have been reduced and tax breaks granted in the cases of Eastern Siberian, Black Sea and Sea of Okhotsk development.

The challenges of oil development in Russia lie primarily in the enormity and remoteness of the Eastern Siberian expanses and their harsh weather. To bring new fields online in this environment would require the construction of significant new infrastructure at great expense – in effect, spelling the end of cheap Russian oil. Sea drilling is another option for Russia, but it is expensive and does not have an established offshore oil industry to serve as a foundation. In 2009, Russia signed a deal with China that will see the China Development Bank loan $25 billion to Rosneft and Transneft (the Russian pipeline monopoly). The funds are intended for the development of oil production in Eastern Siberia, including the construction of a pipeline from the Siberian fields to Chinese refineries. In exchange, Russia is to supply China with crude oil equal to 301,000 barrels per day for the next two decades – 10 percent of China's current oil imports.

LEFT A derrick at the South Russian OAO Gazprom gas field, 250 km from Urengoy, Eastern Siberia, in February 2008. OAO Gazprom, Russia's largest energy producer, claimed that its first-quarter profits leapt 30 percent to a record high owing to rising natural gas prices.

As noted, Russia has profited considerably from the commodity cycle that began in 2000, but it has capitalised on the present at the expense of its future. In many respects, it has played its hand too early and alienated too many other powers, as its newly acquired wealth reawakened nationalist impulses aimed at regaining the country's former imperial glory. Its consolidation of the nation's energy assets into Gazprom and Rosneft was not only an economic decision but also a strategic one. Through this merger, Russia has been able to use its resource exports as political leverage against the European states dependent on the continued flow of oil and gas. (In 2009, Gazprom and Rosneft announced an increase in supply to Asia but confirmed that Europe would remain its primary market.) Europe is very vulnerable to strong-arm tactics, such as threats to shut off or restrict supply, a rather unwelcome reality in light of US efforts in recent years to provocatively push for the expansion of NATO to Russia's borders. This antagonistic policy, pursued under the administration of George W. Bush, has since been moderated under Obama.

All things considered, as the world's largest exporter of both oil and natural gas, Russia will continue to wield considerable geopolitical power in the new multi-polar world. Indeed, it is expected to become the first nation to conduct its oil sales in roubles instead of dollars, thus strengthening the rouble and making it a petrocurrency, and thereby weakening US financial power.

LIBYA

Several African countries depend on oil as a major or predominant percentage of their GDP, foremost among them being Libya and Nigeria. The list is longer, however, with Algeria, Angola and Sudan placed sixteenth, eighteenth and twenty-fourth, respectively, with regard to global proven reserves by country. Several other oil-rich countries, such as the Democratic Republic of Congo, Gabon, Ghana and Equatorial Guinea, have emerged as African emirates as well.

The World Bank ranked Libya as Africa's third-wealthiest state (by GDP per

ABOVE Italian Prime Minister Silvio Berlusconi, French President Nicolas Sarkozy, Russian President Dmitri Medvedev, US President Barack Obama, UN Secretary General Ban Ki-moon and Libyan leader Muammar Qaddafi pose for a 'family photo' on the third day of the G8 summit on 10 July 2009, which was held in L'Aquila, Italy.

capita) in 2008, after the Seychelles and Equatorial Guinea. Yet the country leads in terms of living standards, according to the United Nations Development Programme's 2009 Human Development Index (HDI), in which it ranked fifty-fifth in the world – above the Seychelles (fifty-seventh) and well above Equatorial Guinea (118th). (An oppressive, oil-flush state with great wealth concentrated in the hands of very few, Equatorial Guinea may be said to suffer from the 'resource curse', whereby poor countries with an abundance of natural resources nevertheless experience negative growth, rampant corruption and low quality of life. Other examples are Angola and Chad.)

Libya is emerging from years as a pariah state with an entrenched reputation for sponsoring terrorism. As a result, in the 1980s and 1990s, a series of US and UN sanctions were imposed on the country. In 2003, following Libya's acceptance of responsibility for the bombing of Pan Am Flight 103 over Lockerbie, Scotland, in which 270 people were killed, the UN lifted the sanctions at the urging of the US and the UK. The same year, Libya also agreed to halt its nuclear weapons programme, and Western oil companies began signing contracts with Tripoli.

Relations between the North African state and the West are steadily normalising. German, French, British and Italian heads of state have visited Libya since 2004, and in 2009, Barack Obama became the first US president to meet with Muammar Qaddafi in over two decades. The deeply eccentric Qaddafi – currently Africa's longest-serving leader (since 1969) – nevertheless addressed the Security Council in 2009, in a ninety-minute harangue during which he suggested, among other things, that it should be disbanded as a 'terror council'.

Libya's oil reserves, ranked ninth in the world at nearly 44 billion barrels, are the main reason behind the West's warming to a country that it once shunned (and, in the case of the US in 1986, bombed). Qaddafi has also embarked on a $123 billion five-year-plan to invest in the country's infrastructure, which has sparked a great deal of trade with Western countries as well as with China, Brazil, Turkey and South Korea.

However, since the re-opening of Libya, new drilling in the country's expanses of under-explored terrain has not yielded any remarkable finds. There are indications that the government is leaning more towards increased resource nationalism as well. In 2009, for example, the French petroleum corporation Total conceded to a renegotiation to halve its 75 percent stake in two Libyan oilfields. The firms OMV of Austria and ENI of Italy have also faced new terms when renewing their licences, giving a much greater share of their production to the state. Libya has asserted itself with regard to China, too, blocking the sale in 2009 of a Canadian firm operating in Libya to the China National Petroleum Company, for approximately $443 million. The deal collapsed, and Libya itself purchased the company for around $304 million.

Dealing with Libya has never been a particularly smooth undertaking for foreign oil firms. Libya's duties and royalty schemes are often contradictory, with fees being extracted despite exemptions in the contract. Firms must also endure the inertia of an administration based on Qaddafi's idiosyncratic version of Socialism.

NIGERIA

Nigeria, with just over 36 billion barrels, is tenth in the world in terms of proven reserves. According to the World Bank, the country depends on oil for 95 percent of its export revenue and 85 percent of government revenue. Its oil industry, concentrated in the Niger River Delta, has been ridden with human rights and environmental issues. Nigeria struggles with the said 'resource curse': corruption is endemic, and few Nigerians – least of all residents of the Delta regions – have benefited from the wealth that the oil industry has generated. The Ogoni people of the Delta, in particular, suffered from abuse in the 1990s as security personnel reacted to protests with extreme violence, displacing thousands and destroying villages.

Since 2004, militant Niger Delta activists have targeted oil production in the region. Various amnesties and peace processes with these activists began in 2009, thus reducing the level of violence, but although incidents of civil unrest have been sporadic, they have persisted. (Other armed groups in the region claim to fight for the rights of local peoples and for a larger share of oil income, but merely kidnap foreigners for ransom. In early 2010, three Britons and a Colombian were added to the nearly one hundred expatriate workers abducted to date, of which five have been killed.)

Inter-ethnic conflict, in this country of 140 million people and 250 ethnic groups, often boils over, as in early 2010 among Christians and Muslims in Nigeria's heartland.

ABOVE A boy stands near an abandoned oil well head leaking crude oil into Kegbara Dere, Ogoni Territory, Nigeria, in April 2007. A combination of corruption, violence and civil unrest in the region have led to this kind of environmental damage going unaddressed.

All told, the security situation, plus pipeline damage from theft and sabotage, has cost the government millions of dollars in lost and potential revenue.

In late 2009, Nigeria moved closer to resource nationalism with the introduction of a much-debated Petroleum Bill. This may see international oil companies face higher royalties and taxes, as well as lower stakes in exchange for production rights. However, the passage of the bill has been far from smooth and was compounded by the crisis over the prolonged absence of the country's president, Umaru Yar'Adua, who had been out of the public eye since late November 2009, returning in February 2010 and dying in May. Yet change is possible under new President Goodluck Jonathan. With strong leadership, ensuring that peace negotiations with Niger Delta peoples continue and that poverty and corruption – the real fuel of unrest – are tackled, Nigeria could live up to its potential as one of Africa's key economic forces.

KAZAKHSTAN

Landlocked Kazakhstan, Central Asia's largest and dominant nation, top oil producer and eleventh on the list of proven reserves by country, is the source of over 50 percent of the Caspian Sea region's oil production of approximately 2.8 million barrels per day. By 2015, it is expected to be capable of producing 3 million barrels per day. Kazakhstan was the last of the Soviet republics to declare independence following the demise of the USSR in 1991, and it is now the ninth-largest country in the world with just over 16 million very ethnically diverse people. It has prospered greatly from its copious reserves, and has quickly achieved political stability as a republic under President Nursultan Nazarbayev, in whose hands a great deal of power has been concentrated and who has removed any barriers to perpetual rule. Poverty is still widespread, and economic growth braked significantly in 2009 on account of the global financial crisis. The rule of law is relatively weak, and corruption is a growing problem. Yet the country ranks a global eighty-second in the HDI, grouped with other states enjoying 'high human development'. Since independence, a prosperous entrepreneurial class has risen (although well in the minority), and Kazakhs have enjoyed rapid development and the highest standard of living in Central Asia.

While many international oil companies are active in Kazakhstan, KazMunaiGas – the world's youngest national energy company, intended by design to eventually devolve into a publicly held company – maintains control over the terms of drilling and production. In 2007, the government adopted a law allowing it to break contracts and to force changes in contract terms with foreign oil corporations if it perceived doing so was essential to national security. Resource nationalism is thus alive and well in Kazakhstan.

Because it does not have a sea border, Kazakhstan's oil pipeline network is a crucial factor in getting the country's exports to international markets. A pipeline to the Russian port of Novorossiysk on the Black Sea opened in 2001 and currently has a capacity of around 660,000 barrels per day. Kazakhstan has elected not to rely solely on Russian land transport, however. In addition, it ships oil across the Caspian Sea from the port of Kuryk to Baku, Azerbaijan, where it is then transported via the formidable and controversial Baku–Tbilisi–Ceyhan pipeline to the Black Sea. Another pipeline, which was completed in 2009, crosses Kazakhstan from the Caspian Sea shore into China's Xinjiang Uighur Autonomous Region. With a capacity of 120,000 barrels per day, it is China's first direct-import pipeline. (In late 2009, another pipeline to Xinjiang was inaugurated, originating in Turkmenistan.)

With Kazakhstan a rising energy power, Nazarbayev has nevertheless embarked on an ambitious industrialisation plan aimed at diversifying the country's economy and skills base. Still in its early days, it focuses on such sectors as agriculture, construction, chemicals and metals. However, the energy sector remains a priority, including the development of clean energy, although, as Kazakh oil production continues to expand over the next decade, the country will be dependent on this commodity for some time.

OFFSHORE OIL: THE SLEEPER SOURCE

The price of oil has surged over the decade from 2001 to 2010, rising as much as 900 percent in 2008. However, at the senior levels of major oil companies, the reluctance created by the collapse in oil prices in the 1970s still prevails: badly needed investment to extract oil from new sources, in extreme sub-sea locations, has been unreasonably sluggish.

During this period, as discussed, the geopolitical significance of oil-exporting states increased significantly, and the transference of wealth to them from developed consumer countries was immense. That the West, with its advanced technology, should dedicate its efforts to securing new sources of oil offshore should have been strikingly apparent; it did not, however, demonstrate such vision.

RIGHT The Petrobras P-51 semi-submersible offshore oil platform being built at the Brasfels shipyard in Brazil in August 2008.

BELOW RIGHT Aker Solutions' H-6e, the world's largest drilling rig, transits in the Suez Canal on 8 May 2008, on its way to Norway after being built in Dubai. It is the most advanced kind of semi-submersible platform and is used to drill for oil at depths of 3000 m.

The few exceptions include the UK, which, in 1993 – well before the rise in oil prices – established a 200-mile oil exploration zone around the Falkland Islands, into which British investors committed approximately $530,660 billion (£327 billion) by 2009. Drilling ceased on account of high costs, but current oil prices have renewed interest among UK firms. The sea around the Falklands is thought to contain, potentially, 60 billion barrels of oil. If such estimates are correct, and if oil can be extracted, the UK would be catapulted into the ranks of the world's top oil producers. Argentina, which in 1982 unsuccessfully attempted to retake the islands from Britain, has nevertheless continued to lay claim to them. Renewed hostilities could result should oil should be found in the large quantities projected.

Yet, from the point of view of Britain as well as other states, the key question remains: what is the point of owning a potential source of oil if there is not the wherewithal to extract it? The majority of offshore exploration and production fleets are old, dating back to the investment cycle that followed the 1970s oil spike, and they are in desperate need of replacement. New-generation oil rigs are also a far cry from their predecessors, capable of drilling in water to depths and under conditions that were completely unfeasible in previous decades. New realms of exploration can be opened up through them. Nevertheless, and again with few exceptions, corporations have been reluctant to invest in such mammoth projects for fear that current high oil prices are an aberration and will eventually collapse.

There is also limited capacity worldwide to build massive offshore rigs and finance them, the net effect of which is that there is a major shortage of bigger rigs for future charter by oil firms. Charter rates have thus skyrocketed. Barring a solution, the situation will only worsen, and Western states will miss their slender chance of rebalancing energy geopolitics. They must commit to securing offshore oil with decade-long timeframes wise enough to see through the immediate trauma of the financial crisis and market collapse of recent years.

Rig investment at present is the domain of a relatively small group of specialised hedge funds. Their capital is fully invested, and their positions have embedded leverage – they will be increasingly affected by the financial crisis and ebbing credit flow. Firms with long-term vision will need to commit to finding solutions to design and construction,

AKER KVAERNER

with shortened deliveries and reliable technology. The danger is that the safety of Western energy resources will be entrusted to hedge funds and financiers who, because of the financial crisis, are reluctant to invest in new projects with delayed cash-flow dynamics.

On 20 April 2010, the Transocean-owned, BP-leased Deepwater Horizon oil rig off the Louisiana coast in the Gulf of Mexico exploded during exploratory drilling at a depth of 5,000 ft. Efforts to extinguish the resulting fire were fruitless, and two days later the rig collapsed, sinking to the bottom of the sea. Subsequently, the damaged wellhead began leaking oil directly into the Gulf. The result has been widely considered one of the world's worst ecological disasters. The ongoing gusher spread crude oil slicks throughout the Gulf at a rate estimated at 35,000–60,000 barrels per day.

This has clearly illustrated the engineering challenges of operating in 5,000–10,000 ft of water, as well as the risks of taking safety shortcuts. (In the case of Deepwater Horizon, the focus of this failure was the 350–400 tonne blowout preventer – a valve or series of valves that can seal off an oil well that is being drilled. This should have activated automatically but failed to do so). As oil companies continue to develop technology, they contend with massive expenditure yet under shareholder scrutiny. It is only too easy for shortcuts to be made. To ensure the highest standards of operating safety, it is vital that legislation and strict government regulation are implemented.

The Gulf of Mexico has the highest number of small, privately owned rigs, which have historically made this region resistant to regulation. This was undoubtedly a key contributing factor to the disaster and one often overlooked in the rush to apportion blame to BP. It is well known that rig operational costs are much higher, for example, in the North Sea, not only because of the harsh environment there but also because safety standards are so much higher than in the Gulf. Although BP has been justifiably targeted as responsible, also complicit in the disaster is the US government for its failure to properly regulate the industry in the Gulf, plus Transocean and the rig master, with his right of veto in the event of a possible threat to the well-being of the vessel.

However painful the lessons of the Deepwater Horizon catastrophe, it must be emphasised that this incident must not prevent the further exploration and development of deep-sea oil reserves that are so vital for the world's future. Indeed, Western countries, in addition to reviewing and updating safety protocols, should redouble their efforts to provide serious incentives to companies to develop the appropriate infrastructure for securing future energy supplies. Unless governments, finance and industry move quickly, the lead time in which the offshore-energy 'lifeboat' can be deployed will outpace its ability to ease the stress on global energy supplies.

THE FAILURE OF FORESIGHT

The absence of vision among the developed world's governments and their inaction in the face of a mounting resource crisis can be attributed to a lack of judgement and foresight in four key areas.

First, in the early years of the twenty-first century, the West failed utterly to anticipate the expansion of Asia's economies and particularly that of China to the extent that we are now witnessing.

Second, after years of cheap energy, Western governments have abdicated responsibility for energy policy to the private sector.

Third, in turn, the private sector has retained memories of the 1970s oil spike, when massive infrastructure investment left the oil industry in a very difficult financial position once oil prices fell. When oil again increased in price, energy companies were slow to invest in new exploration and infrastructure projects, particularly with regard to expensive offshore exploration.

Fourth, those entrepreneurs who had noted the opportunities presented by the prevailing mindset attempted to invest in oil rigs for charter to major oil companies. The restrictions they subsequently encountered were due to the world's shipyards being at full capacity, building ships to feed the Asian-driven commodity boom. Even when they were able to find a solution to this obstacle, they then found financing difficult, as the leverage bubble that was driving the Western capital system burst. Unless investments paid an income stream – impossible in the building phase of such a project as oil-rig construction – they could not be financed. This proved to be the case. Moreover, day rates for rig charters have been increasing as the rig shortage has became more acute.

This plethora of failures is sadly consistent with a mature empire, such as the WCSE and particularly its American Empire component. The US has proved unable to provide mechanisms for investment in its own future. In stark contrast, China, as it ascends to empire, has consistently invested in its future access to resources.

The West must address the question of its future energy needs now, during the current deflationary phase in the global economy. However, the catastrophic shortage of money as a result of the financial crisis will undoubtedly inhibit it from doing so. China, unaffected to the same extent by the crisis, can be expected to accelerate its resource gathering. Yet, despite the financial challenges, it is imperative that Western governments encourage the expansion of offshore oil exploration. This geopolitically insulated resource must be secured before the next inflationary phase in the commodity cycle takes hold with a vengeance. During this next phase of the K wave, the price of oil will not only damage the world's economies through inflation but may become the focus of the next global war.

Natural Gas

BELOW RIGHT The burning of excess natural gas at an oilfield in the southern Rumaila area, Iraq, in July 2007. Iraq began production in the southern oilfields of Rumaila to increase total production by 15–18,000 barrels per day.

Natural gas is a fossil fuel burned across the world as pure methane. In its raw form, it contains other gaseous hydrocarbons, such as propane, butane, pentane, plus carbon dioxide, nitrogen and helium, which are filtered out in processing plants for commercial use. Oil drilling releases the gas by reducing the pressure between the reservoir and the surface. It can also be found in reservoirs that do not contain oil, as well as in coal beds.

Gas from oilfields is often burned on-site, as separating gas from crude is necessary to produce export-quality oil. (The image of large gas flares flaming from oil wells, oil rigs or refinery chimneys is a familiar one.) Doing so also prevents conduits from accumulating too much pressure. However, gas flaring, as it is called, has harmful environmental effects, and since the beginning of the century, such organisations as the World Bank have called for it to be minimized. In many countries, cutbacks in the practice have been enforced, but it continues in major oil-producing countries such as Russia, Nigeria, Iran, Iraq and Angola, although reductions there are planned or beginning to occur. Gas flaring constitutes a significant percentage of global greenhouse-gas emissions, sending up to 400 million tonnes of carbon dioxide into the atmosphere annually, equal in emissions to that of all vehicles in the UK, France and Germany in a single year.

Yet natural gas is a major energy resource. Another reason it has been burned off so frequently is that producers have tended to lack the infrastructure required for storing or transporting it to market. Pipelines are not always economical or possible. One solution is liquefied natural gas, which is obtained by converting the gas to liquid form. The higher density of gas in a liquid state facilitates cost-effective transport (mostly by ship) from producers to consumers. Increasingly, motivated by high gas prices, oil producers are monetising this former drilling by-product.

Natural gas usage worldwide continues to grow by an average of 1.6 percent, and is expected to reach a total of 4.3 trillion cubic metres by 2030 – up from 2.9 trillion cubic metres in 2006. It is used in developing and developed countries alike in almost equal measure. Member states of the Organisation for Economic Cooperation and Development (OECD) used just under 1.5 trillion cubic metres of natural gas in 2006, and non-OECD countries used just over the same amount. The US is the world's largest consumer of natural gas, at 653 billion cubic metres in 2007. Russia (an OECD accession candidate) ranked second in global natural gas consumption in 2007 at 481 billion cubic metres. Neither China nor India consumes natural gas in very significant quantities; it comprised 3 and 8 percent, respectively, of these countries' total energy consumption in 2006, but these percentages are expected to rise rapidly over the next two decades.

Natural gas is used for the efficient generation of electricity via gas and steam turbines, either separately or, for greater efficiency, when employed in a combined cycle power plant. Gas is a much cleaner energy source than oil or coal, producing 30 percent less carbon dioxide than the first and 45 percent less than the second.

Given that natural gas is a major by-product of oil exploration, the list of global proven gas reserves by country resembles a reconfigured version of the list of proven oil

Gas Reserves (Cubic Metres), 2009 *(fig. 33)*

Data source: BP Statistical Review of World Energy 2010.

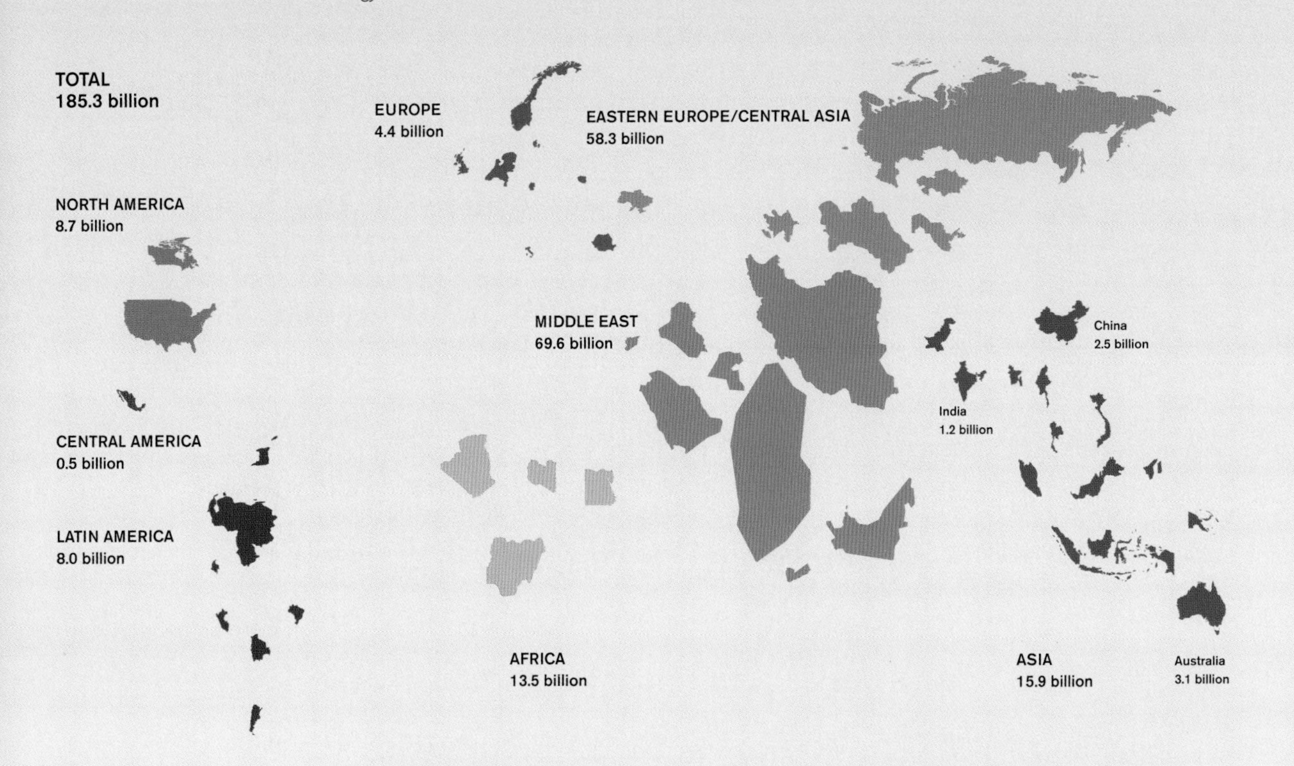

reserves. (Figure 33 shows global oil reserves in 2009.) Russia is first, followed by Iran, Qatar, Saudi Arabia, the United Arab Emirates, the US, Nigeria, Algeria, Venezuela and Iraq (2008 rankings). The Gas Exporting Countries Forum, which has been meeting annually since 2001, groups together many of the world's largest natural gas producers, including Russia, Iran, Qatar, Venezuela and Algeria. In 2008, those five countries controlled nearly two-thirds of the global natural gas reserves and accounted for 42 percent of global production.

As noted earlier, Russia has been unafraid to apply pressure on Europe as one of the continent's major energy suppliers – fully one-quarter of Europe's natural gas supplies derive from Russia. Certain countries, such as Ukraine, Belarus and Germany, have been particularly vulnerable to the tactical use of gas supply as political leverage by Russia. In 2009,the EU signed a deal that it had been working on since 2002, which will see a gas pipeline constructed to transport gas from Turkey to Austria via Bulgaria, Romania and Hungary. The Nabucco pipeline, which is expected to be operational by 2015 and to permit an annual flow of 31 billion cubic metres of gas, will be supplied with gas by Azerbaijan. This is significant in its decisive attempt to diversify Europe's source of gas and alleviate its dependency on Russia.

Metals

The use of metal is a hallmark of our species' mastery over its environment and its technological success. From the Bronze Age to the First and now Second Industrial Revolutions, our consumption of metals has attained a frantic pace, resulting in a substantial inflationary price spiral. The main impetus has been the demand on the metal markets stemming from China's huge construction programme. As Britain impressed the nations of the globe with its achievements during the First Industrial Revolution of the nineteenth century, and the US took industrial development to new heights in the twentieth, so has China awed the world with the scale of its ambitious building.

The elasticity of supply where metals are concerned is greater than that of oil – metals are more readily available and are relatively cheaper to access (see Figure 34). Nevertheless, the effects of Chinese demand created a powerful spike in the price of metal between 2003 and 2008. In accordance with the Kondratiev cycle, we can expect a period of relative price stability (2008–12) followed by one of stagnation, before another inflationary cycle takes hold of this sector.

Assuming that, unlike oil, metals are more evenly distributed across the globe's surface, the relationship between the size of a country's landmass and its industrial power base should quickly reveal which states are net importers and exporters of base metals. Thus it is no surprise that Russia and Canada are the biggest exporters, while the US, Japan, China and Germany are the biggest net importers. Russia, then, has a powerful hold over not only a considerable percentage of the world energy supply but also the supply of industrial metals.

Metal Reserves, 2009 *(fig. 34)*

Data sources: USGS, Mineral Commodity Summaries 2010, estimates for 2009. Uranium: World Energy Council, estimates for 2007.

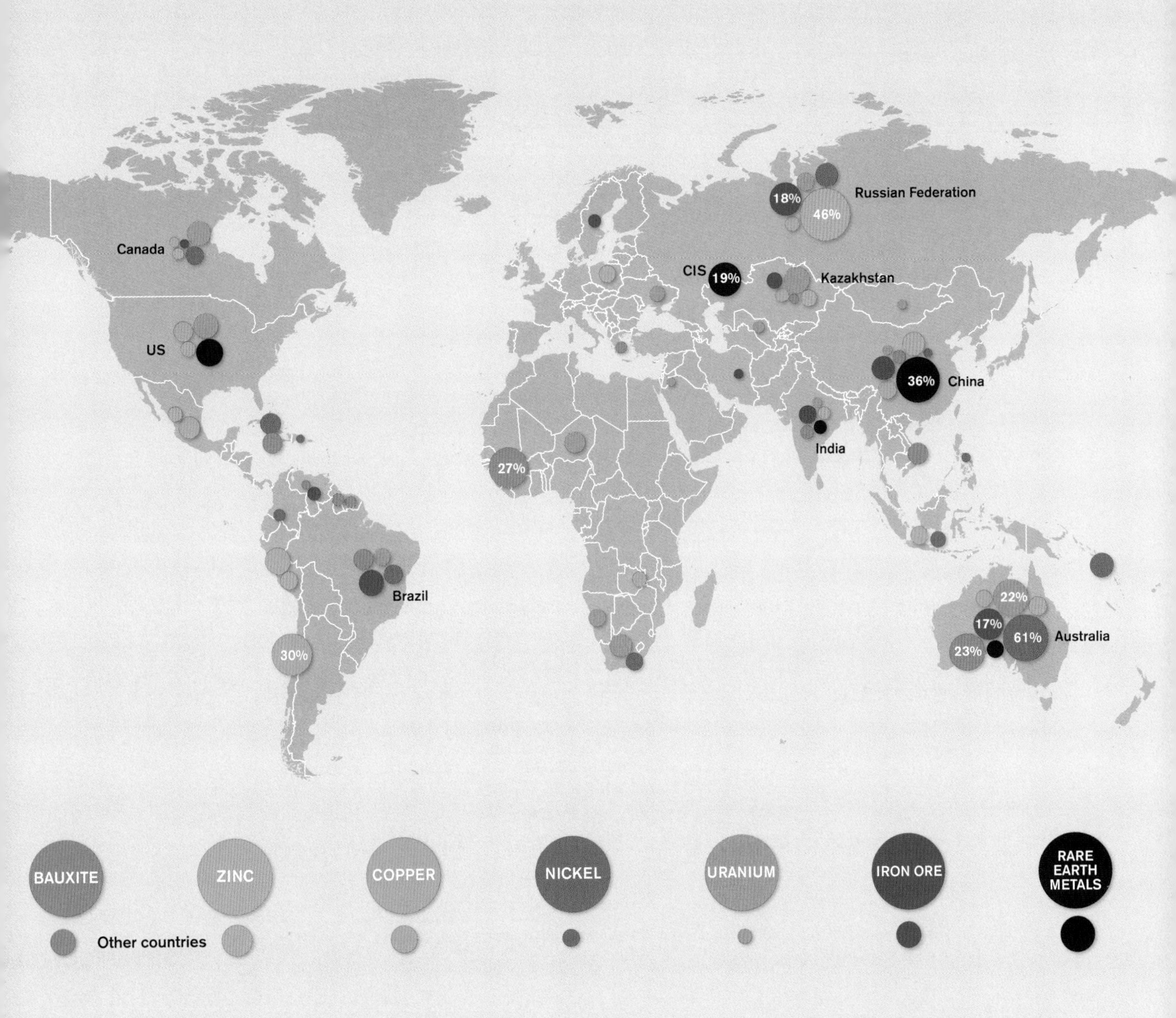

IRON

The widespread availability of iron allowed early humans to enter the eponymous Iron Age. That there have never been any 'iron wars' is a testament to the metal's ubiquity. To extract the metal from the ore, it must be chemically reduced. It is then alloyed with carbon to form steel. Strong, cheap iron accounts for 95 percent of all metal production and is the foundation of our modern industrialised society.

COPPER

The first metal exploited by early humans to make tools was copper, and its importance endures. An ideal conductor, it is mainly used in electrical installations but also in plumbing, heating and ventilation. Copper is readily available globally but is found in vast quantities in Chile, which has thus experienced an economic windfall. China is now the world's biggest importer ahead of the US and Germany, its construction and infrastructure programme pushing its needs even above those major players.

NICKEL

Nickel is another key industrial metal. When alloyed with iron and chromium, it forms stainless steel. As the fifth-most common element after oxygen, iron, silicon and magnesium, nickel nevertheless exists predominantly in the Earth's core. It is relatively evenly distributed across the surface of the globe. Thus, once again, countries with large landmasses, such as Russia and Canada, are the largest nickel exporters. Again, as well, the biggest consumers of nickel are the US, China and Germany.

ZINC

Also key to industry, zinc is valued as a coating to protect iron from oxidisation. As an alloy with aluminium, it also produces high-performance components. Zinc is often found in ores along with lead, copper and minor metals such as cadmium. Extracting it is complex and expensive, requiring huge amounts of power. As a result, it has only been available for industrial use relatively recently, and it is not as widely available as other industrial metals. Canada and Australia are the world's largest exporters.

ALUMINIUM

Found in the Earth's crust in more ample quantity than any other metal, aluminium has nonetheless been commercially produced for only about a century. This non-ferrous metal is noted for its low density and resistance to corrosion, so it has found widespread use in the aerospace industry. Aluminium is also used as a chemical catalyst, and as an additive in chemical engineering. The most common source of the metal is bauxite, which is mined in open pits. To smelt it requires a massive input of power.

URANIUM

Uranium was discovered as an element during the late eighteenth century but its radioactive properties were not appreciated until late in the following century. Once used as a colouring agent for ceramic and glass, from the twentieth century uranium has been applied in only two ways: for destruction and for power production. In the 1930s, the explosive power of nuclear fission – the breaking apart of nuclear fuels, of which uranium was the first – was proven. This discovery led to the first nuclear bombs deriving their power from uranium and from plutonium, which is synthesised from uranium. In recent years, uranium has found a new application in its depleted or non-radioactive form, on account of its density, as military-grade armour-penetrating ammunition. With regard to its use as a source of energy, it is salutary to note that 1 kg of the metal generates as much energy as 3,000 tonnes of coal.

Uranium exists in nature in the form of the isotopes U-235, U-238 and, far less commonly, U-234. Its radioactive effects are for all intents and purposes permanent – U-238 has a half-life of 4.47 billion years, and U-235, 704 million years. The processing of uranium results in coarse 'yellowcake' powder, which contains 70–90 percent uranium oxides and is used in the fuelling of nuclear reactors.

Canada, Kazakhstan and Australia, the world's top uranium miners, produced 63 percent of the world's uranium in 2009. Other top producers include Namibia (9 percent), Niger (7 percent), Russia (7 percent), Uzbekistan (5 percent) and the US (3 percent). Known uranium reserves are sufficient for the next eighty-five years. This suggests that, as with oil, the uranium supply cycle will peak, making the metal a sought-after resource. The importance of this metal raises the geopolitical significance of the uranium-producing countries.

RARE-EARTH METALS

Improved technology has opened up a range of special, technical applications for metals distributed in relatively small quantities across the Earth's surface. Lasers, electronics, hybrid cars, power-plant cooling systems, military armour, precision weapons, navigation systems, wind turbines, mobile phones, portable media players and smartphones all make use, by varying degrees, of these resources. In short, where there is advanced modern technology, there will be a demand for rare-earth metals.

These metals are found with the more common elements, and are derived as by-products of the primary mining of base metals. When the price of many basic resources becomes too high, solutions are found in their substitution with other, cheaper resources. However, in the case of rare-earth metals, it is precisely their unique properties that make them valuable: substitution is not an option.

There are, strictly speaking, seventeen 'rare-earth elements', which are classified together in the periodic table of the elements from lanthanum to lutetium. Other key metals in the development of advanced technology, while not classified as 'rare earth', are nevertheless also not found in great abundance. Examples include molybdenum, which is mined as a by-product of copper and used in steel production and cooling systems. China produces 30 percent of the world's molybdenum supplies. Another, indium, is mined as a by-product of zinc. More than 60 percent of all indium production originates in China, while Japan accounts for 60 percent of the metal's global consumption. A small amount of indium is used to make every liquid-crystal display (LCD) screen. Indeed, this is the primary use of indium today and it accounts for more than 50 percent

of consumption. These LCDs are key components in laptop computers, flat-panel monitors and flat-panel televisions.

Copper indium gallium diselenide (CIGS), meanwhile, is used in the construction of photovoltaic solar cells. As alternative power sources are developed, solar cell technology will only increase demand for CIGS.

Similarly, the power source of choice for the new generation of hybrid fuel-efficient vehicles is the lithium-ion battery, which replaces the current nickel-metal hydride variety. The global rush for this new technology, first and foremost in the US (where the government is pushing the country's auto industry to manufacture fuel-efficient cars), raises a question about lithium supply. (Lithium-ion batteries are also used in mobile telephones and laptop computers, generating even greater demand for the element.) The US's lithium imports currently come from Chile and Argentina, both major producers along with Brazil, China and Australia. Bolivia has, in its vast Uyuni Salar – the world's largest salt desert – an estimated 5.4 million tonnes of lithium, and may yet become a major player if the world moves towards 'peak lithium'. (Early in 2010, the country began to take steps to formally establish a state lithium corporation.)

Both lithium-ion and nickel-metal hydride battery-powered cars, however, also require the genuinely rare-earth metals neodymium, praseodymium, dysprosium and terbium for their generators and motors.

China, Japan and South Korea, with heavy industrial bases, have recognised the importance of rare-earth metals as vital commodities. According to the US Geological Survey, 97 percent of rare-earth metals are produced in China. The scramble to discover other sources has therefore been undertaken vigorously by the US, Canada, Brazil and South Africa, among other countries. China began expanding its mining of rare-earth metals in the late 1990s (mostly in Inner Mongolia), and it has steadily reduced exports of these resources and stockpiled them since 2004. In 2009, the Australian mining company Arafura Resources Ltd (25 percent of which is China-owned) alleged that the Chinese government had drawn up a draft report titled 'Rare Earths Industry Development Plan 2009–2015', which called for a complete ban on exports of terbium, dysprosium, yttrium, thulium and lutetium. It also reportedly aimed to drastically curtail neodymium, europium, cerium and lanthanum exports, which would leave global demand for these elements severely undersatisfied.

If correct, China – which the Geological Survey reports as consuming 60 percent of the world's rare-earth elements – is surely claiming the resources for its own development. And its stranglehold over supplies could make for a very effective economic weapon against Japan and the US. Japan has begun developing a rare-earth metals strategy, which calls for stockpiling and obtaining the resources from other countries. The US Department of Defense has also reportedly advocated protectionist measures with respect to the flow of rare-earth metals into the hi-tech US defence industry, as well as stockpiling less scarce metals, such as platinum, iridium, tantalum, columbium, tungsten, beryllium and cobalt. A bill to this effect was introduced in Congress in March 2010.

LEFT The construction of mobile phones makes use of lithium.

BELOW LEFT Bolivia's Salar de Uyuni, the world's largest lithium salt flat at 4,680 sq. miles, sits at an altitude of almost 12,000 ft. Ringed by the sierras of the high Andes, the Salar's bizarre moonscape can take the best part of a week to traverse.

ABOVE Rioters in Shanghai, China, protest against food shortages in 1948. Such events have been commonplace in China's history throughout the millennia and are deeply scoured into the minds of the nation's leadership.

Food Security

Since the low of the Kondratiev wave in 2001, appreciating oil prices gradually received more attention. By the 2008 peak, concerns about energy security were firmly embedded in the global policy mindset. Metal prices followed oil prices during that period. The last item in the commodities 'basket' to receive attention was the most basic of all: food. In mid-2008, after an eight-year appreciation of food prices (some 100 percent at its peak), it was widely recognised that we were no longer in the era of cheap food. The issue of food security had come of age.

For industrialised societies, the loss of adequate energy sources would grind civilisation as we know it to a halt; but without food, we would simply die. As such, the issue of food security will come to dwarf that of energy security over the coming decades.

Historically, a society's failure to feed its population has ended in one outcome: famine, followed by the desperation of food riots – the driver of revolutions. Chinese history is particularly replete with declining empires, a central feature of which has been financial overextension with regard to defence, resulting in a reduction in labour to work the fields and thus dramatic declines in food production. Nearly four out of the five Chinese empire cycles over the past 4,000 years ended with peasant food revolts, altering the political landscape. As the proverb has it, 'A hungry man is an angry man.'

Over the last half-century, the West has come to take access to food for granted, allowing its consumption to become excessive (witness the increasing tendency in the West towards obesity). Where once ancient societies stored food as insurance against times of scarcity, in modern times we have forgotten this expedient precaution. For example, the United States Department of Agriculture projected US wheat stocks in 2008 at 272 million bushels – the lowest level since 1948.

Since 2001, the rise in oil prices has increased the input costs of agriculture via soaring fertiliser, transport and fuel costs. Biofuel development has been spurred as a result. This has initially drawn on food sources, such as sugar and maize, to produce energy, and in so doing has fed back into food prices. The decline in the price of oil will, no doubt, be accompanied by a corresponding short-term decline in food costs. Yet there are other, more significant variables related to food costs that have been masked by high energy prices and that must be considered in greater depth in order to understand what the future holds for this vital resource. These factors are overlapping and interrelated.

POPULATION GROWTH

As discussed in PAST, the world's ever-growing population poses a serious threat on numerous fronts, but food security must be considered foremost among them. Over the last 100 years, the global population has quadrupled, and it increases daily by 200,000 people. By 2040, it is expected to reach 9 billion. As we have also seen, this population boom is not occurring in the West but in Asia and Africa.

INDUSTRIALISATION

Since the First Industrial Revolution, beginning in Britain and expanding later to the rest of Europe and the US, agricultural outputs increased and populations exploded with the advent of ever more sophisticated technology. Yet this process affected a relatively small percentage of the Earth's population. Then, from the mid-1990s onwards, Asia entered the Second Industrial Revolution, led by China (then with a population of approximately 1.2 billion) and India (953 million). The scale of this second wave of industrialisation dwarfs that of the West. In 2007, the urban population was estimated to have exceeded that of rural areas for the first time in history by researchers working with UN predictions of a 51.3 percent urban population in 2010. Moreover, while in 1996 some 42 percent of the world's population was employed in agriculture, ten years later that number had been reduced to 36 percent.

Affluence and city living tend to drive up the consumption of animal protein. In 1985, global meat consumption was, on average, 20 kg of meat per year per person; in 2002, that figure was a staggering 40 kg. By comparison, per capita meat consumption in North America was found to be 109 kg in 1985 and 123 kg in 2002. (Denmark, however, eclipsed the US with 124.8 kg, the world's highest rate.) Meat consumption per head in Sub-Saharan Africa, on the other hand, only amounted to 14 kg in 1985, and this pattern had not changed significantly by 2002. China, for its part, doubled its per capita meat consumption from 1990 to 2002; in 1961, that figure amounted to 3.6 kg, rising to 52.4 kg by 2002. (In terms of pork alone, China accounts for half of the world's consumption.)

ABOVE A butcher checks carcasses of meat at Smithfield Market in central London in December 2006. Consumption of meat in the UK is estimated at 25–50 percent above the World Health Organisation recommended maximum level.

Given that producing 1 kg of animal protein requires 10 kg of grain, it is clear that demographic drivers exert powerful pressure on both crop and protein prices globally. (It should be noted that animal protein is consumed way out of proportion to human needs in such countries as the US, Argentina, Northern Europe, Australia and, to an extent, South Africa. These 'meat cultures' must take their share of the blame for the intensive use of land and water resources required to derive animal protein, where a more moderated diet – to say nothing of a vegetarian one – would ease the pressure. China's meat-eating population, large as it may be, is far from the only nation contributing to this problem).

Arable Land (Hectares), 2007 *(fig. 35)*

Data source: Food and Agriculture Organization of the United Nations.

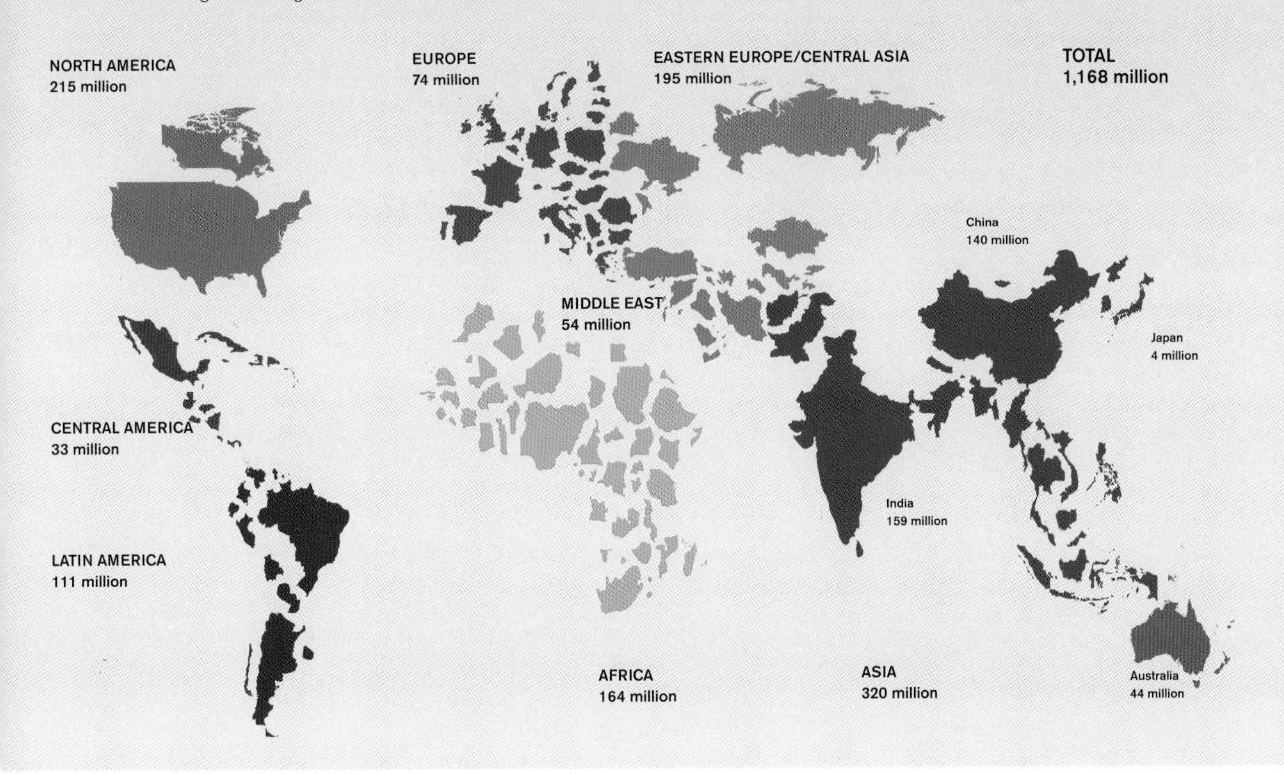

AVAILABILITY OF LAND

Approximately 30 percent of the Earth's surface area is land, with some 37 percent of it – 5 billion hectares – as arable (see Figure 35). However, 13 million hectares of this arable land is lost every year through deforestation and subsequent desertification. Currently, around 1.5 billion hectares, or 11 percent of the total agricultural land on Earth, is currently being used as cropland. The 26 percent (3.5 billion hectares) of uncultivated arable land that remains is situated in, it is estimated, the US, Russia, Brazil and, in largest measure, Africa. Yet the greater part of this land would also need considerable essential investment in transport, power and irrigation to make it productive.

The scale of such a challenge makes it extremely unlikely that an increase in productive arable land can take place quickly enough to meet demand. Even if it could be engineered, climate change is rendering vast swathes of land unproductive. Every year, 12 million hectares (an area just slightly smaller than Greece) with the capacity to grow 20 million tonnes of grain is lost to desertification. One prominent example is the encroachment of the western desert on Chinese agricultural production, to the point where the country is now necessarily a net importer of food. At present rates, China will face a 100 million tonne shortfall by 2030 as its population reaches its apex, and will consequently require an additional 10 million hectares of land in order to satisfy its consumption requirements.

PRODUCTION CAPACITY

Since the 1960s, agriculture has been one of the most neglected areas of investment. Globally, agricultural land has only grown by 10 percent. If yields had not increased over this period, the world would have required an area the size of the US (970 million hectares) to keep pace with the growth of the population during the same period. Yields have been increased using improved crop variety developed through breeding programmes as well as, more recently, genetically modified (GM) crops.

Given the grave situation that we now face, debate over whether or not to use GM crops is looking more and more like a luxury. It is extremely likely that we will only be in a position to grow crops that are resistant to climate change and disease with the aid of this technology. GM crops such as soybeans have already demonstrated benefits in increased yields, particularly when combined with precision farming techniques. For example, no-till methods do not plough the land but instead leave all biomatter but harvestable grains, so as to create a surface mulch that increases water retention over three years by up to 30 percent.

Another global revolutionary trend in agriculture has been the move, over the past decade, away from the subsistence farm operated by successive generations of a family to larger organisations that have the benefits of economies of scale. Countries that encourage this trend will enjoy a significant advantage in the decades to come with respect to food security. China, for example, has 40 percent of the world's farmers but only 9 percent of its arable land (as well as 5 percent of the world's renewable water supply). The country relies on a large number of small farms, each averaging just 0.8 hectares. To date, attempts to modernise the Chinese farming system have failed, although they will at some stage be forced into transition. On the other hand, Argentina has, over the last decade, successfully transformed its family farming industry into one that is dominated by corporate agriculture, a sector that has become a sizable contributor to the country's GDP.

China's precarious food supply situation points to Africa as the one place from which it can achieve food security. As we have seen, China has already developed oil and mineral acquisition programmes; its next investment will focus on agriculture. Africa contains large tracts of undeveloped land, variable political risk, and longitudinal and latitudinal diversification as a hedge against climate change. Development of agriculture across Africa by China could grow local economies while guaranteeing the People's Republic a large food basket for its growing empire.

Food security should be expected to become a major source of conflict across the globe. Emerging nations with expanding and poor populations will feel this pressure keenly, particularly in those areas most affected by climate change. The richer the country, the more able it will be to adapt and invest in new agricultural infrastructure. The Arab countries of the Gulf, for example, have wisely sought to ensure their food security by investing in agriculture overseas, with yields imported into their homelands, although their initial choice of Islamic African states may prove problematic with regard to ever-scarcer water resources.

CLIMATE CHANGE

As I shall discuss in detail in Chapter Ten, it is not a question of whether or not climate change is taking place but of what its effects are, and how severe they will become. The polar ice caps have shrunk, carbon dioxide levels are higher than at any time in the past 650,000 years and the tropical zone is expanding north and south. Current conservative estimates indicate that the temperature this century will rise by 1.8 to 4°C, compared with 0.76°C in the last century. This represents enormous climate change with very serious consequences.

The net effect of changing weather patterns will very likely mean a decline in productivity in areas that, until now, have been optimised for certain types of farming. It will take time for climate patterns to stabilise and for new planting models to be established. During the interval, the world's food supply will be extremely vulnerable. One of the key strategic considerations for any nation will be diversifying food supplies across latitudes and longitudes in order to mitigate unpredictable changes in climate. The problem is summed up by the World Bank, which has expressed its concerns that climate change over the next decade could halve the productivity of Asian and African farming. There is no doubt that changing weather patterns are beginning to have severe consequences for production, with storms, droughts and crop infestations becoming more common as the Earth shifts into a new climactic paradigm.

BELOW **Scientists experiment on the production of genetically modified (GM) strawberries. GM crops are the subject of much controversy, but with the increasing shortage of food, such concerns may become an expensive luxury.**

WATER SUPPLY

RIGHT Water sprinklers saturate the lawn in front of the US Capitol on 5 October 2009 in Washington, DC. The West takes the availability of water completely for granted.

BELOW RIGHT During a severe drought in eastern Kenya, women drag their containers for many miles each day to fill with water at a pastoral delivery point.

Water is essential to the survival of humanity. As we have seen, ancient civilisations were centred on major rivers, which provided a plentiful supply of water for irrigation systems and led to the development of the first intensive farming measures. Arable land is useless without water, and access to a plentiful supply for every nation's needs (for farming, industry, household and personal use, and sewage) will only become more, not less, critical with time. Water has traditionally been in plentiful supply, and our thoughtless sense of on-tap entitlement in the West means that it is not used efficiently. With the increase in global demand for water, and changes in rainfall and irrigation patterns, water rights within countries and between nations will become an increasingly contentious issue.

Where water is plentiful, irrigated land is independent of the prevailing rainfall patterns that make agricultural production predictable (all things being equal, i.e. discounting occurrences of disease and blight). However, many of the world's artesian basins – subsurface water reservoirs – have been declining with increased usage, placing a premium on surface water that flows into floodplains from high mountain ranges. At the same time, this source is also endangered. China and India, for example, depend on Himalayan glacial meltwater during the dry season for much of their water supply; as the glaciers recede, the Yangtze and Yellow Rivers may not be able to provide the water supply needed to sustain critical agricultural programmes. China has embarked on a project, to be implemented over decades at a cost of billions of dollars, to divert water from the wetter south of the country to the more industrialised north, where rivers are drying up. As its main feature, the plan will see the waters of the Yangtze River diverted north, via a series of canals and pumping stations, to the Yellow River, quenching the north as far as Beijing. The project will see some 330,000 people relocated as expanded reservoirs flood their homes.

To sum up, the message is clear, and can be very plainly stated: as the climate changes, an alarming percentage of productive land stands to be adversely affected by reduced access to water.

Dams will become strategic assets as conflicts ignite over river systems that originate in one country and flow into another. Should the upstream country decide to extract more water than the downstream country is prepared to accept, tensions will rise. Like food, energy and metals, the fierce competition for dwindling resources will be played out in the case of water, too.

Chapter Seven

Polarisation: The Road to War

The Tie that Binds

In physics, the term 'polarisation' denotes the condition by which the oscillation of certain types of wave can be oriented on the same plane. Individuals, cultures and empires can be similarly polarised: that is, they can define their values unanimously and cohesively, bonding as a single society and focusing their energies against a perceived threat from a competing system. Competition can bind a group by leading it to establish a common goal.

Polarisation can act as a positive force that is expressed, for example, through community self-improvement or in team-based contexts, such as sports. However, as competition increases between two groups, withdrawal from these amicable relations becomes more pronounced, and a hardening of differences leads to inevitable conflict. This process occurs at the individual level as well as the group level, and most people will have experienced it in one form or another. Polarisation manifesting at the level of nations and empires leads to war, with the collective character becoming more extreme or fundamentalist in its values. In the process, killing other human beings becomes justifiable because they (the opposition) embrace values that are anathema: they are 'the enemy' and no longer viewed as human.

Genghis Khan had a masterly understanding of this mechanism. Prior to his rise, the Mongols had been controlled by the Jin Dynasty, which exploited their tribal enmities to play one tribe off against another. The young Genghis had been cast out from his tribe upon the death of his father. As a man, he first used the process of polarisation to bond the stragglers from his tribe into a new combat-ready force by fighting their traditional enemies, the Tatars. He then proceeded to build up this force until it was able to challenge the tribes that dominated the region, finally using the threat of the Jin to unite all of the tribes and create the Mongol nation. This nation rode out of the steppes and conquered the known world.

The long-term memory of a collective can be highly selective, consigning some parts of its history to oblivion and holding on to others for centuries, furthering the group's sense of identity and purpose. It generally achieves this by feeding on the darker aspects of the collective memory, highlighting the enemy's despicable characteristics and emphasising fear and revenge to ensure that it has a decisive advantage.

The Law of Concentric Competition *(fig. 36)*

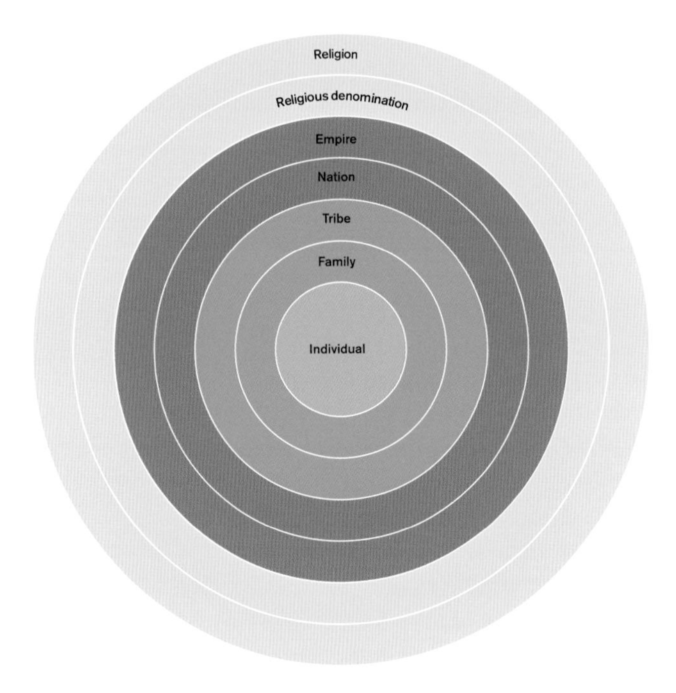

THE LAW OF CONCENTRIC COMPETITION

Humans are social animals, forming family and tribal units rather than existing as lone hunters. This is probably a necessary characteristic of human evolution, owing to the years of nurturing required by our offspring. As a result, tribal behaviour is hardwired into our brains, generating a powerful need to belong to a greater whole – the tribe being among the earliest units of this commonality. Yet the tribe is not simply a group of humans but a complex social structure with unique cultural values that differentiate it from others. Over time, we have developed larger, more complex social structures. However, our social conditioning still drives us to associate with other humans and absorb memes that define us as individuals and the group to which we 'belong'.

The law of concentric competition (see Figure 36) recognises the fractal structure of human organisations and the human need to belong to a social group. The circle describes the smallest unit at its centre as the individual, widening to encompass couple, family, tribe, nation, empire and religion. For polarisation to operate, at least two competing groups are required. Any group inside the circle will compete with another at the same level – from individual to individual, to religion to religion. However, when competition from outside enters the circle, all units inside it may respond by binding together to neutralise the threat. Thus tribes may compete within a nation, but can be united against another nation, as Genghis Khan and many other leaders have realised.

(It is known in popular culture, too: a recurring science-fiction theme, for example, has the warring inhabitants of Earth uniting in the face of an alien invasion.)

In accordance with the law of concentric competition, two memes that are used as vehicles for social polarisation are nationalism and religion. The first is based on the chain of obligation to, and reliance upon, other humans in our largest social structures, beginning with family ties and extending to imperial loyalty. The other transcends the bonds of social responsibility – it trumps all other memes, defining a group at the highest level by the gods that it worships. It lies outside the concentric circle for this reason.

THE MECHANICS OF POLARISATION

Polarisation is driven by an empire's underlying demographic trend, which in turn, as we have seen, underlies the Five Stages of Empire (see Figure 37). Empires in their expansionary stages can be said to be experiencing primary polarisation. The youth, confidence and appetite for risk of the dominant regional power characterises primary polarisation as it ascends to empire. Empires in their contraction phase have passed their peak and can no longer rely on the law of concentric competition to consolidate their national energy. Civil war is one predictable result. When an empire is facing decline, its energy is focused on maintaining the *status quo*. The empire is characterised by conservatism, protectionism and aversion to risk. This is secondary polarisation.

The consciousness of any collective can be shaped and directed as a larger fractal of the individual, with a similar, if not more basic, psychology. Just as an individual displays a range of responses to a threat depending on character, physiology, age and the type of danger, so does a group. Under stable conditions, rational responses predominate. In situations where survival is at stake, very extreme behaviour can result in which the destructive side of human emotional response can be witnessed on a collective scale.

Tribal dynamics are such that the majority of the group behave as an unthinking herd, its direction decided by a few individuals who usually coalesce into two dominant groups battling for control. (As discussed in PAST, the right leadership often emerges from a collective at certain junctures during an empire's cycle to help protect or redeem it.) But the majority are not merely passive; it is the needs of the collective, after all, that produce leaders to act in the best interests. Group dynamics occur on many subtle levels.

CONSTITUENT FACTORS AND THE POLARISATION CYCLE

The process of polarisation is determined by three interlinked factors, as follows.

1. NATIONAL CHARACTER. The entire history of a nation, itself arising from such key factors as climate, topography and geography, gives rise to a value system and culture that defines it.

2. EMPIRE CYCLE. As I have discussed above, the position of an empire along the power curve of the Five Stages of Empire model determines whether it is undergoing primary or secondary polarisation.

Polarisation as a Function of the Empire Cycle *(fig. 37)*

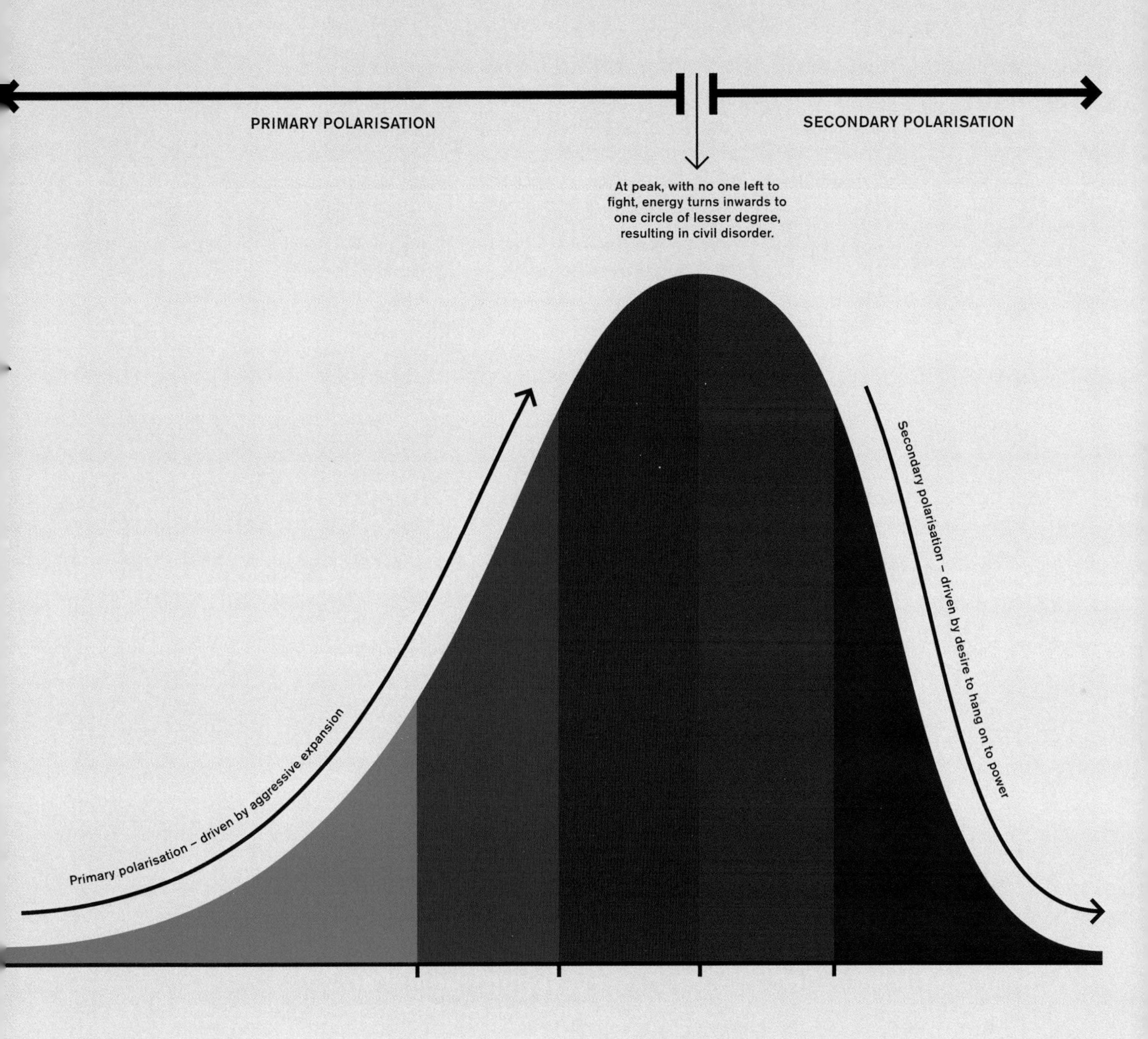

3 **GEOPOLITICAL ENVIRONMENT.** An empire can find itself facing a range of options in its interaction with other powers, from maintaining normal relations to facing a threat to its survival. The resources with which it meets such challenges depend on its stage of empire.

The following factors in turn impact on every stage of the polarisation cycle.

1 **COMMON PURPOSE.** A collective concurs in identifying a single enemy and narrows its focus accordingly. It then mobilises maximum resources to attack the perceived threat. The aggressor (in primary polarisation) plays on collective emotions such as insecurity, anger and the desire for vengeance in order to spur its population. The defender (in secondary polarisation) responds with fear, outrage and protective conservatism.

2 **NEUTRALISATION OF DISSENT.** As polarisation acquires momentum, it gains strength through consensus. The removal of opposition to the leadership of the collective then becomes a necessity. It is enacted through either the force of public opinion drowning out any dissenting voices or – as is so often the case – brutal suppression. This process ensures the collective's focus, and strengthens the power base of the leadership. The more extreme the environment, the greater the likelihood that measures to ensure consensus will be punitive.

3 **WEAKENING OF MODERATING TENDENCIES.** Throughout the polarisation process, two competing systems bring out the worst in each other. Assume a simple case of two polarised protagonists with more moderate elements in the centre who are affiliated to each. At the beginning of the process, the moderates will observe relatively mild aspects of each protagonist. However, as polarisation increases, the worst aspects of each protagonist will become evident to the moderates. A battle for the hearts and minds of the moderates becomes inevitable and they will be forced to choose their allegiance.

4 **WAR.** When one side perceives their differences as intractable, such that eradication of the other is the only solution, war becomes inevitable.

5 **AFTERMATH.** When one side destroys or subsumes the other, the latter's culture is significantly modified or simply disappears. Deadlock is another possibility, leading to mutual exhaustion, eventually forcing peace and coexistence. Exhaustion may, in fact, alter the perception of both protagonists. If they have suffered enough loss or become weakened to the point where they are vulnerable to other (even common) threats, the process of polarisation will be reversed and peace obtained.

Polarisation: The Case of Germany, 1862–1945

It would be misguided to automatically associate the Germany of the 1920s and 1930s exclusively with Adolf Hitler's National Socialist Party. There were myriad other volatile elements in German society at the time who were crushed as Nazi power grew, not least the Communists, as well as moderates of many persuasions. Hitler was the archetypal antihero: his rise began during 'the time of heroes' in the German empire cycle – when it faced economic disaster following the Weimar Republic – but he harnessed the darker forces of the collective psyche, which ultimately led to the nation's destruction.

NATIONAL CHARACTER

BISMARCKIAN ORIGINS

Nineteenth-century Prussia was expansionary and aggressive, defining itself through its force of will and diplomacy. In 1862, the formidable Prussian Minister-President, Otto von Bismarck, made the case for military preparedness in the context of expelling its Austrian rival in the German Confederation and unifying the German states. His army overwhelmed Austrian forces in seven weeks during 1866, gaining Prussia the dominant role among the German principalities. Four years later, Bismarck provoked France, then Europe's biggest power, into declaring war, as part of a strategy to eradicate Napoléon III's influence over the German states and Europe as a whole. The Franco-Prussian War ended in French surrender in 1871.

Following this victory, Germany ascended to empire. Although it had an emperor (Wilhelm II), it was Bismarck – elevated in 1871 to the position of Chancellor of Germany (a post he held until 1890) – who wielded real power. Although he was a skilled diplomat who disliked war in principle, he was a pragmatist and saw no other way to bring about unification. His famous statement that it would be 'blood and iron', not words, that would resolve 'the mighty problems of the age' came to define Germany's national character until 1945.

GERMANY POST-BISMARCK

As Prussia continued to consolidate its military and economic power, it was inevitable that it would find itself in conflict with the other empires in the WCSE, particularly those of Britain and Russia. Despite its aggression, Germany was ignominiously defeated in 1918 and, at the insistence of France, it was forced to sign an armistice that imposed massive and punitive fines upon the Germans. The 1920s was a period of intense social revolution for the German people as the country experimented with democracy. However, when the Great Depression struck in 1929, because of the financial burden that it still carried from the war, Germany suffered disproportionately among the nations of the world. However, its underlying demographics, along with its industrial power, remained. The potential to ascend to empire a second time was therefore present, but the country needed a new infusion of pride and purpose, which in turn required a new brand of leadership and vision.

ABOVE The German Civil Servant Bank goes into bankruptcy in January 1929. This was just one of numerous institutions that failed in the period of hyperinflation during the Weimar Republic.

During the early 1930s, the very survival of the German state was at stake. The nation faced the enormous challenge of total economic collapse together with the Weimar Republic's inability to function. Extreme measures were required if the country was to continue intact. It was against this desperate background that Hitler rose to power. The twelve years of Nazi rule demonstrate clearly how the polarisation cycle operates.

THE POLARISATION CYCLE OF NAZI GERMANY

COMMON PURPOSE

Hitler made his long-term goals starkly obvious when he created National Socialism, an ideology that stood in clear opposition to the WCSE's religious value system. Nazism was not totally atheist – it espoused a loose, opportunistic and exclusive Christianity blended with various occult and atavistic 'Aryan' tribal beliefs, and pseudoscience, crowned with a strictly racist worldview. It coexisted with compliant sectors of Christianity in Germany, but it ruthlessly persecuted those who resisted its approach. The Nazi belief system provided a powerful polarising focus and common purpose for the German people – or at least the members of society whom it considered qualified as German. Removing the inhibitions of the traditional Christian value system, it was no longer God who sat in judgement but the *Führer*.

ABOVE German *Führer* Adolf Hitler (1889–1945) walks up swastika-lined steps accompanied by other party functionaries during a mass National Socialist rally in 1934.

Hitler's was a much more literal 'blood and iron' approach than Bismarck could ever have conceived, and this allowed the nation as a whole (if not every individual) to commit brutal acts without traditional religious remorse – a significant advantage in the ruthless war of expansion upon which the Third Reich embarked. The Nazi value system was an extension of Hitler's own, which had been hardened in the trenches of World War One and during his subsequent harsh life in post-war Germany. He also used the injustice of the Treaty of Versailles, which had inflicted such suffering on the German people, as a focus for German rage, pride and desire for vengeance against the dominant powers of Europe.

Before Germany could engage its enemies 'without', Hitler first had to bond the German people in opposition to the enemies within. The scapegoating of Jews, gypsies, homosexuals, Communists and other 'poisonous' threats to the health and purity of the German *volk* was essential in this context. In Nazi ideology, these 'undesirable elements' were a key factor in Germany's current economic, social and political crisis.

NEUTRALISATION OF DISSENT

Hitler's rise to power was marked by a relentless ongoing search to root out all dissenting elements of society. Moderate voices whose ethos represented old German values, Communists and progressive social reformers were all targeted. Even the *Sturmabteilung*,

Jch bin am Ort
das größte Schwein
und laß mich nur
mit Juden ein!
Jch nehm' als Judenjunge
immer
nur deutsche Mädchen
mit aufs Zimmer!

DEM EUTSCHEN OLKE

a paramilitary wing of the Nazi Party that became overly powerful, was thoroughly purged. Neutralisation of any and all dissenting elements was so effective that, as Germany stormed its way to its own destruction from 1943, no one was left who was able to depose Hitler and attempt to find a less than totally devastating solution to end the war.

LEFT **A Jewish businessman and his Christian girlfriend are forced to wear signs bearing rhymes that discourage Jewish-German integration in Cuxhaven in 1933. The woman's sign translates, 'I am fit for the greatest swine and only get involved with Jews.' The man's reads, 'As a Jew, I only take German girls to my room.'**

BELOW LEFT **The badly damaged *Reichstag* (parliament building) in Berlin in December 1944. The city was devastated as Allied and Russian troops advanced.**

WEAKENING OF MODERATE TENDENCIES

The rhetoric emanating from 1930s Germany escalated year by year. Initially, the European powers observed Germany under Hitler with detachment and even condescension, finding it of interest but perceiving no threat. However, by 1937, the Nazi regime's aggressive and expansionary policies were quite conclusive. Even so, the complacency and arrogance of the old powers was such that they failed to fully take on board the coming danger until it was much too late. Hitler was appeased: France and Britain stood by as he annexed Austria and sent tanks into Czechoslovakia. This merely served to confirm Nazi belief that the old system was weak and ready to be supplanted by a new European paradigm under the boot of the Third Reich. However, appeasement by the moderating element had to be abandoned in the face of the Nazi's invasion of Poland.

WAR AND AFTERMATH

It is not my intention here to go into the history of World War Two, which is widely known, save to say that although casualty figures vary widely, it is certain that tens of millions of lives were lost on the European continent alone. Germany as well as much of Europe was devastated. The aftermath has been positive, however, with modern Germans among the world's most militantly pacifist, and the country leading Europe in prosperity and good governance in the EU.

Religion: Transcending the Circle

The supremacy of religious over nationalist memes might stem, perhaps, from the fact that while the governing body of a tribe or nation controlled one's physical existence, God or the gods were believed to control one's spiritual life and, in particular, access to the eternal afterlife. The latter could motivate individuals and societies to make sacrifices far in excess of those that national bonds could command, including a willingness to lay down one's life in the name of religious belief.

Early tribal values generated the fundamental memes of culture and religion, which then developed over time as societies expanded into nations and then ultimately into empires. Religions were predominantly exclusive: the tribe owned its gods, which in turn empowered it to survive, and the gods being worshipped were the birthright of every member of the tribe.

However, religious structures, whereby a faith was 'inherited', were limited: without population growth, they were not able to readily expand to encompass a larger

ABOVE A golden Buddha sits against the backdrop of a mural at Wat Phrathat Doi Suthep Buddhist temple in Chiang Mai Province, Thailand, in 2009.

collective. As the human population grew, the need arose to bond numerous disparate groups into larger organisational structures. New concentric circles had to be added to the society. Inclusive religions provided the solution. Faith could be adopted, rather than being determined solely by birth. Inclusive religions could poach or convert followers, either through the power of their message or through coercion. Moreover, in the event that a culture that regarded its gods as protectors was defeated by another civilisation, the adoption of the victor's 'stronger' gods would have been a natural consequence of the conquest. The potential of a religious meme to expand and then eventually encompass a vast geographical area added the outermost concentric circle to the structure of human organisation. (Figure 38 shows the current situation with regard to the religions that predominate in different regions of the world.)

Among the first inclusive organised religions was Buddhism, which was initially a reformist movement within Hinduism, originating in the sixth century BC. Five hundred years later, Buddhism assumed its universalist form and spread into Asia, just as Christianity – itself a reformist movement within Judaism – was beginning to arise only decades after the death of Jesus Christ. Islam, arriving in the sixth century AD, posed itself as the fulfilment of Judaism and Christianity. It is interesting to note that the birth of each prophet of the three great inclusive religions – Gautama Buddha, Jesus of Nazareth and Muhammad Ibn 'Abdullah – occurred in roughly 500 year cycles.

THE POWER OF INCLUSIVE RELIGIONS

Greece and Rome were the forerunners of modern secular society, where the interests of the state took priority over those of religion. Jesus had been born during a stable Roman Empire that was entering its mature stage. As it approached its peak of power, its citizens were enjoying a high standard of living that freed them to ponder the spiritual questions of life. Enter the new and expanding religion of Christianity, which offered, as its core message, a far more humane way to live and to treat others, repudiating the often severe practices of the past and describing higher realities than the mundane to which human beings had access. As the empire continued through maturity over the next two centuries, Christian values gained greater traction within Roman society. Constantine, operating with more political than religious inspiration, recognised this trend clearly and consequently made Christianity the official religion of Rome.

By the time Muhammad founded Islam, 300 years after Constantine's fateful decree, he would have been able to observe how Christianity had acted as a bonding

Largest Religious Adherence by Country *(fig. 38)*

Multiple religions are displayed for a country if the margin is less than 20 percent.
Data: CIA Factbook, Pew Forum, others. The definition of adherence varies by country.

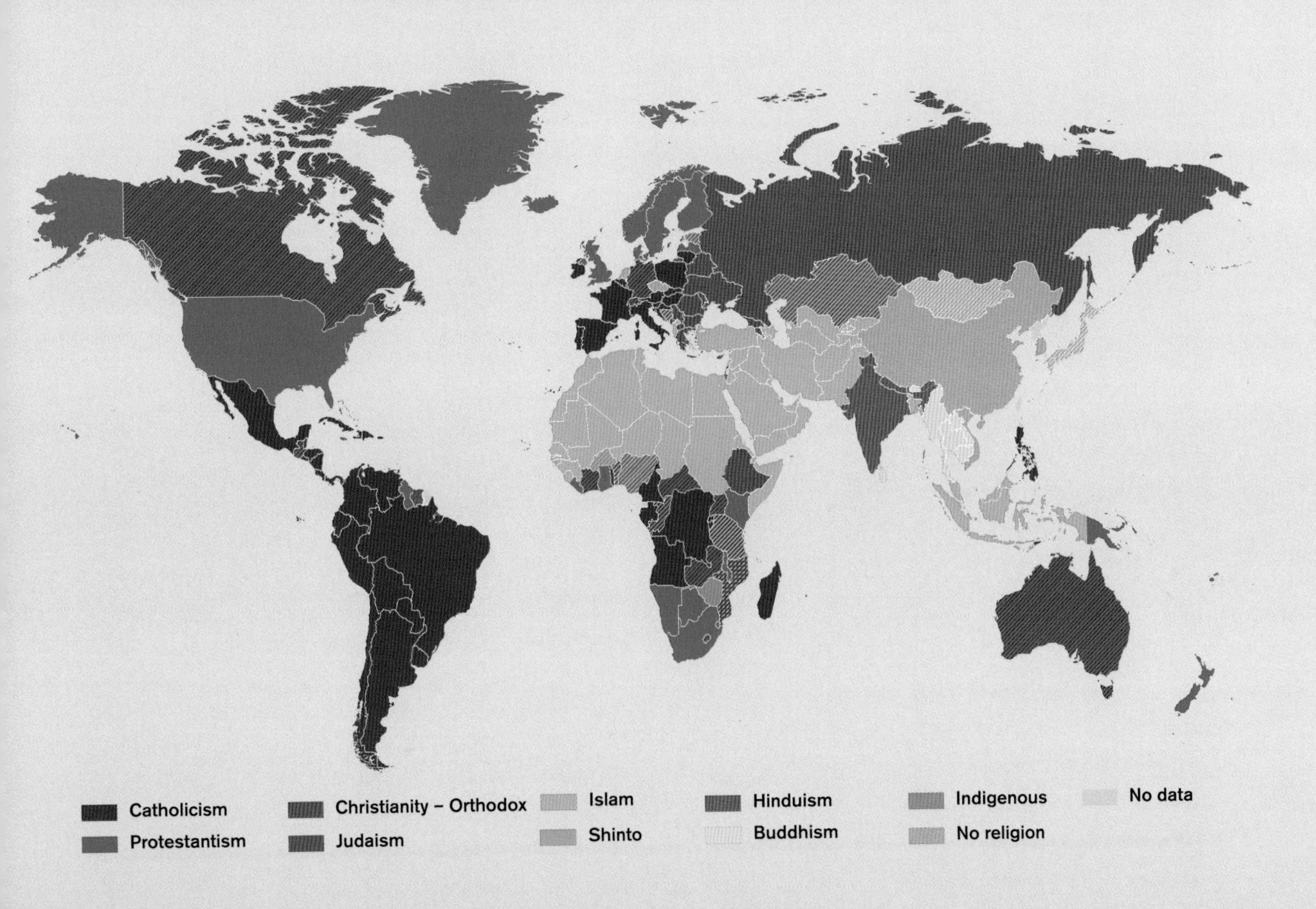

meme even after the collapse of the Western Roman Empire. Unlike Christianity, created under a maturing empire, Islam was founded at the start of a regionalisation stage in Arabia. As such, it was initially an aggressive, expansive force, successful in establishing itself as a regional power – within a century, the first caliphate occupied an area larger than the old Roman Empire. From this point, the Christian and Muslim empires would become locked in a struggle for control of the territories once dominated by Rome. This cycle of conflict directly correlates with today's situation in the Middle East.

Ironically, Christianity began as a religion that espoused tolerance, but its character began to transform into that of a warring, subjugating religion after Constantine. Islam, on the other hand, emerged first as a warring religion but, once it had established itself, its empires would rank among the most tolerant. It would appear that religions acquire a totalising momentum once they have survived one complete cycle of empire, regardless of the position within the cycle in which they are birthed. Once a religion has achieved this status, it can propagate as a powerful meme.

The belief systems that came to dominate the West (Christian) and the Middle East and South Asia (Islamic) transcended any single ethnic group. In this way, they became the larger bonding memes. Countries and empires were either Christian or Muslim first, before their national status came into play. The polarisation of the Christian and Islamic nations generated conflict, which further refined the nature of their respective beliefs (humans are just as easily defined by what they are not as what they are). In essence, these two religions represented at their cores a striving towards individual and collective humanity. Yet they have been co-opted throughout history by extremists and opportunists who have subsumed their fundamentally humanist precepts to encourage intolerance and acquire power.

Christianity and Islam evolved from similar roots contesting the same territory, but their interdependence has often been overlooked in favour of crude interpretations, such as the 'clash of civilisations' line of thought. In reality, influences from Islam and the West have washed back and forth across any attempted imposition of barriers between them. Cross-pollination has long occurred, aided in the current era by the dynamics of global finance, geopolitics and immigration, even as polarisation persists.

Religion was the controlling factor of empires in the West from AD 200, ebbing during the sixteenth century when European nations began the movement towards secular administration. This development released Europe from the constrained thinking that dominated the papacy, unleashing an explosion of creative energy. This was catalysed in part by Eastern influences that had been gaining purchase since the first half of the fourteenth century. Whereas the Islamic world had been a great deal more advanced than Europe throughout the Middle Ages, having absorbed Greek civilisation and innovated far beyond it, the trend now began to reverse, although the West would only begin to eclipse the Islamic empires with the advent of industrialisation, and the conquest and settlement of the New World.

It appears that sub-cycles exist within empire cycles, determining which meme is activated to cohesively bond a society, oscillating between nationalism and religion. The peaks show a greater propensity for the religious meme than do the troughs, where the underlying belief of a collective may reflect a sense of abandonment by its gods.

It is clear from the interrelationship between the Christian and Islamic empires that the rise of Christian empires correlates with the decline of those under Islam, and vice versa (see Figure 39). This is not altogether surprising, given that the Europeans were vying with the Arab and Turkish Islamic empires for occupation of the same territory around the Mediterranean.

One of the legacies of the historically intense competition between the Christian and Muslim empires is the current manifestation of primary polarisation in some Islamic countries in the early stages of regionalisation (e.g. Iran and Pakistan) and secondary polarisation in the West – particularly the US, which is in overextension. (Europe displays little secondary polarisation as its societies are more secular than ever, religion holding little attraction for the present European collective. However, it filters through into the nationalist meme, with its themes of 'protecting the way of life', 'controlling immigration', 'integrating young Muslims into "multicultural" society' etc.)

The Mutual Exclusivity Between the Christian and Islamic Empires *(fig. 39)*

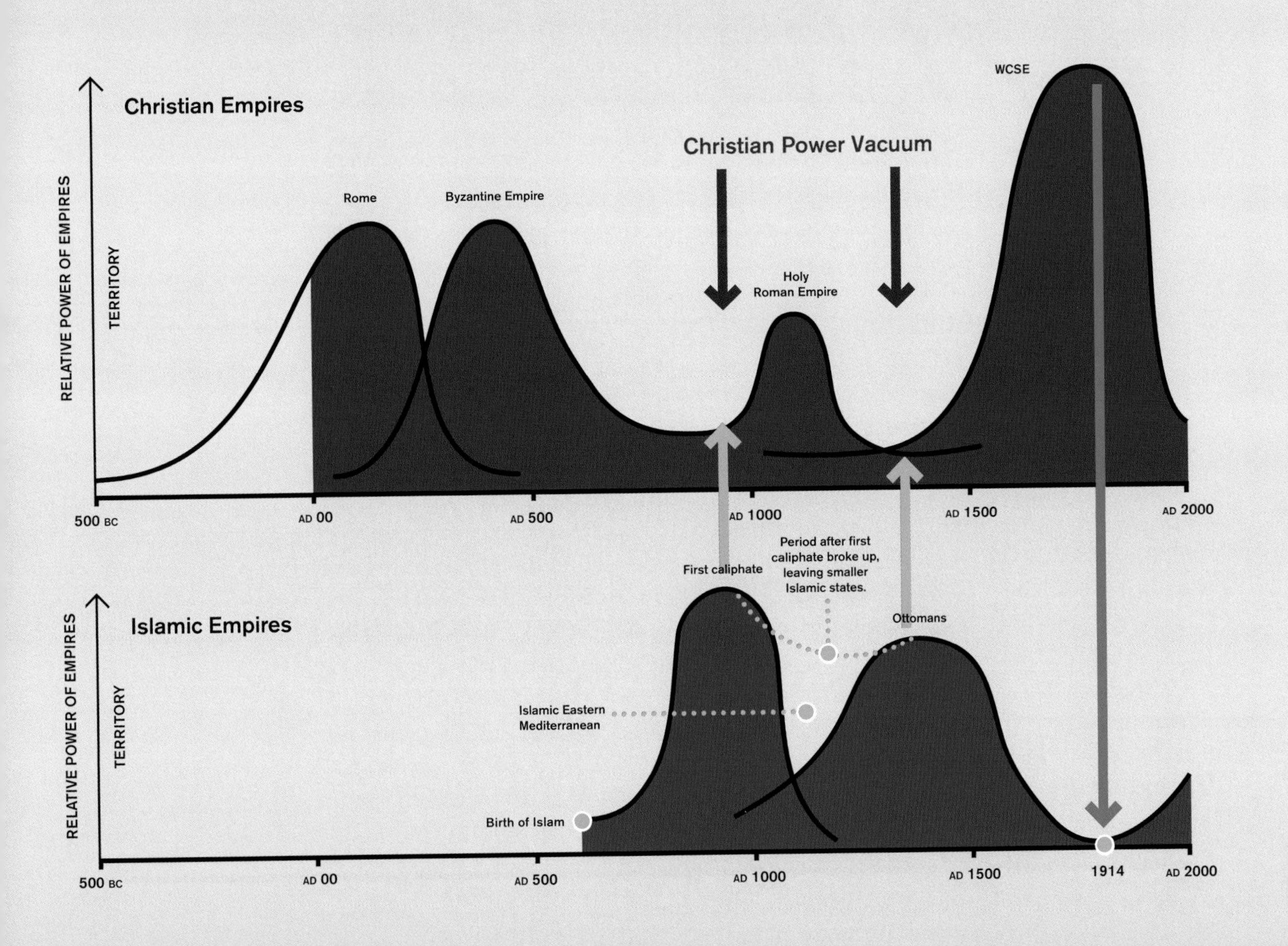

ABOVE **Observant Jews pray at the Western Wall in Jerusalem, Israel, in August 2003. Venerated as the sole remnant of the Second Temple, the site is a place of pilgrimage for Jews in a city that also boasts some of the holiest sites of Islam and Christianity.**

To understand how the powerful dynamics of polarisation operate in the modern world, we should examine embedded prejudices and beliefs in both Christianity and Islam.

THE BIRTH OF CHRISTIANITY AND THE RISE OF THE CHRISTIAN WORLD

What facilitated the growth of Christianity during a period of history when empires and polytheism were synonymous? As discussed, the drive towards monotheistic Christianity within the Roman Empire was likely to have been facilitated by the increasingly high standards of living that came about as the empire matured. Roman technological sophistication resulted in a lesser degree of dependence on fluctuating environmental circumstances than ever before. An efficient supply of water, the availability of diverse foods, durable shelter and clothing, and other advances permitted time to cultivate a more complex spiritual dimension to the lives of Roman citizens. In this context, the teachings of Christ were revolutionary; in addition to the adoption of one god and a messiah, the concept they promulgated of tolerance, forgiveness and love for one's fellow neighbour contrasted with Roman pantheistic codes of deity appeasement. They also opposed more draconian Jewish doctrines based on Mosaic law. (Christianity, it must be remembered, was originally a revision of the

Jewish faith.) This revolution in thinking grew rapidly, leading to a sea change in Western thinking and spirituality that has shaped the world to the present day.

The Jewish people were subjugated harshly under Roman rule, but their faith provided hope through messianic prophecy. Christ's followers believed that he was their long-awaited saviour, and they hoped that he would lead them to freedom. However, this was not his message; Christ believed in a spiritual struggle, not a military one – a society where warfare became obsolete in the face of love and fellowship. The Jewish establishment feared this message would quickly remove their power base, and so moved to persuade the Romans to have him executed in the cause of civil tranquillity. The persecution of Christ was a watershed in Jewish history, leading ultimately to the destruction of their Second Temple in AD 70 and the resumption of Jewish exile.

Belief that Jesus Christ was the Son of God was spread across the Roman world at great personal risk to early Christians. Naturally, Rome resisted this new challenge to its authority and enacted an official policy of suppressing Christianity, resulting in the enduring image of believers being crucified or thrown to the lions. Nevertheless, the religion only grew in strength. By the fourth century, every major city contained large Christian communities, polarised and strengthened by their history of persecution.

The civil wars that wracked the Roman Empire ended with Constantine's victory in AD 312, shortly after which he announced his conversion to Christianity. He wisely used the Christian religion to join the competing elements of the declining empire, adding a powerful new outer concentric circle to bind its adherents to Rome. The message was that the gods of old Rome had lost power, while the Christian religion would renew the empire with fresh vigour. In the ensuing years, Constantine made Christianity the official state religion, thus neatly associating the defence of Rome with the defence of the followers of Christ. The teachings of Christianity became, in this way, associated with a warrior culture that would dominate European history for the next 1,700 years, and provide both an inspiration to and a counter against the spread of Islam.

Proof of the power of a religious meme over that of nationalism is that it can outlive the empire that gave birth to it. Roman civilisation left a huge footprint in terms not only of its culture but also of its Christian values. Moreover, the religion even survived the definitive split of the empire, with Roman Catholic and Orthodox Christianity enduring even to the present day.

The Christian city-states, Pisa, Genoa and Venice, arose in the tenth century and began to flourish during the twelfth and thirteenth centuries between the Second and Third Crusades. Their strength and prosperity endowed Rome with increased security once more, as it became the centre of the Catholic Church. In the north, during the Early Middle Ages, Britain and Ireland provided a bastion for Christianity – and Christian missionaries – as Northern Europe was awash with invasions by Germanic tribes. From the fourth century, Christianity would eventually inculcate the Franks, Goths, Ostrogoths and other tribal peoples of Europe. The Scandinavian peoples were Christianised from the eighth century onwards.

The Byzantine Empire provided a safe haven for the development of Christianity from the collapse of the Western Empire to the start of the eleventh century. It was the global centre of Christian theology, and endured intense factionalism as well as the onslaught of Islam's initial expansion. However, over 400 years before Constantinople's fall in the fifteenth century, the Catholic Church began to surge in power, and challenged the Byzantines for spiritual control of the Christian world. In 1054, the Great Schism cleaved Christianity into Latin (Catholic) and Greek (Orthodox) factions, each regarding the other as heretical and thus excommunicating them. Gradually, Christian power seeped from East to West, and, with the Ottoman conquest of the diminished husk of the Byzantine Empire in 1453, the region east and south of the Italian Peninsula finally came under the complete control of Islam. (The collapse of the Byzantine Empire resulted in an exodus of Greek Christian scholars into Europe, however, which nourished and fostered Renaissance thought.) Rome came to dominate the religious meme that was now binding the growing WCSE.

RIGHT **A car bomb exploded in Iraq's Sadr City in Baghdad on 12 August 2006 as part of the ongoing *Shi'i-Sunni* sectarian violence. Four Iraqis were killed and twenty-six were wounded.**

BELOW RIGHT **Ten years earlier, the remains of the Killyhelvin Hotel in Enniskillen after a car bomb exploded on 14 July 1996, injuring forty.**

Elsewhere in Europe, the Holy Roman Empire was facilitating the development of the WCSE, functioning, in effect, as an outer containing concentric circle to the various kingdoms in constant conflict. As Portugal rose to power, Europe, which was blocked to the east by the Islamic Empire, expanded westwards. As we have seen, sea power proved decisive in accelerating both the growth of the WCSE and the expansion of the Catholic and later Protestant Churches, both in the New World and across the globe. Like Catholicism, Protestantism took on its own close identification, with the state fuelling Dutch and English missionary zeal. (The Puritan pilgrims who landed on the coast of the future state of Massachusetts were, in fact, religious extremists in search of a new, purer world. The echo of their belief system can still be clearly discerned in the present-day US.)

The successful spread of Christianity across the Americas was aided by the near-eradication of the Native Americans through disease and war, but also by the inclusive dynamics of the Christian system. The larger its following became, the more rapidly the religion pervaded the minority tribal regions, as various tribes chose to align themselves with the dominant power bloc. This dynamic would be repeated elsewhere in the world over the coming centuries, in other colonised lands, such as those in Africa.

The legacy of the WCSE is evident in the number of Christians worldwide today: 2.1 billion (33 percent of the global population). However, although Christianity spread rapidly outwards throughout the Mediterranean and into Europe during its early years, it failed to make significant headway east of the Tigris – the limit of the old Roman Empire. Islam later became the dominant religion of the Middle East and made significant inroads into South and Southeast Asia, whereas in the Far East, Buddhism and other inclusive belief systems based on Taoism and Confucianism, which had held sway there for millennia, dominated (although Christianity and Islam were able to establish themselves as minority religions there as well).

Exclusive religions are vulnerable to expanding inclusive religions. When two inclusive religions collide, however, the outcome is unpredictable, and the competition between the two can result in both intolerance and violence. Indeed, not only has inter-religious conflict been rife throughout history, but factions within the same religion

have also often clashed violently – as, for example, witnessed in the bloody struggles between Catholics and Protestants, as well as between *Sunni* and *Shi'i* Muslims.

Currently, in its regionalisation stage, Europe's attachment to the formerly defining religious meme has remained relatively passive since 1914. In addition to the rise and fall of atheist Communism and God-minimising Fascism in the twentieth century, the observance of religion has lost its attraction for most Europeans. Modern-day Europe still respects its religious past and values, but its states are distinctly secular. Moreover, notwithstanding the tilt to the right in many European governments in the twenty-first century, Europe is, on the whole, tolerant of the influx of other religions into its societies. (This, despite the declaration of the former Belgian prime minister, Herman Van Rompuy, in 2004 – five years before becoming the EU's first long-term, full-time president – that Europe's 'universal values' are synonymous with Christianity, and would 'lose vigour' with the admission of Islamic Turkey to the EU.)

The US, on the other hand, mired in decline and already contending with a strong religious culture, is undergoing a powerful phase of secondary polarisation, its influence palpable at every level of society. The US, in fact, is arguably the developed world's most religious civilisation, an assertion that is perhaps borne out by using belief in Creationism as a simple test. Surveys are regularly conducted in the US by a number of professional polling organisations. One 2006 poll found that 55 percent of respondents believed that humans were created by God, with a further 27 percent declaring a belief in evolution as a process 'guided' by God. There have been legal challenges in a number of states for the right to teach 'intelligent design' (which views scientific evolutionary theory as consonant with a divine plan) alongside evolution in American schools. By way of contrast, in 2007, the Committee on Culture, Science and Education of the EU's Council of Europe decried Creationism and intelligent design as 'blatant...fraud' and 'intellectual deception', posing dangers to education.

In the face of expanding Muslim demographics, and the rise of various Islamic states and groups as regional forces, US reaction has been one of powerful, unconscious, defensive polarisation, exacerbating many recent geopolitical zones of friction. The process had begun well in advance of the 9/11 attacks on New York and Virginia, but this epochal event certainly accelerated and coarsened the response. With greater conscious awareness of the context in which events were unfolding and a more judicious application of force, recent US foreign policy disaster might have been avoided. (The wider WCSE has, of course, also been drawn in, notably in the case of the UK, but the repercussions of US policy have affected several other European states too.)

THE BIRTH OF ISLAM AND THE RISE OF THE ISLAMIC WORLD

Islam, in Arabic, means 'submission' – that is, to the will of God. The term in this sense originated with the man who would become known as the Prophet Muhammad. Born *circa* 570 in the Arabian city of Macca, situated in the Hejaz region near the Red Sea, Muhammad's visions in his fortieth year placed him at the heart of a religious revolution. The Arabian Peninsula had not been converted to Christianity and had never been part of

ABOVE **Muslims perform the final walk (*Tawaf al-Wadaa*) around the *Ka'ba* at the Grand Mosque in Mecca on 30 November 2009, during the annual Muslim *Hajj* (pilgrimage) to the Saudi holy city.**

the Roman Empire, although Jews had settled in the area. By this time, however, the example of the Byzantine Empire and its warrior religion would have resonated throughout the whole of the old Persian Empire. This monotheistic power in full flourish would have set a groundbreaking new standard.

In social terms, the revelation of God according to the Prophet Muhammad was based on the same principles as the religion founded by Jesus Christ more than 600 years earlier. It appealed to all social classes, allowing access for the poor to a higher state of being through the power of their own humble existence (and, as Muhammad led an army of Muslims against the Makkans who had persecuted them, access to worldly power through victory in combat too). A key difference between the two religions, which still has ramifications more than 1,300 years later, is that from the earliest days of the WCSE, the Church coexisted with various independent states; Islam, on the other hand, which unlike Christianity grew outside the context of a state or empire, was conceptualised as an *ummah* ('nation'), or a totalising form of social organisation that supersedes the state. (As with all religions, this concept is based on an ideal that has been imperfectly realised. Nor, as we shall see later, was Islam immune to schisms.)

When Muhammad began receiving what he called divine visions and reciting what would become the *Qur'an*, he accumulated followers from among his relatives and from the dispossessed. His reformist ideas challenged the order of Makkan society and aroused the enmity of the ruling Quraysh tribe. (Among other things, Muhammad decried the worship of other gods as well as the attachment to worldly wealth. Neither message would have been particularly welcome in this polytheistic, prosperous society.) In 622, Muhammad and his followers fled under threat of assassination to the nearby oasis of Yathrib, also in the Hejaz. (It soon became known as Madinat al-Nabi – 'City of the Prophet', or simply Madinah (Medina). This emigration from Mecca is known as the *Hijra* and marks the beginning of the Muslim calendar.) Medina became the first Muslim stronghold. A Jewish community had established itself there, and the concept of a single God would not have jarred local sensibilities as in Mecca.

RIGHT **Palestinian elder Haj Khalil Ali Zuwara in February 2006. The eighty-two-year-old spends his days reading from the *Qur'an*, the Islamic holy book, in a mosque in the West Bank town of Bethlehem. The ever-fragile balance between Christian and Muslim residents of the town where, according to tradition, Jesus was born, is threatened by the rise of the militant Islamic group Hamas.**

Once Muhammad had consolidated power in Medina, the Muslims began to attack the Makkan caravans in retaliation for their having confiscated their wealth. (Medina was agricultural, whereas Mecca was not. Thus Muhammad and his followers were initially at a loss for resources.) A series of battles ensued, most of which resulted in victory for the Muslims. (The eighth-century biographer of Muhammad, al-Waqidi, refers to this period between the Hijra and Muhammad's death in 632 as *al-maghazi* – 'the raids'.) War was also declared on the prosperous Jewish tribes of the former Yathrib, which found themselves participants in these political circumstances. These tribes are believed to have aligned themselves with the more established Makkans, with whom they had more in common, against the new, upstart Muslims, and they fell foul of Muhammad and his followers as a result. In a series of raids, the Muslims either expelled or decimated the Jews of the oases in and around Medina.

This seminal conflict between Muhammad and the Arabian Jews, which arose out of a specific political conflict of the day, was ultimately hardened over the centuries as a key tenet in the polarisation of both Jews against Muslims and Muslims against Jews – although it must be noted that the *Qur'an* also contains numerous injunctions against intolerance of both Jews and Christians.

As is so often the case, Muslim victory encouraged other tribes to ally themselves with Muhammad's people, or to convert and become Muslims themselves. Islam began as an inclusive warrior religion that had the capacity to attract ethnic sub-groups to form a greater collective with shared beliefs, a model that has served as an integral geopolitical element up to the present day. As with all polarisation, the meme also incorporated ways of identifying its followers, such as the commands to make ablutions and pray five times a day, for men to grow their facial hair and so on.

Muhammad also introduced the concept of *jihad*, a term that remains highly contentious in any discussion today of Muslim extremism. Often misunderstood as meaning 'holy war', the word in fact translates as 'striving'. Recalling that the development of early Islam unfolded during a time of strife in western Arabia between factions of polytheists, believers of the new faith called Islam and local Jewish tribes, and considering that material survival was at stake for the disenfranchised Muslims in the face of competing tribes with greater wealth, power and status, certainly this 'striving' had a military component. Yet *jihad* is also understood as a greater striving,

وأوحينا إلى موسى إذ استسقىه قومه أن اضرب بعصاك الحجر فانبجست
منه اثنتا عشرة عينا قد علم كل أناس مشربهم وظللنا عليهم
الغمم وأنزلنا عليهم المن والسلوى كلوا من طيبت ما رزقنكم
وما ظلمونا ولكن كانوا أنفسهم يظلمون ﴿١٦٠﴾ وإذ قيل لهم اسكنوا
هذه القرية وكلوا منها حيث شئتم وقولوا حطة وادخلوا الباب
سجدا نغفر لكم خطيئتكم سنزيد المحسنين ﴿١٦١﴾ فبدل الذين
ظلموا منهم قولا غير الذى قيل لهم فأرسلنا عليهم رجزا من
السماء بما كانوا يظلمون ﴿١٦٢﴾ وسلهم عن القرية التى كانت
حاضرة البحر إذ يعدون فى السبت إذ تأتيهم حيتانهم يوم سبتهم
شرعا ويوم لا يسبتون لا تأتيهم كذلك نبلوهم بما كانوا يفسقون ﴿١٦٣﴾
وإذ قالت أمة منهم لم تعظون قوما الله مهلكهم أو معذبهم عذابا
شديدا قالوا معذرة إلى ربكم ولعلهم يتقون ﴿١٦٤﴾ فلما نسوا ما ذكروا به
أنجينا الذين ينهون عن السوء وأخذنا الذين ظلموا بعذاب بئيس بما
كانوا يفسقون ﴿١٦٥﴾ فلما عتوا عن ما نهوا عنه قلنا لهم كونوا قردة
﴿١٦٦﴾ وإذ تأذن ربك ليبعثن عليهم إلى يوم القيمة من يسومهم

العذاب إن ربك لسريع العقاب وإنه لغفور رحيم ﴿١٦٧﴾ وقطعنهم
فى الأرض أمما منهم الصلحون ومنهم دون ذلك وبلونهم بالحسنت
والسيئات لعلهم يرجعون ﴿١٦٨﴾ فخلف من بعدهم خلف ورثوا
الكتب يأخذون عرض هذا الأدنى ويقولون سيغفر لنا
وإن يأتهم عرض مثله يأخذوه ألم يؤخذ عليهم ميثق الكتب
أن لا يقولوا على الله إلا الحق ودرسوا ما فيه والدار الآخرة خير للذين
يتقون أفلا تعقلون ﴿١٦٩﴾ والذين يمسكون بالكتب وأقاموا
الصلوة إنا لا نضيع أجر المصلحين ﴿١٧٠﴾ وإذ نتقنا الجبل فوقهم كأنه
ظلة وظنوا أنه واقع بهم خذوا ما ءاتينكم بقوة واذكروا ما فيه
لعلكم تتقون ﴿١٧١﴾ وإذ أخذ ربك من بنى ءادم من ظهورهم ذريتهم
وأشهدهم على أنفسهم ألست بربكم قالوا بلى شهدنا أن تقولوا يوم
القيمة إنا كنا عن هذا غفلين ﴿١٧٢﴾ أو تقولوا إنما أشرك ءاباؤنا من
قبل وكنا ذرية من بعدهم أفتهلكنا بما فعل المبطلون ﴿١٧٣﴾ وكذلك
نفصل الآيت ولعلهم يرجعون ﴿١٧٤﴾ واتل عليهم نبأ الذى ءاتينه
ءايتنا فانسلخ منها فأتبعه الشيطن فكان من الغاوين ﴿١٧٥﴾ ولو شئنا

حزب ١٨

that is, for spiritual development amid the conflicting demands of worldly circumstances – a concept found in virtually every major religion. *Jihad* in the *Qur'an* obtains plausible readings for both interpretations and they both have currency among Muslims today. Critics of Muslim 'apologists' accuse them of preferring the second interpretation of *jihad* at the expense of the first; and certainly, the fact that Islamists of the *al-Qa'ida*, *Taliban* or *Hizbullah* ilk believe in armed *jihad* as a major part of striving cannot be dismissed. Yet the nuances of the concept cannot be ignored or waved away as inconvenient, and its interpretations are many: Muslims constitute 23 percent of the world population, and by virtue of their heterogeneous origins and traditions, they do not all understand *jihad* (or any other key tenet of Islam) in the same way.

The *Qur'an* contains strong passages, akin to those found in the Old Testament, mandating the total and brutal subjugation of people belonging to a different faith; yet it also contains many injunctions against intolerance, and finds accommodation with people of other religions as long as they do not declare themselves against Islam. Armed *jihad* is also widely understood in defensive, not aggressive, terms – yet another shade of nuance that has caused much debate today.

Under the prophet's successors, however, conquest was a primary driver: less than 150 years after Muhammad's death, the caliphate that Muslims established extended from Portugal to the Chinese frontier. (The concept of the caliphate is a uniquely Muslim one, whereby political authority of the *ummah* is concentrated in the caliph (from *khalifa* or 'successor'), who leads the Islamic world bound by Islamic law – the *shari'a* ('path').) Islam proved unique in its capacity to inspire its followers to fight without fear, in the belief that a greater, divine realm awaited them after death. Muslim armies were thus highly effective against opposing forces whose soldiers had less commitment to dying on the battlefield.

Having conquered such a vast amount of territory so quickly, the Muslim victors avidly absorbed the learning, skills and technology of the civilisations now under their rule. Islam matured from a group of tough desert oasis-dwellers into one of the most cosmopolitan, open-minded and moderate empires to which history has been witness.

The list of caliphates and independent Muslim dynasties ruling southern Europe, North and parts of East Africa, Asia Minor, the Middle East, Persia, Central Asia, India and parts of Southeast Asia is a long and varied one, stretching over centuries from the first caliphs after Muhammad to the modern era. It culminates in the demise of the Ottoman Empire at the end of World War One and the abolition of the caliphate by Mustafa Kemal Atatürk's Republic of Turkey in 1924.

The first caliphates were entirely Arab, but their non-Arab Muslim subjects gradually accumulated more power within them. Once the borders of a new empire had stabilised in maturity, the energy that would normally fuel further conquest turned inwards, and sectarian differences arose – a normal dynamic, according to the Five Stages of Empire model. Independent Muslim dynasties tore into the integrity of the various Islamic empires, which began to tip into decline around the beginning of the second millennium, just as the WCSE was rising.

SUNNIS AND SHI'IS

One of the most important developments in the history of Islam, which continues to have hugely significant ramifications today, was the schism that arose upon the death of Muhammad over the question of who would succeed the prophet as the caliph of the 100,000-strong *ummah*. Abu Bakr, Muhammad's father-in-law and close companion, was elected caliph. However, a faction of Muslims espoused the view that Muhammad's cousin and son-in-law, 'Ali, was the rightful successor. The first group became known as the *Sunni* (i.e. followers of the *Sunnah* – the body of ideas and behaviour modelled after the prophet's own). The second group was called *Sh'iat 'Ali*, or 'Partisans of 'Ali' (the term '*Shi'i*' is a contraction thereof). These very different contentions would prove more serious a rift than those between the Catholic and the Orthodox Churches in the early days of Christianity, and as deep as the initial polarisation between the Catholics and Protestants in the later Christian era. Both candidates for succession were worthy: Abu Bakr is said in some traditions to have been the very first male convert to Islam and stood by Muhammad throughout the prophet's life and trials. 'Ali, who had distinguished himself since childhood as one of Muhammad's most loyal followers and became his beloved lieutenant, is said in other traditions to have also been Islam's first male adherent. The prophet had publicly praised both men in words that could well have indicated approval for succession.

Abu Bakr's election prevailed, and 'Ali stepped aside as two more caliphs went on to rule in succession. When the last, 'Uthman, was assassinated, 'Ali finally assumed the role and held on to it despite the civil war that ensued. He was assassinated in 661, and his sons Hassan and Hussein – the last of Muhammad's bloodline who had been alive during the prophet's lifetime – were later also killed, in further politically motivated attacks. ('Ali and his sons, particularly Hussein, are deeply revered by *Shi'is* as martyrs and cornerstones of the faith after Muhammad.) From then on, the Islamic Empire was exclusively a *Sunni*-dominated affair until the rise of the *Shi'i* Fatimid Dynasty in 909, which contested the rule of the 'Abbasid caliphate and went on to control North Africa, Sicily, Palestine and the Levant, and western Arabia. *Sunni* domination resumed with the *Shi'i* Fatimid Dynasty's decline in 1171, and endured until the last of the Ottomans – barring one critical exception. With the ascension of the Turkic Safavid Dynasty in the early sixteenth century, formerly *Sunni* Persia was converted to *Shi'ism* and sandwiched between the *Sunni* Ottomans and the *Sunni* Mughals in India. Persia, now Iran, remained *Shi'i*, and is, in the present day, the only *Shi'i* power in the Islamic world (although countries such as Syria and Pakistan have significant *Shi'i* populations, and *Shi'is* in fact constitute a majority in Iraq, Bahrain, Lebanon and Azerbaijan).

Both *Shi'is* and *Sunnis* believe in the *Qur'an* as the unmitigated revelation by God to Muhammad. *Sunnis* tend towards a more legalistic and administrative bent, and believe in the selection of successive caliphs according to elections among an elite most qualified to rule the *ummah*. *Shi'is*, however, believed that the caliphate should have remained within the family and descendants of the prophet, retaining the spiritual truth of Islam's origins. They revere as holy men the imams descended from the *Ahl al-bayt* ('People of the House'), or Muhammedan lineage. The *Shi'i* imamate has no correlate in the *Sunni* belief system. Moreover, *Shi'ism* cultivates a simpler, more direct, more flexibly defined experience of Islam's spiritual grace, and it maintains that every age brings its appropriate *imam*. (Numerous *Shi'i* sects exist, extolling the righteousness of various imams.) *Sunni* beliefs represent a more conservative approach, given to more fixed interpretations of Islam.

Modern-day Iran saw Ayatollah Khomeini and the Islamic Republic come to power in 1979 as a response against the inept, oppressive and corrupt Pahlavi regime, which was modernist, secular and pro-Western. Khomeinist Iran, although politically oppressive, has maintained a zeal akin to the energy that drove the first expansion of Islam (although it has been very much influenced by Western revolutionary models), and it boasts a population of 72 million, of which two-thirds is younger than thirty years old. Despite the reformist wave of protests following the 2009 presidential elections, which were nevertheless significant, the 1979 revolution's hardliners and their ideological descendants remain in charge.

RIGHT **Hundreds of thousands of people attend the funeral of Ayatollah Khomeini at Behesht-e Zahra cemetery in Tehran, Iran, on 6 June 1989. His death closed a turbulent chapter in Iran's history, but opened a long and still uncertain phase in its evolution as a nation.**

That an aggressive *Shi'i* state is constantly jockeying for regional supremacy has long kept the *Sunni* states of the Middle East nervous. Their anxiety has not been soothed by the opportunities in Iraq that the 2003 US invasion opened up to Iranian influence, by Iran's progressive pursuit of nuclear weapons and by its ability to foment unrest in Lebanon and Palestine. The conservative *Sunni* states of the Arabian Gulf have sought to maintain the *status quo*, and have developed alliances with the West as a hedge against a second phase of Iran's Islamic Revolution. The weakness of the *Sunni* states, particularly Saudi Arabia, lies in the narrow wealth holdings of their royal families. Unless clear, effective policies are implemented to alleviate the disenfranchised poor in these states, the risk of a fundamentalist Islamic revolution led by Iran remains high.

THE EXPANSION OF ISLAM TODAY

Once again, the phenomenon of surging demographics must be considered, this time in the case of the Islamic world. At 1.57 billion people, adherents to Islam comprise one-quarter of the world's population (see Figure 40). The densest concentration of Muslims is in the Middle East and North Africa, where they constitute approximately 95 percent of those regions' inhabitants. Further east, India and Indonesia together account for over 23 percent of the global Muslim population. (The Asia Pacific region contains nearly 62 percent of the world's Muslims, compared with the Middle East and North Africa at just over 20 percent, and Sub-Saharan Africa at just over 15 percent. Muslim numbers in Europe and the Americas combined amount to around 3 percent of the world total.) The population of the Islamic world is growing at 1.8 percent per annum, compared with the world rate of 1.12 percent; the meme of Islam is thus expansionary. A 2007 study by the Carnegie Endowment for Peace found that Islam topped the list of the fastest growing faiths at 1.8 percent, followed by Bahá'í (1.7 percent), Sikhism (1.6 percent), Jainism (1.6 percent), Hinduism (1.5 percent) and Christianity (1.3 percent). Birth rates were found to be the main factor for religious growth in all cases.

The Islamic world is in a primary phase of expansion and polarisation, in which inward demographic pressure causes an expansion outwards, triggering the defensive

Muslim Population, 2009 *(fig. 40)*

Data Source: Pew Forum

AUX
LIBERTICIDES !
NO

response of secondary polarisation in countries that resist or attempt to curtail its expansion. This is a long-term trend, comparable to more conventional empire cycles, which will unfold over centuries.

Yet, as much as the West must acknowledge the demographic upswing of Islam, there is no 'Islamic Empire' with which to contend, only individual states or blocs informed by Islam, and dealt with like any other state within the framework of international relations. Moreover, Islam does not conceptually or philosophically stand in opposition to the West. However, in the light of terrorist attacks by the extreme fringes of Islam, Europe is understandably anxious, and Western denunciation of fundamentalist Islamic practices has been vocal and often overbearing as a result. Western Muslims have also become a focus for varying degrees of secondary polarisation, with, for example, the Islamic dress controversy in France since 2004 (still ongoing), and also in Germany and the Netherlands; the Danish cartoon furore in 2005; the Swiss ban on minarets in 2009; and other instances of reactions to the cultural shifts instigated by the growing minority presence of Muslims in Europe. Elements on the left and right alike in Europe have attempted to assert national values of secularism or Christianity.

LEFT Thousands of Muslims, mostly women, demonstrate on 14 February 2003 in Lyon against the proposed banning of the wearing of Islamic headscarves in schools. Despite widespread demonstrations, the law was passed. The tensions between secularist and religious elements in Europe remain.

Muslims in the West, meanwhile, are negotiating complex questions of identity, including the assimilationist demands of Western societies; perceptions of them as a fifth column or as a 'new' ethnic minority; feelings of alienation and rootlessness, which have been exploited by radical Islamists; and simply living as fully participant citizens of secularised states while simultaneously maintaining their beliefs and cultural traditions – which the vast majority, of course, do.

These social changes and concerns are indeed significant, and the geopolitical implications of the violent radical Islamist agenda must, of course, be confronted; but there is no projection of 'Islamic power' in the conventional sense of an empire. The preoccupation by so many in the non-Muslim West over the 'threat' of Islam is, and has been, a distraction from the growing challenges that the West faces from China, which really does have the power to change its way of life over the coming decades.

ISLAMISM

From its splendid height under the Umayyads and the early 'Abbasids, the caliphate dissolved into various dynasties during the later 'Abbasid reign, regaining some degree of coherence under the Fatimids (one of the rare eras in which *Sunni*s and *Shi'*is coexisted peacefully under a single rule). It is important to note that there was never any centralised state or religious authority for all of Islam, but a series of empires and dominions accompanied by a widely varying degree of interpretations regarding religious practice.

The Ottoman Empire, arising at the beginning of the thirteenth century, would become the longest lasting of the Islamic empires, eventually dominating North Africa and parts of the East African coast, Greece, the Balkans, great swathes of Eastern Europe and most of the cosmopolitan Middle East. (Islam had begun to spread into Southeast Asia 100 years earlier via Muslim sea traders, and it made significant inroads into that region long before the arrival of Christianity.)

By the nineteenth century, Ottoman power was faltering, but Britain and France prolonged its life well past the extent that might normally have been expected; they saw in the empire a useful buffer against Russian ambitions. At the end of World War One, the British and French partitioned the Ottoman Empire, and the last vestiges of Islamic imperialism were dissolved. At the empire's nadir, the WCSE was thus at its height. However, ironically, World War One also marked the beginning of the WCSE's decline, and, although the territory of the old Ottoman Empire was now occupied, the European hold over it was tenuous. It would last less than half a century longer.

Meanwhile, political Islam – Islamism – developed into a considerable force. Islamists view Islam as not just a religion but also a set of values to be implemented in every aspect of modern society. It was embodied in movements such as the Muslim Brotherhood, the *Sunni* organisation founded in 1928 in Egypt by schoolteacher Hassan al-Banna. Widely acknowledged as the originator of modern-day Islamism, it espoused a revivalist course for Islam that stood in opposition to Western imperialism, emphasising pan-Islamic unity and the supremacy of *shari'a* law. Its heirs are seen to be such groups as Hamas and *Hizbullah*, which also make use of religion-infused radical politics – although with differing aims and through armed struggle. In contrast, the Muslim Brotherhood (where not outlawed) has by and large stayed within legal boundaries.

Hamas and *Hizbullah* fight for local and nationalist agendas, as do latter-day Muslim Brotherhood branches, despite their pan-Islamic rhetoric. Radical supra-national Islamist groups like *al-Qa'ida*, on the other hand, adhere to a considerably more fundamentalist and conservative interpretation of Islam that holds the implementation of *shari'a* law as paramount. Without any social agenda, nurtured in a complex equation by globalised rootlessness, modern revolutionary politics and Westernisation, these offshoots are far from manifestations of traditional Islam.

The 1948 creation of Israel in Palestine, the heartland of the Middle East, served to focus radical Islamist polarisation against the empires of the WCSE deemed complicit in the triumph of Zionism. (Hostility towards Jews, Christianity and *Shi'*ism – embodied by Iran – are also typical of radical *Sunni* Islamist groups in the *al-Qa'ida–Taliban* mould.)

Ironically, on several occasions in the twentieth century, Islamist forces were ideologically allied with the West. Western (as well as Western-friendly Arab/Muslim state) support of the Afghan *mujahideen* against the Soviets during the latter's invasion of Afghanistan is the most readily summoned example, but there have been many other instances of Western-Islamist alliance. These include Bolshevik support of Islamist factions against the 'imperialist' West in Dutch-ruled Indonesia in 1920 and in Azerbaijan during the same period; Fascist Spain under Francisco Franco recruiting Muslim Moroccan fighters to attack Republicans; US-backed anti-Socialist Saudi opposition to Nasserist Egypt; and instances in Algeria, Sudan, Yemen and Oman.

However, the ill-conceived US invasions of Afghanistan and Iraq since 2001 and 2003, respectively, have resulted in the death of a great many innocent civilians and have as a result radicalised a new generation of Muslims. The insurgencies in these countries have maintained low-level resistance effectively, and thwarted US-led attempts to stabilise these regions – a significant indication of strong Islamist motivation

ABOVE The flag of the future Jewish state is raised during morning parade at a training base of the fledgling Israeli Defence Forces on 27 April 1948, less than three weeks before Israel's independence, in what was still the British Mandate for Palestine.

that, facing occupation, is greater than that of the West. The perception of Western failure and decline are bound up with the US-prosecuted wars in the Middle East, thereby fuelling Islamist vigour.

The US cannot afford to deploy enough troops to win the First Gulf War outright. Instead, it is engaged in protracted deal-making between factions that are still likely to descend into sectarian civil war once the US withdraws. This further risks a more decisive drawing of majority-*Shi'i* Iraq into Iran's sphere of influence. The Afghanistan conflict, meanwhile, is failing to secure stability for ordinary Afghans because the *Taliban* remains a force in the region and has spilled over into neighbouring Pakistan, threatening to destabilise that already-insecure country. The situation has radicalised Pakistani Muslims on the Afghan frontier, and polarised Muslim sentiment against the West. The worst-case scenario has it fuelling an Islamic revolution that could sweep the country, making Pakistan, with its nuclear capability, potentially one of the world's most dangerous states.

THE NEW WAVE OF POLARISATION

Since the birth of Muhammad, Christian and Muslim forces have contested the Mediterranean region. With the decline of Western power and the ending of the British and French Mandates, Islamic states have sought to occupy the power vacuum, as the

ABOVE Afghan anti-Soviet resistance fighters in the east of the country in 1980. The Afghans repulsed the Red Army's 1979–89 invasion at a huge human cost and with the material aid of the Western world, above all the US. Estimates indicate that around 1.5 million Afghans were killed in the fight against the Soviet invaders.

first caliphate did after the decline of the Roman Empire. The Israel–Palestine conflict is but a microcosm of this wider dynamic.

Since the *al-Qa'ida* attacks of 11 September 2001, the world has been witness to the struggle between the primary polarisation of radical, supra-national Islamism and the secondary polarisation of Christian-led American neo-Conservatives, each side warring for the allegiance of moderates on both sides of the divide. History teaches that such conflicts can become protracted, with no limits to the violence and atrocities that they can unleash. Once momentum builds on both sides, it is very difficult to reduce tensions, and, despite the ebbing since 2001 of both sides (*al-Qa'ida*'s capabilities have diminished and US neo-Conservatives hold less direct sway in the post-Bush era), they are far from spent forces – each remains dangerous and has inspired numerous successors of one kind or another. As such, there remains a pressing need to prevent this conflict from spreading further across the globe.

The history of the Middle East during the last century yields a real-time example of polarisation in action and insight into the potential mechanisms of reversing the process. It can be divided into five stages, as described below.

STAGE ONE: OCCUPATION, 1918–45

THE BIRTH OF ZIONISM

Jews in the WCSE had been tolerated during some eras and persecuted during others. As the super-empire moved into its peak prior to World War One, persecution was particularly vicious in Russia and Eastern Europe. In response, Zionism, an international movement, was founded in 1897 with the sole purpose of securing sanctuary for the Jewish people by establishing an independent Jewish state in Palestine. This was the spiritual home of the Jews, but was at the time under Ottoman rule and populated by a majority of Arab Muslims and Christians.

THE COLLAPSE OF THE OTTOMAN EMPIRE

After 1918, the British and French Mandates over the formerly Ottoman Middle East heralded the return of Christian imperial rule to the region (short-lived though it would be). This process marked the low point of the empire cycle for the Islamic Middle East and the high point for the WCSE. Both the Arabs and the Jews in the region supported the Triple Entente during the war, with Zionist Jews hoping to obtain Western backing for a state and the Arabs seeking to realise their newly nationalistic aspirations.

The British, attempting to balance these dynamics, fared badly by making conflicting promises to both peoples, in the form of the 1915–16 Husayn–McMahon correspondence to the Hashemites and the 1917 Balfour Declaration to the Zionists. Violent riots targeting Jews ensued, followed by decades of further Arab resentment and Zionist paramilitary terrorism. Jewish immigration to Palestine surged, although tensions precluded the new arrivals from integrating effectively with the local Arab population. The British response was to restrict the number of Jewish immigrants, which nevertheless continued illegally, expanding once again following Hitler's rise to power in Germany.

Meanwhile, as has been discussed, the Muslim Brotherhood was founded in Egypt in 1928 in response to the country's occupation by the British. The organisation opposed violence much as Gandhi did in India, and it expressed its aims politically, weaving itself into the fabric of Egyptian society.

This period marked the early stages of regionalisation in the Middle East, resulting in a succession of regional wars. When the West withdrew following the end of World War Two, it left behind a string of newly created and/or independent states, including the Iraq of today – encompassing Kurdish, *Sunni* and *Shi'i* zones – and the 'island' of Israel.

STAGE TWO: THE REBIRTH OF THE MIDDLE EAST, 1945–79

ISRAEL

As Israel came into existence, supported in great measure by Jewish Zionists in the US and by American politicians, tension arose between the US's support for Israel and its need for strategic access to Middle Eastern oil and political alliances. This has lasted for over sixty years and is one of the great influences on Middle Eastern geopolitics.

The original borders of the Jewish state were expanded in both offensive and defensive wars (most fatefully in 1967 by the annexation of the West Bank and Gaza

PEOPLE
OF AMERICA

Strip, as well as of the Golan Heights), and then contracted when Israel handed the Sinai back to Egypt in 1979 in exchange for the peace that endures tenuously to this day.

US support for Israel stayed strong, particularly during the Cold War. Britain offered more limited support, also judiciously maintaining ties with Arab states. The Soviets generally supported Arab states. Israeli victories in the wars of 1956, 1967 and 1973 were achieved against overwhelming odds, backed by superior US-supplied military equipment. Israel's conventional dominance was reinforced when it became a nuclear power in, it is believed, the late 1960s.

LEFT **Palestinian refugees from Gaza leave the West Bank, occupied by Israeli troops, to go to Jordan in January 1968. After the Israeli army started a lightning war in Syria, Sinai and Jordan in June 1967, the problem of Palestinian refugees increased without precedent in the Arab countries around Israel.**

BELOW LEFT **Protestors burn the Stars and Stripes during an anti-US demonstration outside the US embassy in Tehran in November 1979, after the flag was seized by revolutionary students.**

IRAN

The reformist-minded Mohammad Mossadegh was named Prime Minister of Iran in 1951 but, as his secular government moved to nationalise oil production, it was overthrown in a coup sponsored by the CIA and MI6. Mossadegh remained under house arrest until his death in 1967. The Shah, meanwhile, concentrated all power in Iran on himself and ruinously dominated the country for twenty-five years as a compliant Western ally who ensured the steady supply of oil to the US and Europe. Widespread resentment against the Shah's poor administration of the country and his feared and brutal SAVAK secret police provided fertile breeding ground for revolution, which arrived by way of Ayatollah Khomeini in 1979. The Islamic Revolution set Iran decisively against the West, and particularly the US.

STAGE THREE: ISLAMIC REVOLUTION, 1979–89

The populist revolt against the Shah of Iran began in 1978 with a small demonstration by religious students, who at the time were violently repressed. The widespread anger that ensued set off waves of similar protests. Despite the thousands of casualties, one in three Iranians would ultimately take to the streets. The despised Shah was forced to flee, and Khomeini's theocracy was instituted. A stalwart of American foreign policy in the region had, in the space of a year, turned into the US's greatest enemy.

The rage that Iranians had built up against the Shah was now refocused by Khomeini against the American 'Great Satan'. With their supreme leader's blessing, a group of students from the Islamic associations of Tehran's main universities stormed the US embassy and held fifty-three Americans hostage for 444 days, demanding the return of the Shah for trial. The incident was a clear challenge to US power in the region, compounded by President Jimmy Carter's humiliating failure to rescue the hostages. It acted as a major contributor to Islamist polarisation, further alienating the US, which had been completely outflanked.

THE SOVIET INVASION OF AFGHANISTAN

As the Cold War heated up, the US consolidated strategic gains through the Egypt–Israel peace accord, and closer military ties to Saudi Arabia and Iraq. In turn, the USSR provided billions in economic and military aid to the Communist People's Democratic Party of Afghanistan, which had overthrown the republic established by Mohammad Daoud.

One month after the Islamic Revolution in Iran, Muslim militants abducted and killed the US ambassador in Kabul, marking the rise of radical Islam in the country. When the government's conflict with the *mujahideen* intensified, the USSR saw an opportunity and invaded the country. In response, Britain, the US and Pakistan deployed covert operatives to train and lead the *mujahideen*, supplying them with weapons, such as Stinger missiles, that endowed them with a strategic advantage. (As noted, this support was a prime example of common cause between Islamists and the West – albeit a short lived one that would have serious future consequences. The war was a proving ground for Osama bin Laden and other Islamist terrorists. Indeed, the *mujahideen* would continue its armed *jihad* even after the Soviet withdrawal – as *al-Qa'ida*, against the perceived Western occupation of sacred Muslim lands.)

LEFT **Iraqi soldiers keep a close eye on Iranian prisoners captured on the front line, southeast of Tigris, in March 1985 during the Iraq–Iran War. The conflict was initiated in September 1980 after Iraqi troops and tanks crashed across the Iranian border in a dawn thrust to encircle Abadan, an oil refinery city.**

The long guerrilla war sapped the strength of the USSR and would ultimately contribute to the collapse of the Berlin Wall ten years later.

THE IRAN–IRAQ WAR

In 1980, seeking to exploit the chaos of the Islamic Revolution, Saddam Hussein invaded Iran with tacit US approval (and, later, support). The eight-year war that followed – for Iraq had sorely underestimated Iran – was one of the bloodiest and most consequential conflicts in the history of the Middle East, with an estimated 1.5 million casualties on both sides. Hundreds of thousands more combatants were severely wounded.

OTHER INCIDENTS

This period saw numerous other pivotal incidents that helped to usher the Middle East to where it is today, and accelerated polarisation on all sides. Among them were Israel's successful pre-emptive air strike against Iraq's nuclear programme; the beginning of the internecine fifteen-year Lebanese Civil War in 1975; the Israeli invasion of Lebanon to expel the Palestine Liberation Organization (PLO) in 1982; the 1983 bombing of the US Marine barracks in Beirut, Lebanon, by the Islamic *jihad* terrorist organisation (possibly a *Hizbullah* alias); and the 1988 attack by a US missile cruiser on an Iranian airliner, killing all 290 passengers.

STAGE FOUR: VICTORIES AND DEFEATS, 1990–2001

THE FIRST GULF WAR

Financially crippled by the war against Iran and indebted to Saudi Arabia and Kuwait, Saddam Hussein used reduced oil prices and an old territorial dispute as pretexts to invade Kuwait in 1990 with the aim of refilling Iraq's coffers. However, he did not expect the US (which its ambassador in Baghdad implied would remain neutral) to seek to liberate Kuwait. The US was, in fact, taking advantage of its role as the world's only superpower in the last days of the USSR to create a broad-based Western-Arab coalition and showcase the military technology that it had developed during the Cold War.

The war ended little more than six months later. It was a watershed for US foreign policy: having joined forces with the Gulf states against a common threat, and

ABOVE Iraqi girls, wearing *keffiyeh*, march in front of a painting of Iraq's then-president, Saddam Hussein, in Tikrit, north of Baghdad, during celebrations for his sixty-fourth birthday on 28 April 2001.

then persuaded Israel not to retaliate against Iraq for firing eight Scud missiles at it in an attempt to draw it into the war, the US had greatly improved its profile in the Arab Middle East.

THE ISRAEL–PALESTINE CONFLICT

During this period, the prospect of an independent Palestinian state came closer to reality than ever before. The promising but largely symbolic Madrid Conference in 1991 was followed by the 1993 Oslo Accord that created the Palestinian Authority. Criticisms of both conferences were numerous, and the possibility of a separate state did not progress beyond 1993 in any meaningful way. This laid the ground for the rise of the Islamist group Hamas as a political force in Gaza over the coming decade before winning an election in 2006 (unrecognised by Israel). The first *intifada* (Palestinian resistance) wound down around 1993, but a second was brewing and would erupt in 2000. Suicide bombing became a more widely employed tactic by radical Islamist groups after 1993. The issue of Palestine remained a primary focus for the Islamic world in its dealings with the West, and was also a hotly brandished recruitment tool for radical groups.

BOMBINGS AGAINST US TARGETS

The 1993 car bombing of the World Trade Center in New York was attributed to a group that had trained with *al-Qa'ida* in Afghanistan, and whose ringleader was the nephew of Khalid Sheikh Mohammed. He would later be cited as the 'mastermind' behind the 2001 destruction of the same buildings. The attack was motivated primarily as a result of America's support for Israel.

In 1998, suicide truck bombings occurred simultaneously at the US's embassies in Nairobi and Dar es Salaam, killing hundreds. The operations were tracked to the radical Islamist group Egyptian Islamic Jihad (EIJ), in concert with Osama bin Laden and Ayman al-Zawahiri. They marked the first time that the names of *al-Qa'ida*'s number-one and number-two leaders came to public prominence in the West. (The attacks are believed to have been acts of retaliation against the US for its role in the arrest of EIJ members two months earlier.) The US under Bill Clinton responded by attacking targets in Sudan and Afghanistan with cruise missiles. A Sudanese pharmaceuticals factory that the US claimed (ultimately unpersuasively) was producing chemical weapons was destroyed. The Afghan targets were terrorist training camps. Although the response was, all told, woefully insufficient, the US came to pay closer attention to bin Laden's activities – but not, as the attacks in New York and Virginia three years later would make clear, close enough.

BELOW **Palestinian boys throw stones at an Israeli tank during clashes in the West Bank city of Nablus in August 2003. Tension between the Palestinian populace and the Israeli authorities continues unabated.**

In 2000, a small seacraft laden with explosives rammed into the hull of the US navy destroyer USS *Cole* as it refuelled in Aden in the Yemen, killing nineteen servicemen and injuring thirty-nine. *Al-Qa'ida* claimed responsibility.

RIGHT **The front page of the *Daily News* of 27 February 1993 headlines the 680 kg bomb set by *al-Qa'ida* conspirators that was intended to knock the North Tower into the South Tower, bringing both down. It failed to do so but killed six people and injured 1,042.**

THE RISE OF GEORGE W. BUSH

With his controversial election victory in 2001, Bush ushered in a neo-Conservative agenda and personnel that pushed for unrestrained US geopolitical dominance, advocating widespread implantation of US-style democracy and free-market values, and laying the theoretical groundwork for the disastrous invasions of Afghanistan and Iraq. The Bush regime embodied the values typical of secondary polarisation, reasserting its narrow convictions and nurtured by fundamentalist Christian beliefs.

STAGE FIVE: SECONDARY POLARISATION, THE US AFTER 9/11

The attacks of 9/11 constituted a stunning success for radical Islamists, but, as noted, it was an avoidable failure of the US security services. The World Trade Center was destroyed and nearly 3,000 people killed, yet arguably the greatest damage was the psychological blow to a nation that had believed itself invulnerable to domestic attack. The US under threat resorted to an extreme response, which in turn made it easier for Islamist terrorists to justify their methods and recruit more fighters. The Bush neo-Conservatives now had a focus for their agenda, positing an 'Axis of Evil' that included North Korea, Iran, Iraq and, subsequently, Cuba and Syria, and justifying a range of controversial measures. A prominent example is the Guantánamo Bay detention camp, to which terror suspects were sent, and which operated outside US legal jurisdiction and without adherence (until 2006) to the Geneva Conventions. Another is the concept of 'extraordinary rendition', whereby prisoners could be transported clandestinely to states that practised torture. As a result of such policies, the US under Bush lost any claim it had to moral supremacy.

WAR IN AFGHANISTAN

The US invasion of Afghanistan was conducted with minimal land forces, and, with reliance on covert operatives, airpower and the Northern Alliance, a patchwork collection of armed groups united against the ruling *Taliban* – *al-Qa'ida*'s protectors. World sympathy accompanied the campaign to overthrow the *Taliban*, but initial success was not followed by the necessary dedication of resources towards building a sustainable Afghan state or economy once the *Taliban* had been removed from power. The *Taliban* regrouped as an insurgency that has since inflicted over 1,700 casualties (as of mid-2010). The Afghan government under Hamid Karzai is far from a model of stability and good administration, bin Laden remains elusive and the Pakistani border areas with Afghanistan have become inflamed, threatening that country's internal cohesion as well. The net effect for the US has been huge expenditure, failed strategy and a sapping of will from its staunchest allies.

SPORTS ★ ★ ★ ★ FINAL

DAILY NEWS

40¢ NEW YORK'S HOMETOWN NEWSPAPER Saturday, February 27, 1993

NEW YORK'S DAY OF TERROR

KEN MURRAY

Trade Center bomb kills at least 5, wounds hundreds; link to Bosnia airlift feared

14 PAGES OF COVERAGE BEGIN ON PAGE 2

THE SECOND GULF WAR AND ITS AFTERMATH

ABOVE **US military police guard *Taliban* and *al-Qa'ida* detainees in a holding area at Camp X-Ray at the naval base in Guantánamo Bay, Cuba, in January 2002. Despite promises by Barack Obama to close the facility, in May 2009 the US Senate blocked funds for the transfer of prisoners. As of July 2010, 176 detainees remain.**

The 2003 invasion of Iraq was undertaken by the US with the cooperation of the UK on what proved to be the false pretext that Saddam Hussein was developing weapons of mass destruction. Initially, it was a military success. As a sustained campaign, however, it utterly squandered the goodwill of most of the population – particularly the Iraqi *Sunni* Muslims, who formerly made up the ruling class (where Kurds and the majority *Shi'is* had long suffered repression). In terms of winning the hearts and minds of most of the Islamic world, moreover, the war in Iraq was a complete disaster. Poorly equipped and undertrained personnel, who were outmatched by an unexpectedly robust and complex insurgency, failed to understand or connect with the local people or to improve their lives. Scandals such as the Abu Ghraib torture incidents, and the arrests and killing of innocent civilians, severely weakened the US's moral position in Iraq and the Arab Middle East, and once again nurtured radical Islamist prejudices and self-justification. Plans for reconstruction were insufficiently well developed where they existed, and ineffective troop deployment levels doomed the campaign early on. While some tangible progress was made – notably the troop surge in 2007 – reconstruction proceeded slowly and the threat of sectarian civil war remained a troublesome possibility as the US prepared to withdraw the majority of its troops. By 2010, the war in Iraq had

ABOVE In the *Shi'i* Muslim suburb of Sadr City, Baghdad, in May 2004, Iraqi boys play in front of a mural depicting the Statue of Liberty and a painting copied from a widely published photograph taken in the US-run Abu Ghraib prison, which shows a hooded Iraqi prisoner.

yielded well over 4,000 US casualties. Its duration, like that of the Boer War in the case of Britain, signalled to detractors of the US that it was vulnerable.

ISRAEL, LEBANON AND PALESTINE

The 2006 Israel–*Hizbullah* War began when the Islamist group fired on an Israeli patrol across Israel's border with Lebanon, killing three soldiers and capturing a further two. Israel responded with a massive onslaught on South Lebanon, in a war that lasted thirty-four days and killed 1,000 Lebanese civilians, uprooting as many as 1 million more (in addition to displacing around 400,000 Israeli citizens). The Lebanese infrastructure was severely damaged, inflicting high costs on its economy. Yet Israel could not claim victory: *Hizbullah* had been engaged and incurred losses in terms of both fighters and weapons, but Iran and Syria would continue to sustain the organisation. Worst of all, *Hizbullah* was widely viewed as a resisting force that had challenged the Israelis yet again (the first confrontation had been in 2000, when Israel was forced to withdraw from South Lebanon). The group accordingly acquired a good measure of respect across Lebanon and even the wider Arab world, even grudgingly among its enemies; Israel, on the other hand, was seen as having engaged in combat operations, indifferent to the suffering of innocent civilians. Nor was this expensive and futile war popular in Israel. The propaganda victory for *Hizbullah* was decisive, and it hardened both Islamist and ordinary Muslim sentiment against Israel.

In 2008, a truce between Israel and Hamas in Gaza ruptured, and hostilities escalated to the point where Israel launched a blistering assault on Gaza, using what many commentators expressed to be disproportionate force by air, land and sea. Casualties have been said to number around 1,400, with a great deal of infrastructural damage sustained. While the conduct of Hamas was irresponsible and in violation of combat statutes, the extremity of Israel's response undoubtedly tainted its reputation abroad. In May 2010, Israeli commandos raided a ship in international waters – part of a convoy bringing aid to resource-starved Gaza – and killed several passengers, prompting further outrage and condemnation around the world.

BELOW **The Israeli navy intercepts Turkish passenger ship the *Mavi Marmara* in international waters in the early hours of 31 May 2010. Part of a six-vessel 'Freedom Flotilla', the ship proved to be carrying humanitarian aid to Gaza. Israeli forces killed a reported nine people, with many more being missing or injured. The event was a public relations disaster for the Israeli Government.**

OTHER INCIDENTS

In March 2004, an *al-Qa'ida*-inspired terrorist cell detonated a series of bombs in the Madrid commuter train system, killing 191 people. This was an attack by militant Islamists against Spain, then a member of the Bush alliance in the invasion of Iraq. Popular backlash against the government of José María Aznar affected the outcome of the general election held only three days later – José Luis Rodríguez Zapatero's Socialist Workers' Party was swept into office, and, indeed, Spain immediately pulled its troops

out of Iraq. The bombings, therefore, must be considered to be the second time since 9/11 that an act of terrorism produced a strategic success for Islamist extremists.

Another multi-pronged public transport bombing attack, in London in July 2005, was perpetrated by British nationals of Pakistani and Jamaican descent, demonstrating the domestic threat from homegrown Islamist terrorism. In contrast with Spain, the London bombings hardened British opinion against Muslim terrorists, adding continued support to Prime Minister Tony Blair's polices in Iraq and later Afghanistan. It also set the scene for a wave of secondary polarisation that expressed itself through new legislation that gave the government sweeping powers in the name of anti-terrorism.

There have been additional terrorist attacks since 2001 attributable to Islamist groups, many of them to *al-Qa'ida* – in India, Indonesia, the Philippines, Kenya, Iraq, Israel, Morocco, Russia and elsewhere. Notable incidents include the foiled multiple-transatlantic airliner plot in London in 2006, the assassination of Pakistani Prime Minister Benazir Bhutto in 2007, the Mumbai attacks of 2008 and the 2009 attempted explosion onboard an airliner in the US.

Religion and the Nascent ASE

Unlike the nations of the WCSE, in which a single religion – Christianity – was the dominant thread of belief binding them into a super-empire, the states that comprise the rising ASE each have their own distinct religious beliefs. This religious plurality may come to activate potential polarising mechanisms within the super-empire as it expands. Inclusive religions, such as Buddhism and Taoism, have been harnessed in order to polarise some Asian populations to serve a greater national purpose, much as Rome and the European powers did before them. Japan, India and China are prime examples.

JAPAN

Shinto ('Way of the Gods') is the basis of Japanese spiritual belief, with 119 million adherents today. It does not resemble what most monotheists would understand as a faith, namely with canonical sacred texts and prophetic teachings, a supreme deity, fixed observances and so on. It can be understood as a way of life, integrating the concept of a divine force – the *kami*, believed incarnated in an array of 8 million deities or spirits – with various ritual practices, a code of ethics, folkloric and shamanistic beliefs, and ancestor and nature worship. Over the approximately one-and-a-half millennia since its formal development, the Shinto belief system has incorporated Buddhist, Taoist and Confucian elements, a syncretism tolerated for most of Shinto's history and later forbidden as 'impure' (a rigorous definition of 'purity' being among the main features of the Shinto path). The religion also encompasses various concepts that have analogues in Hinduism as well as, arguably, the esoteric component of the great monotheistic faiths (e.g. the Kabbalah and Sufism).

Shinto has served as the Japanese state religion in a fundamental way: the Imperial Family was accorded an exalted role within it as the embodiment of the soul of

Japan, descendants of the Sun Goddess Amaterasu. (The divinity of the Emperor was formally renounced following Japan's defeat during World War Two.) The origins of Shinto as a formal religion are inseparable from the beginnings of Japanese civilisation, as one of its core concerns was the development of a strong and efficient state apparatus with which to organise society.

Shrines – public and private – are a focal point of Shinto, and they are ubiquitous throughout Japan, with many of them occupying a high place in Japanese culture as national treasures. They act as conduits for the *kami*, and accordingly prayers, various kinds of rituals and festivals are conducted on shrine grounds. (The names of newborns are also registered at local shrines, each child another addition to the national 'family'.)

Shinto, particularly when considered to have absorbed so many Buddhist, Taoist and Confucian beliefs, would appear to be benign, and as such a very unlikely force for religious polarisation. However, the dynamics that saw it assume such a role can be traced to the events of 1853, a seminal year in Japanese history, when Commodore Perry compelled Japan to commence full trade relations with the US at cannon-point.

The institution of the feudal Shogunate was replaced in 1868 by the beginnings of a modern industrial state, led by the reformist Meiji Emperor. Within sixty years, Japan would become Asia's first superpower. Despite the Meiji Emperor's embrace of Western ways and the political imperative for society to follow suit, the response of the Japanese majority to deeper involvement with the Europeans and Americans was fear. To unify the populace, polarise them around the emperor and implement rapid industrialisation, Shinto was enforced as the religion of imperial Japan. The Shinto Worship Bureau was created to oversee the purging of Buddhism and Confucianism from Japanese spiritual life. Shrines were declared state property and emperor worship was encouraged.

The country grew wealthier and more sophisticated as it developed into a regional power, and the population expanded accordingly, propelling Japan further along its empire cycle. The country's first overseas military action took place in 1882, when it deployed troops in Korea to force reparations for an uprising in which the Japanese legation had been attacked.

By 1890, Shinto had become fused with nationalism in the expression of national sentiment and goals, and it was officially the state religion: all other faiths required government approval. Shinto as the cornerstone of national unity was reinforced over wars with Russia and China, and intensified during World War Two.

In contemporary Japan, Shinto traditions have retained currency; the religion is now culturally ingrained within Japanese society rather than enforced, and denuded, of course, of emperor worship. It has also engendered numerous and mostly marginal sects, which blend various Judeo-Christian, Buddhist and/or Taoist beliefs. Shinto is distinctly Japanese and can be considered an exclusive religion with limited potential for expansion. Today, Japan is in the overextension phase of empire, making primary polarisation highly unlikely for a long time, in any form. However, as a response to the ongoing challenge from China, it can be expected to manifest secondary polarisation in the form of nationalism, which will simultaneously bolster the country's Shinto values.

INDIA

One of the world's oldest religions, Hinduism claims the third-largest number of adherents today after Christianity and Islam – approximately 1 billion, with nearly 830 million living in India. Its origins are varied, deriving from ancient Vedic traditions that date back to the second millennium BC, contemporaneous with the very beginnings of Indian civilisation, centred on the Indus River. Hinduism today comprises a diverse number of sects and sub-sects, with a body of sacred literature at their core. There is no single system of unified Hindu creed, but a plurality of spiritual practices and beliefs that find expression in countless ways. Certain tenets are adhered to by all Hindus, however, including the supremacy of *dharma* (perhaps best expressed as 'right living', or a code for virtuous existence) and *karma* (discussed below).

The Indian caste system, which still predominates, is based on an ancient system of class division, with the 'purest' caste ruling the lowlier ones. This has been widely believed to be Hindu in origin, but this assumption has been challenged with the suggestion that it is more a social construct than a religious injunction. There has also been a great deal of fluidity in the application of caste law over the centuries, with the practice being more lenient in some periods and more stringently enforced in others.

The law of *karma* is a constant in Hinduism (and a concept, moreover, that found a home in the later-arising Buddhist religion as well). *Karma* is a complex spiritual process that can be simply expressed as the view that our actions in this life have direct consequences in the next – that human beings can make choices that determine their station in the cycle of continuous rebirth. One interpretation of *karma* is that it holds out hope to alleviate the frustrations arising from low status in a rigidly hierarchical society. The belief that a person can 'do better' in the next reincarnation thus acts as a safety valve, allowing the downtrodden to accept their lot and behave virtuously. (The concept is not all that different from, say, the notion of heavenly reward in Christianity, which was often used to console slaves in the Americas, among others.)

There are mechanisms of conversion in Hinduism through marriage and revivalism, and, as India grows in economic power, this process can be expected to accelerate. Essentially, however, it is an exclusivist religion into which a person must be born and is deeply entwined with the fate of India.

The seeping of bigoted nationalism into the generally tolerant Hindu religion was politically driven, baptised in blood during the 1947 birth of the Union (later Republic) of India and the Dominion (later Islamic Republic) of Pakistan. Competing Muslim and Hindu independence movements on the eve of the dissolution of the British Raj degenerated into religious polarisation, and the rivalry soon escalated into all-out war. Partition, the division of the British colony into the two new states, resulted in the deaths of as many as 1 million people on both sides of the new border, as Muslims moved into Pakistan and East Pakistan (later Bangladesh), and Hindus crossed into India. This insidious Hindu-Muslim polarisation has persisted ever since, finding an additional focus in the hotly contested region of Kashmir – in 2001–2002, tensions over this northwestern state currently under India's jurisdiction ignited the threat of a nuclear exchange. This alarming situation can be analysed as the manifestation of primary,

ABOVE Hindu pilgrims at sunrise at the ritual bathing site at Sangam, Allahabad, the confluence of the Ganges, Yamuna and mythical Saraswati Rivers during the Ardh *Kumbh Mela* (Pitcher) festival in January 2007. Devotees believe that bathing in the Ganges at this time washes away their sins and paves the path to salvation.

expansive polarisation, and, although relations have calmed somewhat in recent years, the two countries exist in a precarious state of potential conflict.

CHINA

With regard to religion in its most commonly understood sense – the worship of a higher power or powers – the People's Republic of China can most likely claim the lowest percentage of the world's population that can be classified as 'religious'. Chinese spiritual beliefs have always been diverse. Westerners attempting to understand China would do better to consider cultural practices than organised religion, which does not play a role in the lives of most Chinese. This is not to say that China is devoid of religious currents. Taoist, Confucian and Buddhist traditions remain. Christianity – which has had a presence in China for over 1,000 years – is growing in popularity (in both officially registered and 'underground' forms). Islam is practised by the Uighurs of the western Xinjiang Province, Buddhism and the ancient Bön religion are native to Tibet, and new religious movements, such as the widely persecuted Falun Gong, have emerged. Yet China, which has a long history strongly anchored in Taoism, Confucianism and Buddhism, and a more recent history dominated by atheist Communism, cannot be readily polarised along religious lines.

Whereas in AD 845 Emperor Wuzong of the Tang Dynasty believed that Buddhism undermined the health of the empire, the succeeding Song Dynasty promoted the integration of Buddhism with Confucianism and Taoism. The last years of the Tang Dynasty saw the ruthless suppression of Buddhism along with other religions, but the neo-Confucianists of the tenth-century Song era studied Buddhist and Taoist teachings very carefully, and began a process of integration that would endure for centuries. Buddhism infused neo-Confucianism with a more structured cosmology, and Taoism a spiritual practice. Confucianism dates from the sixth century BC. A philosophical code emphasising ethical, moral and social values that facilitated the governance of China's huge population, Confucianism provided a means of ordering family, village and higher social units effectively.

As we have seen, even the most benign of belief systems can become ripe for polarisation in the context of expanding demographics or pervasive threat. This was the case in the nineteenth and twentieth centuries, when China was subjected to the humiliation of European power and witnessed an ever-growing influx of Christian missionaries rapidly accumulating converts. The result was the Taiping and Boxer Rebellions, which were borne of two very different movements that each made use of religious energy for its political ends.

The former was a civil war lasting from 1850 to 1864, spearheaded by a convert to Christianity, Hong Xiuquan. He developed a messiah complex, claiming to be the brother of Jesus, and he believed that his mission was to overthrow the weak, compromised and corrupt Qing Dynasty and establish his brand of Christianity in China. He founded the breakaway Taiping Heavenly Kingdom, with Nanjing as its capital. The Qing army eventually defeated the Taiping with military guidance and troops from France and Britain. Some 20 million people were killed.

The Boxer Rebellion, also anti-colonialist, aimed to stamp out Christianity and reaffirm Chinese values and traditions. The Society of Righteous and Harmonious Fists, a sect founded in Shandong Province, practised martial arts and dedicated themselves to purifying China of foreign influence. Their training was impressive, although they believed they could become impervious to bullets. By 1900, the Boxers (as they were called by Westerners) had amassed enough power to constitute a real threat. In May of that year, they had seized Tianjin and menaced Beijing. The hated European and Japanese legations in Beijing were forced to evacuate to a fortified compound. Around 1,000 civilians and military personnel, along with 3,000 Chinese Christian converts, came under Boxer siege there. (Approximately 20,000 Christians, missionaries and Chinese converts alike were massacred in Taiyuan in Shanxi Province.) A battle ensued, with the defenders sustaining many casualties, until international reinforcements arrived in August and quashed the rebellion.

In effect, these events marked the beginning of a new cycle of Chinese expansion, and a resurgence of national pride that paved the way for the later victory of the Communists under Mao (who often hailed the rebellions as revolutionary forerunners).

Any primary polarisation to come in China will be under the influence of a nationalist, not a religious, meme. However, the possibility of the Christian and Islamic

worlds bonding against a 'godless' nation in any confrontation with China is not inconceivable (Afghanistan would be a precedent). In the face of the threat from a common 'enemy', Chinese expansion may yet bring about a truce between Islam and Christianity.

ABOVE **Members of the banned Chinese Falung Gong spiritual movement stage a mock trial for Chinese President Jiang Zemin in Yokohama, southwest of Tokyo, in August 2003. They found him guilty of unfair prosecution of the movement and of torturing hundreds of its members to death.**

CHINA AND JAPAN: THE RISE OF ASIAN NATIONALISM

The barely concealed hostility between China and Japan – reminiscent in many ways of that between nineteenth-century Britain and France – could provide fertile ground for nationalist polarisation. As neighbours, the countries are competing for the same regional resources. Moreover, the memory of Japanese atrocities in China (and throughout Asia) during World War Two is frequently invoked by the Chinese.

Japan, which fervently made pacifism an immovable tenet in the years after 1945, has since begun to rethink this abstention from the notion of an official military. Under Prime Ministers Junichiro Koizumi and Shinzo Abe (2001–2008), the country began to move towards greater emphasis on military preparedness against potential threats from North Korea and China, including talk of reformulating the country's constitution to allow Japan to take its place in a mutual military alliance with the US.

Prime minister until June 2010, Yukio Hatoyama was elected from opposition and pledged to end Japan's sclerotic and secretive governing style. His outlook did not mirror his predecessors' nationalist views, and he attempted to cultivate a more pro-Asia stance that emphasised independence from the US, while still maintaining close relations with it. Nevertheless, the reasons for his resignation after only nine months in office were revealing: he had, he said, failed to rid Okinawa of the controversial US base there, the Marine Corps Air Station Futenma. Hatoyama had agreed instead with Washington that the base would be relocated. Deep public disillusionment over this perceived inability to take an assertive stand with the US, as well as dissatisfaction within Hatoyama's own party, prompted his decision to step down. (A finances scandal earlier in his administration also played a role.) However, the US will need to pay close attention to the currents stirring in Japan, where a close relationship with the country's former occupier is still desired, but a yearning for more independence is growing. Neither country can afford the risk of anti-US polarisation among the Japanese populace.

There has been a strong focus on China's massive growth, but, in comparison, Japan's economy – although unable to grow – is still ten times as large. As part of the ASE, Japan is now newly energised. However, there is no doubt that China's primary polarisation is inducing strong secondary polarisation in Japan, and that this will fuel the rivalry and ultimately again risk conflict in the region.

Chapter Eight

Global Military Balance

The Nature of War

Although most of the human population would passionately condemn it, war has always been a defining characteristic of our existence. As we have seen, history has been driven by cycles in which societies have grown and resource competition with neighbours has intensified. War is the point where collective competition gives way to conflict.

All wars are destructive, but not all wars are equal, because the position of the combatant within the Five Stages of Empire will define the nature of each conflict. In broad terms, an empire will be embroiled in three types of war that create distinct categories of conflict. These are wars of expansion, civil wars and wars of contraction (see Figure 41). However, as a precursor to a war of contraction, a pilot war may demonstrate signs of weakness in an established power and act as a pivotal catalyst for a subsequent major war. Each type of war has its own unique challenges and energy.

WARS OF EXPANSION

Wars of expansion are typically those of a regional power or ascending empire, driven by primary polarisation to confront another power with the ultimate purpose of increasing territory and acquiring resources. In this process, the expanding power uses its growing population as risk capital. Typically, these are wars of choice, and they take place away from the territory of the challenger. If the war is against a declining empire of similar technological status and the conflict becomes inconclusive and destructive, there is usually the choice of withdrawal for the aggressor, with the possibility of making another attempt in the future with less threat to the homeland.

Expanding empires and nations will only make a challenge if they perceive that their potential opponent is weakening and in decline. Thus, in a predatory attack, the relative power between two nations or empires is key. As the balance of power shifts towards the younger, expanding system, the risks of conflict increase dramatically. This occurs because of the expansion of the challenger, but the trigger is a perceived sign of weakness in the established power.

However, there is one slight variation to the predatory model. This occurs when neighbouring territories of similar power and resources become locked in a struggle for the same territory into which they both need to expand. Thus, if the war is against another empire in a similar phase of expansion (as was the case with the Romans and Cathaginians), withdrawal is not a viable option, because the opposition would then

Wars and the Empire Cycle *(fig. 41)*

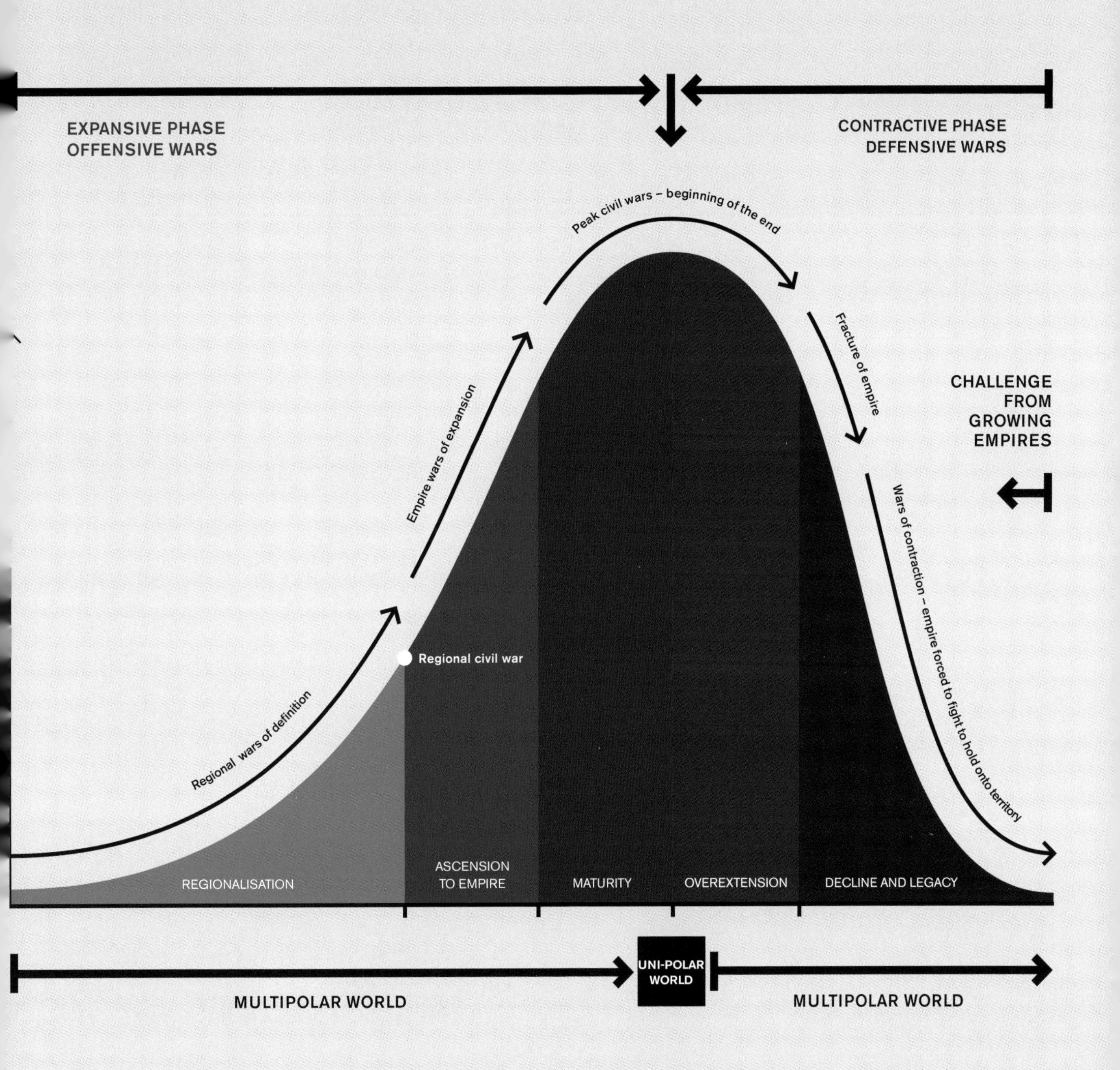

seek to absorb the diminished aggressor. In such cases, the result is a long attritional struggle, as played out in the Punic Wars, the Hundred Years' War and, later, the Anglo-French Wars.

RIGHT A Boer picket on Spion Kop, Ladysmith, South Africa, on 1 January 1900. The battle of Spion Kop was fought on 23–24 January during the campaign to relieve Ladysmith and it resulted in an ignominious British defeat.

BOTTOM RIGHT Republican Forces in the aftermath of the Battle of Brunete against the Nationalists during the Spanish Civil War in 1937.

PILOT WARS

Pilot wars are minor confrontations that take place at the periphery of an empire's sphere of influence. Very often they represent either a regional attempt to break away from the larger empire in order to gain independence, or a bid by a neighbouring state to seize territory from what is perceived to be a declining empire. Pilot wars tend to occur in the period of overextension and decline. Although at the time they do not appear to be a major challenge, they are in fact of critical importance to the world's perception of the failing empire. How a minor, peripheral war is conducted allows potential aggressors to assess their chances in a direct challenge.

Democracies that are the product of a mature and declining empire are particularly prone to being misunderstood and are sometimes viewed by dictatorships as soft targets, which are weak and unwilling to act in defence of their assets. In reality, they are slow to engage in combat but resolute in their purpose once at war. This error of judgement was made by Germany in 1914 and 1939, and by Japan in 1941, and it was a common belief in the Kremlin for most of the Cold War.

The significance of pilot wars can be seen in two key examples. The first is the Second Anglo-Boer War (1899–1902), in which a small group of Boer guerillas in South Africa held out to the British Empire. Britain's failure to act decisively encouraged Germany's belief that it could mount a challenge to this established imperial power. A second example was Germany's contravention of the Versailles Treaty in 1936, when it marched its soldiers back into the demilitarised zone. This was an action that the treaty clearly stated would be interpreted as an act of war, but it was met with British and French passivity. Hitler was emboldened by the lack of response. Four months later, he supported the Franco faction in the spread of Fascism in Spain and continued with Germany's own expansionary policies, which led to World War Two.

Pilot wars are a test of an empire's intention, national vigour and military capability, and shortcomings in any of these areas can encourage a predatory empire or regime to aspire to dominance. Strategically, it is preferable to fight a pilot war properly – for example, to respond to a low threshold challenge such as Hitler marching into the Rhineland – than to fight the all-out war that will inevitably follow.

CIVIL WARS

Civil wars display a particular dynamic and take place at or near the peak of a regional cycle of an empire. Essentially, with no external battles remaining, the last of the expansive energy turns inwards as two internal value systems fight for control.

The clearest examples of peak civil wars are the two world wars for the WCSE. Without a unifying political system, the competition between the constituent empires of this super-empire could only be resolved through conflict.

Lesser-degree peak wars may also occur at the end of, or consolidation of, the regional phase of ascension to empire. The English Civil War of 1642–45 between the Royalists and Parliamentarians was just such a conflict. Not only did it decide the very nature of national governance in favour of a constitutional democracy but it was also a part of the process of establishing Protestantism as the dominant religion of Britain, thereby determining the core value system that would power the country to empire.

When an empire is governed by an absolute ruler, civil war may be the only means of securing power change. However, in a democracy, the democratic process may facilitate change without such destructive conflict. A prime example occurred at the peak of the British Empire with the Second Reform Act of 1867, when Queen Victoria relieved the social pressure of the growing Republican movement by presiding over the evolution of the constitutional monarchy, thereby creating the two-party system and resulting in modern democracy as we know it today.

LEFT **During the Battle of Naseby of 14 June 1645, the Puritan Army under Fairfax and Cromwell won a decisive victory over the Royalists under Charles I and Prince Rupert.**

BOTTOM LEFT **Tourists look at French painter Paul Philippoteaux's newly restored panorama, *The Battle of Gettysburg*, at the new Gettysburg Museum on 27 September 2008 in Pennsylvania. The painting portrays one of the bloodiest battles ever on US soil, in which some 51,000 people lost their lives in three days.**

The American Civil War proved to be the final war in the regionalisation process of the united states with the victorious Northern States dictating the social, industrial and military values that would dominate the newly united States as it ascended to empire.

WARS OF CONTRACTION

Wars of contraction take place when an empire is in decline, as other rising empires and regional powers seek to make their challenge. Typically, the citizens of the empire in decline have grown soft, having enjoyed the material comforts of the mature phase of empire and in sharp comparison to the tough challengers who have lived in a more difficult social environment. The established empire is usually highly dependent on technology to maintain its military edge against an often more numerous and less capable adversary. However, having invested enormous sums of money to develop and maintain a powerful and advanced military complex, such an empire will be vulnerable to one thing – military innovation.

Having invested psychologically in the current military structure, a declining empire can be blind to the benefits of adopting new weapons. The British Royal Navy, for example, was particularly resistant to the introduction of the submarine because the vessel was perceived at the time to threaten the supremacy of the traditional surface-based navy. Furthermore, even if new innovations are recognised as being desirable, two types of financial issue tend to arise: the years of economic investment make it difficult for a nation to scrap the earlier systems, and financial overextension renders new investments problematic. The British found themselves in this situation yet again when they developed the dreadnought, which forced them to dispose of their old battle fleet and build a new one. In the ensuing fourteen-year-long naval arms race, Britain was matched almost ship-for-ship by Germany.

Decline is a difficult and confusing time, as the old empire fights a series of wars that destroy its economic and military power, often through a long process of attrition.

In the cases of both Rome and the WCSE, this took place over a considerable period of time. The declining empire believes that it is fighting the barbarian hordes as the very protector of civilisation. This presumption is accompanied by a consistent underestimation of the competence, capacity for innovation and threat posed by the new challenger, which in fact accelerates the decline.

ABOVE The crew of a German UC-1 class submarine on deck. Introduced in 1915, vessels of this class were employed mainly in mine-laying duties, carrying up to twelve mines. German submarines sank 1,845,000 tonnes of Allied and neutral shipping between February and April 1917.

The Economics and Politics of Warfare in the Empire Cycle

Wars cost money and, as the complexity of military technology increases, so does the expense. Within the cycle of empire, the strength of the expanding empire is derived from its ability to focus its resources on its military machine. In contrast, the problem for the empire in decline is that its resources are not just committed to its military infrastructure but also to social programmes, which its population has come to rely on as part of the fabric of the empire. In most cases, as the empire's borders contract, conflicts move into the heartlands, further reducing offensive capability as the defender has no choice but to use its resources in a war for its very survival.

Different societies have found different solutions to the problem of balancing the need to maintain an economically productive population in order to pay for war, while also maintaining a standing army. The limitation of an agrarian society, for example, is that men are needed to work the land, which restricts the total deployment of the army to about 15 percent of the population for short periods only.

The Mongols moved their whole nation with their army, meaning that men could continue to perform their herding roles and maintain the economies of the tribe. When required in battle, they could all be deployed into the army. Thus, the activity of war cost their tribes no more economic resources than their usual existence, with all the upsides of captured land and wealth. The advantages of this system for the wandering Mongols gave them two centuries of domination over the Asian continent.

Throughout the history of the WCSE, economic power and demographics dominated each constituent empire's ability to wage war. As military costs grew, each empire had to find new financial mechanisms to pay for their armies, such as the city bonds pioneered by Italian cities during the Renaissance. Potential defeat could come both on the battlefield and in the bankruptcy of a nation. As the peak civil wars of the WCSE in the twentieth century engulfed the world, the total war advocated by German military theorist Carl von Clausewitz became a reality. Then, as the destructiveness of weapons increased to include nuclear weapons, economic war came of age in the Cold War, when two very different economic systems fought for supremacy.

For an economic war to unfold at the level that the Cold War required, both sides have to be involved in an all-out arms race while having sufficient collective inhibitions to refrain from actually using the weapons. This legacy continues into the twenty-first century, where the favoured path for China is to use its economic power to reduce the West financially without firing a single shot, much as the US did with the USSR. However, this may only work up to a point because as China's power increases, it may not have the same reservations about going to war and risking millions of lives – for three key reasons. First, it has not experienced a recent total war; second, it has a vast population; and third, and most importantly, it has a higher percentage of men (56 percent) than women in the population, which results in a more aggressive outlook and therefore a lower risk threshold.

Nuclear weapons are attractive to rising powers not just because of their awesome power but also because they represent a cost-effective way of building a military complex without the need for large, expensive standing armies. This lesson was apparent in the relative ways in which the Soviets and the West spent their money: the Soviets on a huge standing army, the West on its nuclear umbrella. The money that the West saved in using this approach could then be used to develop the enhanced technologies that ultimately enabled it to dominate the opposition.

LEFT **Prussian general Carl Von Clausewitz (1780–1831). He is noted for his book *Vom Kriege* ('*On War*'), which advocated the total destruction of an enemy's forces as one of the strategic targets of warfare, and viewing war as an extension of political policy rather than as an end in itself.**

Not only economics but also politics can affect the outcome of conflict. In theory, once the strategic objectives have been determined and given to an army, generals need the freedom to act independently. This was borne out in the Prussian model of a highly skilled general staff that was so effective that it influenced political-military interactions for over a century. In the past, when communications between the administration and army were slow, this was a natural mechanism. Indeed, orders for sailing warships had to be very broad, and massive discretion was given to captains.

However, with faster and, more recently, almost instantaneous communications, politicians can directly interfere with military operations. It is also now almost impossible to contain news, so it reaches the public very quickly and their response feeds back into the perspective of the politicians. This has resulted in a new decision-making process in the UK and US that is more likely to be led by militarily uneducated politicians. Two examples illustrate the effect of this. The Battle of Goose Green in the Falklands War had no military relevance whatsoever, but it gave Margaret Thatcher the public-relations victory that she needed to press home her assault. More recently, Donald Rumsfeld's decision on a troop deployment limited to some 180,000 men in Iraq overruled his generals' advice, and proved disastrous for the US. This politically dominated military decision-making process in the West is a great weakness from which the Chinese system does not suffer.

The Wars of the WCSE

THE WARS OF EXPANSION OF THE WCSE, AD 1000–1900

An analysis of the cycles of war during the WCSE's history (see Figure 42) can provide an understanding of modern-day geopolitics and its challenges. This period covers almost nine centuries, from the beginning of the second millennium to the start of the twenty-first century, and it divides into two distinct categories. The first is the wars between the European empires and the second is the wars of imperial expansion, fought against the militarily less sophisticated cultures of the Americas, Africa and the Far East.

Originally, there were four European powers contending for primacy: the Catholic Spanish and French, and the Protestant English and Dutch. Over a period of some 200 years, they fought each other for dominance of the WCSE. For the first century, the Catholic powers were more successful, but in the second, the Protestant nations were in the ascendency. The wars between these nations were fought with similar technologies and fighting skills, but in the long run the winner was always the nation with the largest resource base, fed by its growing colonial empires.

The wars of imperial acquisition were fought simultaneously with the wars between the European empires. With the powerful Ottoman Empire straddling the land routes to the Far East, the European powers had no option but to find new trade connections by sea. They were advantaged by their geographical locations, with easy access to the Atlantic Ocean, and their strong seafaring traditions.

In the process of this expansion, they discovered the Americas. It is a common misconception that it was the advanced military technology of the Europeans that enabled their acquisition of the Americas. However, how a few European adventurers, remote from reinforcements, conquered the natives of the US with only one-fiftieth of the manpower of the indigenous population was not by military supremacy.

The European powers decimated the natives of the Americas by means of a weapon that they did not even know they possessed – 'germ warfare', the rapid spread of European diseases to which the locals had no immunity. Smallpox was the first to ravage the social structure of the Incas. What opposition was left fell prey to the political manipulation of various factions, gunpowder and the superior mobility of Western sea power. This pattern was repeated by the British and French across the whole of the Americas: over some 135 years the indigenous population is estimated to have declined by 95 percent from its levels prior to the arrival of the Spanish. Without the epidemics caused by the arrival of the Europeans, the Americas may have remained independent for many decades before Western technology and demographic expansion allowed colonisation of the continent.

This pattern of Western expansion was later replicated in Africa and Asia. There it was achieved without the benefit of 'germ warfare', but by this time the industrialised military gave the West overwhelming power of conquest.

The Wars of the WCSE *(fig. 42)*

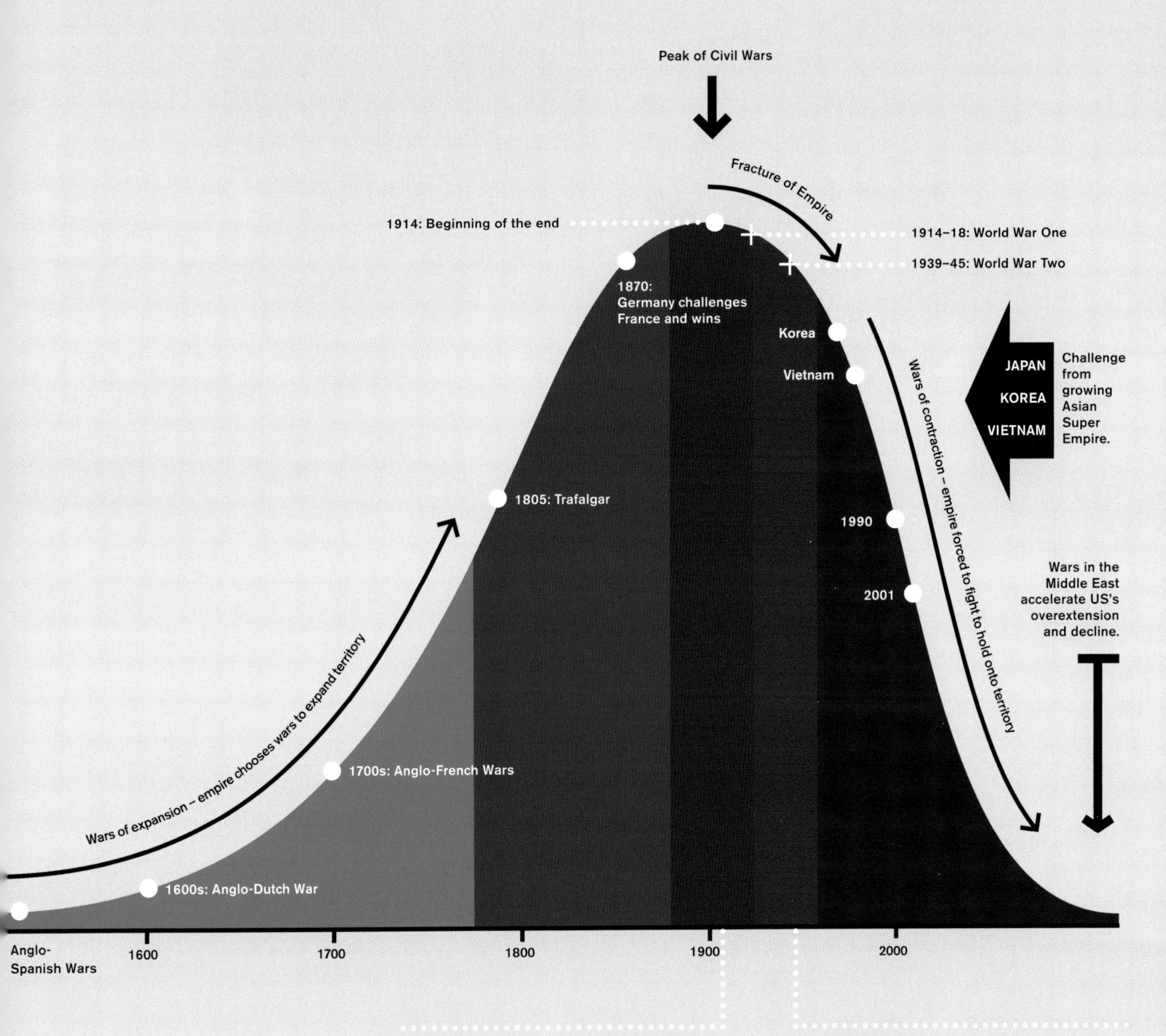

1914: Germany challenges for control of WCSE against Britain and fails.

1917: Russia drops out of WCSE to ultimately challenge for power.

1939: Germany has also polarised itself by becoming Godless as Nationalists Socialists. Once more attacks the core nations of WCSE.

1941: Japan as the leading Asian Power challenge WCSE for power and fail.

1945: Japan surrenders and joins West again.

1945: Germany gives up all aspirations of power and joins Europe.

1945–1990: WCSE newly polarised against atheist communism, Cold War fought by proxy essentially economic in nature, ends with bankrupt USSR and US facing decline.

European Political Alliances, 1914 *(fig. 43)*

European political alliances in the period leading up to World War One.

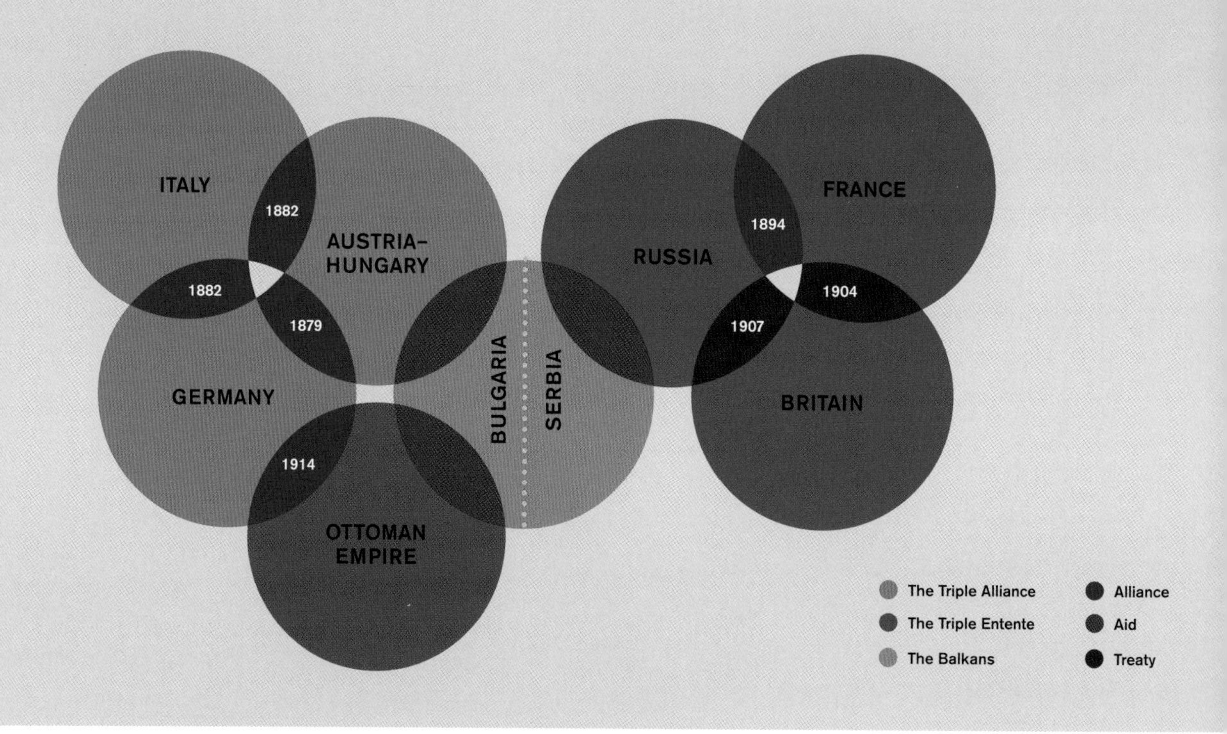

THE PEAK WARS OF THE WCSE, 1914–45

By the late nineteenth century, consistent with the peak of the power curve of the mature phase of the WCSE, two newly industrialised nations that had been growing in power for three decades reached a point where they felt able to challenge Britain for leadership of the super-empire.

The first challenger was Germany, which had risen to prominence under Bismarck's leadership. It had harnessed its chemical engineering industrial wealth to build a powerful, modernised army, and had then taken advantage of the dreadnought revolution in battleships to build a fleet that could challenge British naval supremacy.

The second challenger was the US, which by the start of the nineteenth century, with British agreement, had expanded its sphere of influence into the Pacific and down into South America. The US harboured expansionary ambitions similar to those of Germany, but these would not be realised until much later in the century.

THE FIRST PEAK WAR, 1914–18

As has been discussed, the pilot war that preceded this great clash was the 1899–1902 Boer War in South Africa, which persuaded the Germans that Britain was a failing power ready to be challenged. The process of polarisation to the first peak war took ten years to unfold, embracing the 1904 speech of the German emperor declaring himself admiral of the seas, the 1906 Algeciras Conference through the 1908 annexation of Bosnia by

Austria, and finally the 1911 Agadir Incident. When war broke out, it was in a totally new form. Its destructive power was on a level never before seen and it depleted all of the resources of the combatants in what has come to be seen as the ultimate attritional war.

Although this war is known as World War One, it was, at its most basic, a European conflict (see Figure 43). Its wider spread was due to the appropriation of German colonial assets that, without control of the sea lanes, could not be defended. Britain and her allies very quickly dominated this part of the conflict. However, the intensity and duration of the war, fought along a strip of land running north to south through Europe from the English Channel to the Mediterranean and along the German-Russian Front, was without compare in human history. We are familiar with the roll call of the most protracted battles of this terrible war – the Somme, Passchendaele, Ypres. Before the deadlock was broken, this conflict cost the lives of some 8–9 million soldiers with a further 20–22 million wounded.

However, as destructive as the war was, it still followed the convention of being fought out on a battlefield or, in this case, a battle zone, outside which the civilian populations were predominantly safe (although the battle on the Eastern Front did result in a much higher level of German and Russian civilian casualties). The war was

BELOW A British machine-gun crew at the Battle of Menin Road Ridge on 21 September 1917. The gun's introduction as a defensive weapon transformed warfare. The futility of conventional massed infantry attacks against entrenched defensive positions was illustrated by the first day of the Somme, when the British suffered a record number of single-day casualties – 60,000 – the great majority as a result of machine-gun fire.

ABOVE Soldiers tend to the dead on the battlefield of Passchendaele. Referred to as one of the major battles of World War One, this was in fact a series of operations that took place between July and November 1917. The name has become synonymous with the misery of grinding attrition warfare that is fought in thick mud.

also fought in the sea lanes, where Germany employed its U-boats in an attempt to strangle Britain, whose geographic advantage yet allowed it to seal the majority of U-boats behind barrages of mines, and combine merchant ships into protected convoys.

Two major strategic consequences resulted from World War One. Although initially resistant to involvement, the US's successful intervention enabled it to take a seat alongside the old powers on the world stage. In addition, catalysed by the terrible casualties that it sustained fighting to withstand Germany on the Eastern Front, Russia fractured into revolution in 1917. Under Communism, the Soviets refuted Russia's Christian values in favour of atheism, delineating themselves as no longer a part of the WCSE.

THE SECOND PEAK WAR, 1939–45

The primary effect of the first peak war was to weaken all of the major powers, save the US. Although the other nations of the WCSE tried to return to their pre-1914 position, the truth was that decline had set in. The economic overcompensations that led to the 1929 economic crash set the scene for the rise of the Nazis in Germany and the rekindling of their desire for leadership of the WCSE. This new threat from the Nazis, like the one from the Soviets during the Cold War, was an atheist challenge. As such, it was not just about who would control Europe and the world, but also about the belief system that would dominate, as the values of Communism and Fascism were in stark contrast with the Christian values of the established powers. The pilot war that encouraged Hitler to continue his Fascist expansion was the Spanish Civil War, when the powers of Britain and France stood by without intervening while Franco's Fascists defeated the Communist forces.

BELOW An RAF Supermarine Spitfire Mark I fighter overshoots its target and flies beneath a German Heinkel HE-111 bomber after attacking it on 19 May 1941. The picture was taken from the forward gun turret of the Heinkel.

World War Two (the second peak war) was even more destructive than the first, resulting in the death of millions of civilians and the destruction of vast swathes of continental Europe. In stark contrast, the US's location allowed it to safely continue to develop its industrial economy without being exposed to any conflict within its home territory.

This war was in every way a world war. Initially, from 1939 to 1941, it was an expanded rerun of the 1914 war, with more powerful weapons, and faster moving offensives and campaigns. However, there was one major difference – airpower allowed the battle zone to spread out to encompass the whole of Europe. This brought the centres of civilian population directly into the line of fire.

Once again, Germany sought to challenge the sea power of Britain, relying on its U-boats to sever Britain's links with its empire. With the use of magnetic mines together with new U-boat tactics, Germany almost succeeded. The only factors that saved Britain

from the same fate as Europe were its navy, the English Channel and the RAF's control of the air. This was the first example of the strategic importance of airpower dictating the outcome of a war. Victory in the Battle of Britain ensured that, in time, Britain would become the base from which Europe could be liberated. The importance of air power and strategic bombing was the beginning of a trend that has grown stronger ever since.

LEFT **Black smoke pours out of US warships on fire in Pearl Harbor, Oahu, after a surprise attack by the Japanese, which prompted the US to enter World War Two.**

In the Far East, Japan, the first empire of the rising ASE, had been busy expanding into the Korean Peninsula and Mainland China since 1931, taking advantage of the opportunities created by the Chinese Civil War (a war of regionalisation). With growing confidence, and inspired by Germany's successes in Europe, the Japanese sought to expand their empire to take over that of the British Empire in the Far East.

However, to achieve their goals with what they thought would be relative impunity, the Japanese planned to apply a stunning blow at Pearl Harbor that would render the US ineffective in the Pacific. This was, in effect, the first war of contraction of the WCSE superimposed onto the second peak war. The combination resulted in a war encompassing all of the oceans, and both the European and Asian landmasses. Only sub-equatorial Africa and the US were left untouched.

Throughout this conflict, the Germans were vigorous in their prosecution of their war aims, and it is questionable whether Britain, even with the support of the US, could have won without the equally vigorous Soviet response. Unconstrained by any moral or Christian values, the Soviets attacked the Germans in one of the most ferocious land wars in history.

The conclusion of this war resulted in the almost total destruction of central Europe and Japan, and large parts of Britain – the heartland of the WCSE was in ruins. Not only had the nations of the WCSE expended their financial resources on waging war, but they had depleted their populations and, in addition, they now had to rebuild themselves. It was an inevitable outcome that Britain and France would divest themselves of the majority of their colonial assets during the next decade. Thus, these two peak wars, spread over thirty-one years, left the WCSE exhausted, and without American protection it would have most likely collapsed soon afterwards. However, the now dominant American Empire was to compensate for the decline of the old powers in the WCSE. For the first time in their history, they were bonded together in the new NATO alliance forged to counter the Soviet threat.

In terms of the development of military technology and tactics, the creativity that was unleashed in the two peak wars produced the greatest military evolution that the world has seen. It left the Europeans with an abiding desire never to go to war with each other again, and with aspirations to a greater Europe founded on democratic principles. However, the Communist countries of the Soviet Union and China, with their low regard for human life, took a very different view. They determined that if they had to go to war to achieve their goals, they would do so.

THE WARS OF DECLINE OF THE WCSE

Because the WCSE had all but dominated the globe, it would face multiple challengers during its decline. The Cold War can be seen as the inevitable fallout, as the US and the USSR – the two residual world superpowers – squared up for domination. However, there was one vital inhibition on the Cold War combatants: they remembered the carnage of the two world wars that they had fought. Although they strove to maintain their influence, when frictions grew to breaking point over key incidents, there remained a reluctance on both sides to engage in another war. It is probably fair to say that this saved us from nuclear oblivion. This collective restraint is worryingly not present in the rising powers of China and India to the same degree, and certainly not in other states with nuclear capabilities, such as North Korea and Pakistan.

The challenges post-1945 to the ailing WCSE came first from the Soviet Empire and then from the rising ASE, not forgetting the threat from Islamic extremists.

THE COLD WAR, 1945–90

By the end of 1945, the world was divided into the West led by the US (within the context of NATO) and the Communist Bloc. In essence, this was a battle between two philosophies – the old Christian belief system and the new atheist Communism. Initially, the Western nuclear umbrella held back the potential storm of the immense Soviet army, the largest ever built. In time, as both sides gained nuclear weapons, the concept of MAD prevented the two sides from spiralling into Armageddon.

However, the confrontation of the two superpowers found expression in a series of low-intensity proxy wars, and crises such as the Berlin Airlift, the Cuban Missile Crisis and the Angolan War. With the arms race between the two running at breakneck speed, but capped by the threat of MAD, the Cold War, in effect, became an economic war. The two sides had contrasting characteristics: the West was a Capitalist, consumer society, and the East was a Communist, commodity producer. In the end, the Cold War was to be decided by the commodity cycle: as the prices of commodities appreciated in the mid-1970s, the Communists became richer, and the Capitalist commodity importers became weaker. The situation was worsened for the US by its failing involvement in Vietnam. The Soviet Bloc had maximum influence during this period.

After the 1975 commodity peak, the situation reversed, and year by year the West became richer and was able to build a new generation of weapons, epitomised by the Star Wars programme inspired by President Ronald Reagan. As a result of this initiative, by the end of the 1980s, the US had outclassed its rival's capabilities. It had also used diplomacy to strengthen its ties with Saudi Arabia and joined forces to depress the oil price as a direct challenge to Soviet revenues. Fearful of growing American influence in the Middle East, Russia made the mistake of annexing Afghanistan. The subsequent protracted war destroyed the credibility of Russia's military and sapped the confidence of the Communist population, at a critical time in the declining power of the Soviets.

This was a crucial moment, at which there was perhaps the greatest risk of a Soviet invasion of Europe. However, the Soviet perspective on the West was dramatically challenged by Britain's success in the Falklands. This war fits the classic profile of a pilot

ABOVE Anti-missile rockets pass through Red Square, Moscow, during a Russian Revolution celebration parade on 10 November 1964.

war, in which a small regional power (Argentina) perceives an opportunity to attack a fallen empire on the assumption that it would not, and could not, counter the invasion. If Britain had accepted this act of aggression, the world might well be a different place. Instead, driven by the willpower of Margaret Thatcher, British forces that were optimised only for combat in the North Atlantic sailed 9,000 miles to mount an amphibious invasion to retake sovereign territory. The operation carried considerable risks, and it is doubtful that any other nation would have attempted it under similar circumstances. The effectiveness and toughness of both the military and political bodies of Britain shocked the Russians. Faced with MAD and the apparent commitment of the West to be the first to use nuclear weapons in the event of a Soviet incursion into Europe, the Soviets were deadlocked and imploded internally.

The legacy of the Cold War was a US that had, through a decade of military investment, built the most advanced and capable war machine in history, the effectiveness of which was soon to be tested and proved in the First Gulf War. But there was a price to pay for such advances, and the end of the Cold War marked the peak of American power. From that point it slipped into overextension.

ABOVE British ship HMS *Belfast* fires a salvo from its six-inch guns against concentrations of enemy troops on the west coast of Korea in March 1951. The vessel was originally a Royal Navy light cruiser serving during World War Two and the Korean War. Now permanently moored in London, it is used as a museum ship. Because Korea is a peninsula, naval and amphibious forces were key in the Korean War.

THE ASIAN CHALLENGE TO THE WCSE, 1903–PRESENT

At the beginning of the second decade of the twenty-first century, it is easy to view the challenge presented by Asia as a recent phenomenon. However, the first skirmish mounted by what we can now see as the nascent ASE took place more than a hundred years ago, with the Japanese naval victory over Russia in 1904. This success emboldened Japan to believe that it could challenge Britain and the US in the Far East in 1941, at a time when Britain seemed to be losing the war with Germany. This miscalculation led to the total defeat of Japan at the hands of the Americans.

With Japan then under reconstruction, the next phase of the Asian challenge was cloaked as a threat from the Eastern Bloc. First to attack were the North Koreans with Chinese support, seeking to directly challenge American power and its support of South Korea. These were delicate times, as each side fought to define the new rules of proxy warfare under the threshold of nuclear weapons. There were almost two years of stalemate before peace was signed.

The Vietnam War followed. This was the result of the collapse of French colonial power as well as the US's belief that it was imperative to act against Communism in the region. Once more, China under Mao had a vital role to play in supporting the North Vietnamese, but was limited again by its inability to use airpower to interrupt the supply chain (as was also the case in Korea) for fear of escalation to a nuclear conflict. Instead, it forced an attritional war that did not suit the US's strengths.

Once more, at its essence, this was an Asian-Western proxy war that also ended in American failure. More recently, the Chinese support of North Korea ended with another American defeat as North Korea developed nuclear weapons against American wishes, thereby reducing American prestige and influence.

The pattern of Asian military challenges against the West is clear. For those who believe that China will be the great historical exception in not turning its economic power into military power, this picture is far from reassuring.

Indeed China, India and Japan are currently involved in a naval arms race – a hallmark of nascent empires and a direct challenge to American global power. Just as Britain's warships once cruised off Asian coastlines, the tables could be turned and sooner than we realise we could see Asian vessels along the Atlantic seaboard.

THE ISLAMIC CHALLENGE TO THE WCSE, 1945–PRESENT

During the twentieth century, the Islamic nations, with the exception of Turkey, did not develop into strong industrial powers. As such, the Islamic world has only been able to present a limited direct challenge to Western interests, such as in the Israeli–Arab proxy wars, the Iranian Revolution, the Gulf Wars and the Afghanistan War. In the main, its challenge has been through asymmetric confrontations by 'freedom fighting' groups. However, the acquisition of nuclear weapons that began with Pakistan, via clandestine methods, now risks being spread to the wider Islamic world. Nuclear capability would provide a shortcut for the Islamic nations to gain a degree of parity with the West, certainly to the point where they could resist outside interference. This is an extremely significant matter that could empower a collective Islamic challenge to the ailing WCSE.

War by Land, Sea and Air

The relative strength of empires is based on whether their power derives from land, sea or, more recently, air, and this is of critical importance in understanding today's global military balance.

THE IMPORTANCE OF SEA POWER

The foundation of the WCSE was its sea power, beginning with the Portuguese and Spanish, through to the dominance of the British navy in the second half of the eighteenth century. In response, Napoléon, whose power was based on his army, sought to build a continental trading system across the landmass of Europe to support his empire. However, the resources that Britain could accrue via a global sea empire were far greater than the continental system could provide, and this simple dynamic curtailed Napoléon's ambitions. Similarly, in the American Civil War, the North overwhelmed the South by using its industrial capacity to build a huge navy, which it then used to blockade the ports of the succeeding states. This restricted the South's ability to export agricultural products, particularly its cotton, and to import vital war supplies, effectively choking its economy and undermining its ability to wage war. This was the genesis of the American tradition of maintaining a powerful navy.

World War Two augmented the US's naval strength – to defeat Japan, America had to construct a carrier-based navy on an unprecedented scale. By the end of the war, the US's fleet of large and escort carriers numbered close to a hundred. To place this in perspective, when the war with Germany ended, Britain sent its full force of eight fleet carriers to join the US in the Pacific. This made up only a squadron within the larger fleet. The US had utilised its economic power and taken advantage of its unique geographical position in straddling both the Pacific and the Atlantic to move forces through the Panama Canal, as and when needed, to project power across the globe.

The beginning of the Cold War saw the US, now the world's greatest naval power, pitted against the Soviets with the largest standing land army the world had ever seen. Although the Soviets sought to constrict NATO sea power via submarines, much as the Germans had done previously, they failed to build aircraft carriers on the scale of the Americans. Like France and Germany before them, they were heavily constrained by their geography, which prevented them from having ready access to the world's oceans. Instead, they had to transit tight choke points. This gave the West a significant advantage.

One threatening new weapon developed during the Cold War was the nuclear missile submarine. Whereas formerly submarines and surface ships had been limited to placing a stranglehold on a country by constricting the enemy's trade routes, this submarine was capable of launching its weapons far inland. At first their inaccuracy made them only suitable for use in retaliation, but by the end of the Cold War, American missiles launched from these vessels had a staggering accuracy that could destroy the enemy's missiles in their silos. This completely altered the balance of MAD.

Today, at the start of the twenty-first century, the US navy is the only global force of its magnitude and strength. The US has twelve supercarriers that are capable of striking many hundreds of targets a day. These act as the core of US power projection around the globe. Indeed, Bill Clinton was quoted as saying that the first thing a US president asks in a crisis is, 'Where is my nearest carrier?' In addition, the US has ten assault vessels, around which are based marine expeditionary units. These are capable of deploying US marines directly and rapidly into the world's trouble spots with a fully integrated support force.

These high-value ships act as the focus for taskforces that are protected by screens of cruisers, destroyers, frigates and hunter-killer submarines. In addition, there is a fleet of some fourteen nuclear missile submarines, which act as part of the US's strategic deterrent. In total, the US deploys a navy of some 242 ships and 3,700 maritime combat aircraft that, given the time to mobilise and transit to a given hot spot, could destroy any combination of enemies at sea. At its height of power, Britain had a navy that could take on the combined forces of the two navies below it in strength; the US is in the unique position of being able to take on all of the worlds navies simultaneously with a good chance of winning. The US has invested enough in technology and operating experience to remain the world's most powerful navy for the next ten to fifteen years.

ABOVE A Trident nuclear ballistic missile submarine of the US Air Force, *Henry M. Jackson*, glides along the surface of the Pacific Ocean off the Californian coast, US.

OVER American naval dominance is facilitated by the awesome USS *George Washington*, a nuclear-powered supercarrier, the sixth ship in the Nimitz class and the fourth US navy ship to be named after George Washington, the first US president. In April 2004 the carrier was in the Arabian Gulf to support Operation Iraqi Freedom.

However, like the British navy of the early twentieth century, the US fleet has probably lost the ability to control low-threshold risks, which would act as a deterrent to potential opponents. This limitation has resulted in the growth of piracy off the Horn of Africa, and this signals that the US's dominance of the sea lanes is far from total.

In addition, the new Asian powers of China and India are clearly committed to building their navies. The need to counter the African pirates (who are forcing many vessels to travel round the Cape to avoid the risk of encountering them) has given China a pretext to offer its services with the intention of testing and proving its navy's long-distance deployment capabilities. China's desire to project power is now supported by its growing submarine fleet and the development of large fleet aircraft carriers. However, it has only limited access to the world's oceans and is currently pursuing diplomatic routes to increase its overseas bases for its fleet. The growth of the Chinese and Indian navies provides a sobering warning that their long-term intention may well be to challenge for global power.

However, there are considerable obstacles to overcome in attempting to build a modern navy that could challenge that of the US. From the time of the Greeks, warship design has been at the pinnacle of a nation's industrial complex, in both the technology

舰
八一

employed and the ability to meet the substantial production costs.

The skills and traditions of a navy take time to build into a combat-effective unit. Japan has provided the most impressive example in this respect – in only a few decades, with construction and training support from the Royal Navy, it created a very effective fighting force. Fortunately for the West, China does not have the benefit of such an accelerator for its ambitions and has still to develop its navy on a number of fronts. It faces significant challenges in evolving its predominantly Soviet military infrastructure into one that is on a par with that of the US, Japan and the European allies. However, the Chinese capability gap is expected to narrow during the next two decades due to the country's expanding wealth base (much as Japan's did prior to Pearl Harbor). This inevitably raises the threat of a regional confrontation around the Taiwan Strait. In a bid to counter this threat, the US has increased its cooperation with India.

LEFT **Chinese sailors guard missile destroyer *115 Shenyang* at the port of Qingdao on 21 April 2009. The vessel attended an international fleet review on 23 April 2009 to celebrate the sixtieth anniversary of the founding of the People's Liberation Army Navy.**

The threat of a naval challenge from China is very real and should not be underestimated. Britain and France need to maintain their carriers and nuclear submarine force, which could, within the next decade, be playing a vital role in maintaining the integrity of the Atlantic against Chinese incursion.

CONVENTIONAL LAND WARFARE

The American Civil War introduced a new type of industrialised war, in which the increased destructive power of the weaponry, and the use of railroads to supply the front lines, combined to strongly favour the northern states. This resulted in the deaths of more Americans than all of the subsequent wars from then to the present day combined. Some 618,000–700,000 died, amounting to a staggering 1.5 percent of the then population.

Unfortunately, the lessons of this war were lost on the insular European generals when embarking on World War One, and the result was deadlock and horrendous casualty rates. It was only in 1918, when the *kaiser* launched his last offensive with a million men along a 50-mile front, that a new era of prosecuting land war was ushered in. Germany's final assault ran out of steam after penetrating only 40 miles. The British counter-attacked with a new combined arms capability sweeping all before them. This approach integrated light machine guns to counter the bigger fixed defensive machine guns that had previously dominated no-man's-land; a creeping artillery barrage in front of the advancing infantry; and a combination of tanks and aircraft that provided both reconnaissance and strafing of the enemy positions. In addition, the British solved the problem of supply between the railheads and the advancing troops by using mechanised transport, which maintained the momentum of the advance. This integrated technique returned mobility to the battlefield and forced the Germans to sue for peace.

Ironically, it was the visionary attempts at modernising the British army with the creation of fully mechanised tank brigades by the British commander Major General J. F. C. Fuller and his subordinate, Captain Liddell Hart, that attracted the attention of the German generals when they sought to rebuild their own army. Indeed, history is replete with examples of challenging empires that are eager to embrace new technology in order

ABOVE Two US Army 3rd Infantry Division M1/A1 Abrahms tanks roll deeper into Iraqi territory on 23 March 2003, south of the city of An Najaf. US and British forces assaulted Iraq from land, sea and air as part of Operation Iraqi Freedom. Three versions of the M1 Abrams have been deployed – the M1, M1A1 and M1A2 – including better armament, protection and electronics. These improvements, as well as periodic upgrades to older tanks, have allowed this long-serving vehicle to remain in front-line service as the world's foremost battle tank. The M1A3 is currently under development.

to gain advantage against an older empire, which in contrast is resistant to change as a result of its complacency and faith in its established track record.

Consistent with this pattern, it was the die-hard cavalry and infantry traditionalists in the British and French armies who resisted attempts to mechanise because they were fully invested in the old style of warfare. The Germans, however, quickly recognised the importance of mechanised warfare and built their new army around the concept of *Blitzkrieg*, where concentrated armoured columns thrust like daggers into enemy lines supported by mobile artillery and close-support ground-attack aircraft. As a result, World War Two was a conflict of great mobility, ranging across the entire European continent.

The land war on the Eastern Front between the Germans and the Russians was a staggering struggle in terms of the area over which it was fought, its ferocity, and the destruction and casualties that it caused. Although the Allied forces fought intensively from the D-Day landings as they drove east to Berlin, it was, in reality, a much smaller part of the conflict than that playing out in the East. By the end of 1945, the Soviets had the largest and most capable offensive land army the world had ever seen, and, although the USSR was restricted to being a land power, like the Germans whom it had helped to destroy, it represented a formidable threat to the West and remained so until the late 1980s.

Land war of the twenty-first century deploys weapons that are considerably more accurate and destructive than those of both world wars. The US army is currently 1.1 million strong (including regulars, National Guards and reserves) and, although struggling to optimise itself as a counter-insurgency force, it is unsurpassed in its conventional land-war capabilities. Most importantly, it has continued to innovate and is the foremost exponent of networked warfare, connecting its individual soldiers and weapons platforms to higher command structures with two-way links. This approach provides its generals with an unprecedented perspective of the battlefield and control over the combat process.

The trend towards increased firepower on the battlefield has made the lives of infantrymen ever more precarious but has been compensated for by increasing the average spacing between infantry in the field. However, this highlights the greatest weakness of the modern US army – the casualties that it sustains while at war and their effect on the political will of the nation. This Achilles' heel is consistent with an empire that is declining from its peak and placing a high value on the life of its citizens, and it is in stark contrast with challenging nations that are more prepared to sustain casualties in achieving their objectives. The prime examples are the recent counter-insurgency wars in the Middle East. The casualties that are inflicted here are much greater than the loss of American life, but the loss of every US soldier has a noticeable effect on the desire of the American public to continue with the war.

One of the unspoken trends in modern warfare is the decline in US and British deaths due to increasingly effective combat medical support, but conversely the ratio of wounded to killed has increased dramatically. In World War Two, the ratio was 2:3 wounded to killed. By Vietnam, the number had reversed to 3:2, and survival rates continue to improve. This has, to some extent, mitigated political concern about the ability of the US to fight the wars in Iraq and Afghanistan on a sustained basis.

The US can now go to war with its air force against any enemy in the world and expect very few casualties. Ideally it would like to replicate that dynamic in its army. With this aim, improved protection and survival mechanisms are the current focus of development to reduce the vulnerability of the infantry. The Future Soldier Program, for example, provides for enhanced personnel armour, weapons and situational awareness. Sixty years after the Germans used the remotely controlled mini-tank, Goliath, to clear minefields, the next generation of remotely controlled land vehicles will soon be forthcoming. In any future conflict with China, the US will need to maintain the advantage of its advanced technology to counter China's larger numbers, and as such will seek to develop greater reliance on remote vehicles and robotic combat units.

Throughout history, armies have been optimised to fight the latest war. For example, the French, at phenomenal cost, built one of the largest armies in the world in the 1930s, entrenched behind the all-but impenetrable Maginot Line. However, the Germans used the *Blitzkrieg* to circumvent it, bringing about the surrender of France in just six weeks. Recently, the US army, which was optimised for the conflict of the Cold War, has ended up fighting a counter-insurgency war, yet the risk is that the next threat could come in the form of a conventional war. The most likely enemy could be the Chinese, although their

army's capability is limited by technology akin to that of the old Soviet Bloc. They will need to make substantial progress to be able to match the US army on the open field – unless, that is, the US finds its resources so overstretched that it is overwhelmed on multiple fronts.

ABOVE **An AH-64 Apache, the primary attack helicopter of the US. This features double- and triple-redundant systems to improve survivability for the aircraft and crew in combat, as well as to improve crash survivability for the pilots. It is also in operation in the armies of Britain, Israel, Japan, Greece and the Netherlands.**

The nature of any potential future war is dependent on the world's geography. It is very unlikely that Russia could rebuild its army to a capability that would replicate the Cold War threat, and as such a major European war seems relatively unlikely. The most likely combatants are an alliance of India, the US and/or Japan against China, and it is significant that there are no major land bridges connecting the potential combatants, thus limiting the risk of a massive land conflict. Instead, like the Pacific War before it, any war in this region is going to be highly dependent on amphibious forces. At present, although the Chinese are working on their capacity to invade Taiwan, it is currently only the US that could replicate to some degree its World War Two pan-Pacific fighting ability. The development of amphibious capability of the members of the ASE will be a key indicator for coming regional friction.

One alternative to this scenario could come from the Pakistan/India border. As friction builds between the expanding power of Islamic Pakistan and rising Hindu nationalism in India, the clash of the two could provide the locus of the next Asian ground war, long before China becomes the world's next aggressor.

AIR POWER

The twentieth century saw the first use of the air as a strategic battle space. Initially it was used as a reconnaissance tool. However, by 1918, bombers were making an appearance and fighters were being used to strafe machine-gun nests and trenches in coordination with ground-attack forces.

The British and Americans were quick to see the potential of air power in World War One. In the inter-war years, each had proponents of the strategic capabilities of bombers who theorised that nations could be brought to their knees by air power alone.

When the Germans sought to build a military machine to dominate Europe, air power was a key component of their plan, and they were the first to begin to use it to its full potential in the Battle of Britain. This pivotal battle saw the *Luftwaffe* offensives being countered successfully by the integrated air-defence system of the RAF, which defended with far fewer fighters, and over home territory with tremendous courage.

When the Germans started the strategic bombing of civilian population centres, the consequences for them would be immense. The next four years saw an exponential advance in the capability of Allied heavy bomber aircraft and long-range escort fighters that allowed Britain and the US to destroy the industrial heartland of Germany. Strategic bombing had come of age. This was a watershed in the psychology of modern war, in which civilian populations inevitably became targets. It was a huge departure from the old chivalrous codes that had dominated Europe for hundreds of years, whereby battles were fought on the field between the militaries. As if to reinforce this new mode of war, the delivery of nuclear weapons on Hiroshima and Nagasaki confirmed that the combat zone now encompassed all of the combatants' territory. The destructive power of modern warfare had been increased exponentially.

At the start of the Cold War, the bomber acted as the key delivery system for nuclear weapons, but there was always the risk that these planes could be brought down by air defences. Thus, both sides sought to develop Germany's rocket designs to the point where missiles could be used to deliver nuclear weapons from one continent to another. It was within this context that the space race was born.

Although missiles became the primary delivery mechanism from land silos and submarines, bombers were not abandoned. Indeed, because of the need to launch before incoming missiles hit the air bases, these remained the first stage of strategic escalation.

As radar and missile defences became increasingly effective, the Americans sought to develop a new stealth technology that would make attack aircraft invisible to radar. The solution incorporated a combination of angled surfaces, radar-absorbent materials and active electronic countermeasures that minimised the return signal from the airframe. At $2 billion each, these planes were so expensive that the US could only build twenty-two of them. At this price, only the US could afford such a weapon, but, in terms of stealth and warplane design, the investment has now placed the US at least a generation ahead of any competition.

The plane's ability to deliver nuclear weapons unseen, to any part of the world, has once more made the bomber a vital part of the nuclear and conventional arsenal. Indeed, the capability of this aircraft is such that had it been deployed in the Cold War, it

would have upset the strategic balance and the concept of MAD in one stroke. Weapons development has also undergone a similar enhancement, to the point where each weapon can be specifically targeted with precision accuracy. The First Gulf War showed the awesome capability of precision weapons in their debut, but eighteen years later the destructive capability of the US Air Force (USAF) has increased many times over.

A vital but often unrecognised element of airpower is the transport command, which is responsible for in-air refuelling, which is used to extend the combat range of both fighters and bombers. This is also responsible for transport and support of the army when operating overseas. The USAF deploys a vast fleet unmatched by any other nation, entirely consistent with the goal of being able to project and support military power globally.

Concurrent with the development of its bombing capability, the USAF has also built a force of the most capable fighters ever seen. This means that the US is able both to ensure the safety of its own airspace from attack from any other nation (although clearly not from the kind of attack launched upon it on 9/11) and to dominate any airspace in the world that it chooses, for decades to come.

However, throughout history, there have always been shifts for and against the defender versus the attacker, and no technology will remain dominant. This means that the US will need to continue to innovate to maintain its advantage. Such innovation is going to prove costly at a time of budgetary constraint.

LEFT **A USAF F-22 Raptor, a fifth-generation fighter aircraft that uses stealth technology. It was designed primarily for air superiority, but it has other capabilities that include ground attack, electronic warfare and signals intelligence. Manufacturer Lockheed Martin correctly claims that its combination of stealth, speed, agility, precision and situational awareness, combined with air-to-air and air-to-ground combat capabilities, makes it the best overall fighter in the world.**

SPACE POWER

Throughout history, maintaining the highest ground has always been strategically advantageous. Thus, it was natural that space would become the next battlefield. High-altitude rockets made their debut in 1944 through the German invention of the V2, which was used to deliver one-tonne explosive warheads on London, silently and without countermeasure. Britain, the US and Russia rushed to acquire this technology as Germany was overrun, and their captured scientists were soon put to work on each side of the Iron Curtain in the ensuing space race.

The Russians were first in orbit with the launch of *Sputnik-1*, but it was American innovation that led the field in manned orbital missions and reaching the moon. This civilian technology was quickly translated by both sides into warhead-carrying missiles that were first powered by unstable liquid fuels, and later solid, stable propellants that allowed them to stand in missile silos ready to launch at a moment's notice. It was the Americans who built warheads that could hit targets with the greatest accuracy, initially from land-launched missiles and later, in a quantum leap, from submarine-launched missiles. Thousands of these missiles were built, consistent with the concept of MAD.

President Ronald Reagan proposed the Strategic Defence Initiative or, as it became commonly known, the Star Wars programme, in March 1983. The goal was to use advanced space technology to create a robust defence of the US homeland against a massed missile attack, thus breaking the MAD doctrine forever. The modern public remember the programme as an elaborate dream, but it did have one immediate consequence, which was to force an economically weakening USSR into a new technological arms race against the growing economic power of the US, accelerating the collapse of the Communist Bloc. The challenges were huge, but the technology that was initiated subsequently spread throughout the US military, and is largely responsible for its current relative advantage. Two decades later, some of the 'Star Wars' concepts have been deployed in a limited capacity as a missile shield over the US.

Although its defensive capacity currently only provides the US with limited protection from a few missiles, perhaps launched from a rogue state like Iran or North Korea, this is the first stage of a race to build a larger shield that only the US is running at present. As a result, it is possible to imagine that some time in the next two decades, it could become invulnerable to missile attack. This would be the first time since the peak of the British Empire when a nation could claim such security. As such, it is little wonder that the Russians and the Chinese view this development with great concern, and are dedicated to preventing this outcome.

It should be said that, in the hands of the wrong country, such weapons could be disastrous. However, in the hands of a balanced benevolent state, such immunity from attack, coupled with the ability to attack with impunity, might be the one mechanism that prevents the world from launching into a future global conflict. In the context of the Five Stages of Empire model, such invulnerability would be potentially disastrous for world peace if owned by an empire in the phase of expansion, but a very different and more positive proposition for an empire such as the US, which is in the declining phase of empire – provided that it does not seek to expand its power.

Today, space is a vital medium for sustaining both our modern society and our military infrastructure. The defence of vital satellites must be on the development agenda of all of these nations.

Asymmetric Warfare

Asymmetric warfare is a conflict between two opponents with very different military resource bases. The result is a non-conventional war. Asymmetric warfare is driven by a minority force that, through conflict with its opposition, seeks to spread their perception of their enemy throughout the general population, thereby increasing their own ranks. As such, these conflicts are as much about the shaping of popular opinion as they are about tangible military results. Because of this, it is essential for the defending party to keep innocent civilian casualties to an absolute minimum, otherwise they risk swelling the ranks of the insurgents with new recruits inspired by the desire for revenge.

From the perspective of the hegemonic empire, there are two distinct types of asymmetric warfare depending on where in the Five Stages of Empire the conflict occurs. First, there are the asymmetric wars of expansion, and then as an empire declines there are the asymmetric wars of contraction (see Figure 44). These two have very different characteristics and present different risks for the empire.

The Nature of Asymmetric Warfare and Empire *(fig. 44)*

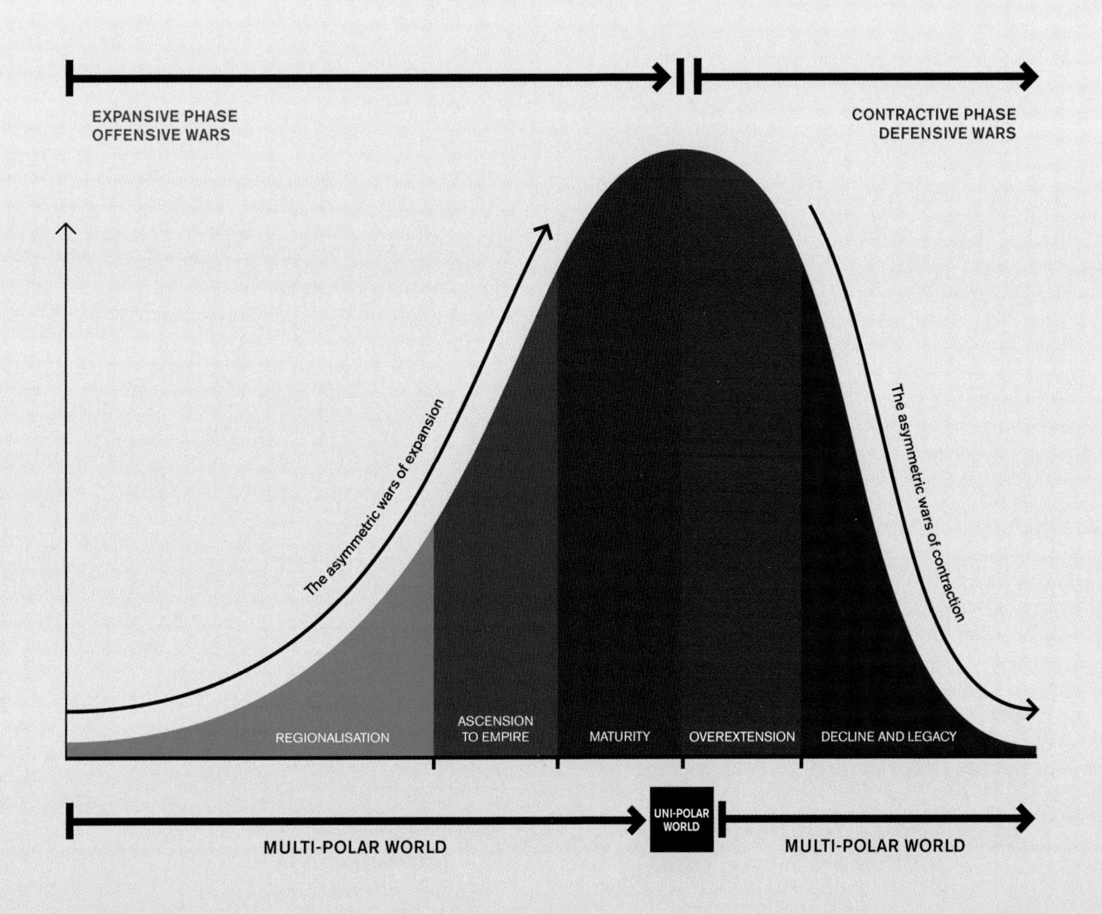

ASYMMETRIC WARS OF EXPANSION

An expanding empire seeks to move into territories that are resource rich and yet relatively undefended in what could best be described as colonial wars of acquisition. This strategy was typified by the WCSE's expansion into the Americas and Africa from the seventeenth century onwards. The drivers of such expansion, as we have seen, were primarily demographic, but they were facilitated by innovations in technology that allowed for conquest and assimilation and, at times, the annihilation of less militarily sophisticated societies. To the threatened societies, the invaders must have seemed overwhelming: their weapons were more powerful and they had a growing populace that was able to quickly replace any losses suffered. The expanding empire's commitment to expansion meant that it was simply a matter of time before it succeeded, and its very confidence in the outcome must have been sensed by its opposition, often hastening the resulting capitulation.

ASYMMETRIC WARS OF CONTRACTION

In contrast, asymmetric wars of contraction take place when a group or nation wants to assert its independence from the empire, or indeed begin to challenge what is perceived as a declining empire. This is part of the fracturing of an empire, and also the beginning of challenges from growing neighbouring regional powers. For the conflict to be classified as asymmetric, the challenge must come from a group that lacks the ability to arm and deploy a coherent military structure. The only option for the challenger is to wage war from within society: a guerrilla or insurgency-based conflict.

The conflict may seem to be unbalanced, with the technology of the empire's established military fighting an apparently rag-tag group of insurgents, but appearances can be deceptive, and such struggles are often far more protracted than one would expect, suggesting that there are key factors at work in their favour, as described below.

INTENTION

The first key factor is the intention to win the conflict. On one side is an empire whose collective toughness has been softened by decades of high living standards and security, and which consequently places a high value on human life. In effect, all the empire wishes to do is to maintain the *status quo*. Thus it is militarily complacent and will probably fail to take the bold steps that are required to mobilise its full resources to quickly and effectively suppress the challenge.

On the other side are the insurgents who are usually part of a demographically expanding population in the process of rediscovering their historic sense of nation or religion. They live in an altogether tougher environment that places a lower value on individual rights, and they have less to lose by pursuing a conflict that they think might help to improve their position. Their strength and commitment to their cause is often underestimated by their seemingly more powerful foe, whose arrogance is founded in the hubris of empire. This is a misjudgement made by many throughout history, including the Greeks of the Macedonians, the Indians of the Assyrians, the Russians of the Afghans and, today, the US of the Middle East.

The higher levels of intention possessed by the minority group give them an advantage that means that they can pose a serious challenge to the waning empire.

TIME

The second key factor is time. Again, this greatly favours the insurgents because, in all probability, they will outlive the political imperatives of the empire to continue their struggle. Indeed, as the empire's decline deepens, the insurgents will continue to expand demographically, giving them a long-term advantage.

LOCATION

While an empire is conducting an expeditionary war, the insurgents will be fighting on their home territory. Without the option of withdrawal, the insurgents may be more willing to exceed the boundaries of acceptable behaviour and to inflict more extensive civilian casualties on their own people. For them, this is preferable to losing the war, their values and their homeland.

Previous empires were not restricted by the moral ethos of protecting the civilian population, so this resulted in widespread destruction. However, today the values of the WCSE impose considerable restrictions on the persecution of civilians, which generates unique problems in the resolution of counter-insurgency warfare. The most relevant historic examples that illuminate today's asymmetric conflicts were those fought by the WCSE from the beginning of the twentieth century.

THE ASYMMETRIC WARS OF THE BRITISH EMPIRE

THE BOER WAR

The first asymmetric conflict of decline fought by the British was the Boer War of 1899–1902. Transvaal State, which the Boers had founded, contained huge quantities of gold, and the gold rush attracted a vast number of fortune-seeking British immigrants, thereby changing the demographic balance of the population and giving Britain the pretext to seek annexation.

To Britain's surprise, the Boers fought hard for their independence and held out for three years before they finally surrendered. Although the British Empire was ultimately the victor, this conflict altered the world's perception of the country, forcing it into a new set of strategic alliances with Japan, France and Russia, and encouraging Germany on its path to the 1914 war.

MALAYA

A half-century later, the British Empire was in terminal decline and had to fight a string of asymmetric wars of contraction during the Malayan Emergency of 1948–60 and the Aden Emergency of 1963–67, and in Northern Ireland. In Malaya, it embarked on a campaign of hearts and minds designed to win the support of the general population, most specifically the 3.3 million disenfranchised Chinese who were the main supporters of the Communist insurgents, the Malayan National Liberation Army (MNLA).

It did this by relocating some 0.5 million people to fortified villages that had superior living standards with defended perimeters supported by rapid reaction forces; by providing medical aid to the population; via the judicious use of firepower; with the development of good intelligence sources; and, finally, via the restriction of the food supply to the insurgents by carefully monitoring the general population as they left and returned to their villages after working in the fields. (In contrast, a decade later, the US failed to win over the Vietnamese and, indeed, by using excessive firepower, converted many a neutral Vietnamese to the side of the North Vietnamese Army (NVA) opposition, thus hastening their defeat.) Of special note in the Malayan success is that it took some 40,000 troops and police to suppress an insurgency of some 7,000 guerrillas. This underlines the point that it takes a disproportionate amount of manpower to win a counter-insurgency war.

RIGHT The second wave of combat helicopters of the 1st Air Cavalry Division fly over a radio telephone operator and his commander on an isolated landing zone during Operation Pershing, a search-and-destroy mission on the Bong Son Plain and An Lao Valley of South Vietnam, during the Vietnam War. It is estimated that of 2.59 million Americans who served in the conflict, more than 58,000 were killed and 304,000 wounded.

NORTHERN IRELAND

The struggle of the Catholics in Northern Ireland against Protestant rule dates back centuries, but in its modern form stemmed from the proclamation by the Irish Republican Army (IRA) in 1916 that in 1922 resulted in an independent Ireland. Decades later, when Britain found its power weakened by the loss of the empire, the old Protestant/Catholic conflict in the British province of Northern Ireland escalated. This spread to bombings on mainland Britain, forcing a long and protracted insurgency campaign that was, in large part, funded by Irish Americans.

Over the decades, Britain developed a strategy that was heavily dependent on intelligence, allowing the application of force in a precise way that minimised the negative effects of public opinion. A key lesson of the conflict was that by occupying moderate ground as terrorist actions become more extreme, Britain succeeded in isolating the IRA from the general population, thereby weakening its support.

The conflict was brought to a conclusion by the Good Friday Agreement in 2005. This allowed for the inclusion of the IRA in the political structure of the province, bringing the former terrorists into the mainstream. Had they continued to engage in armed conflict, it would have damaged their credibility as politicians. The lesson here is that if you cannot annihilate your enemy, ultimately they will have to be included in a political process – the terrorist must be transformed into a politician.

Interestingly, this conflict was resolved at the same time as the Islamist threat to the West began to emerge, almost as if the last Christian religious war was being supplanted by a greater threat from another religion.

THE ASYMMETRIC WARS OF THE US

VIETNAM

As we have seen, when the empires of Britain and France collapsed, it was up to the US to fill the power vacuum. Thus it inherited the conflict in former French Indochina that became known as the Vietnam War. The US entered the war to prevent a Communist takeover of South Vietnam, as part of its wider strategy of containing Communism's spread. Unlike the Malayan insurgents who were isolated, the NVA had the support of

ABOVE A member of the US's 24th Marine Expeditionary Unit has a very close call after *Taliban* fighters open fire near Garmser in Helmand Province, Afghanistan, on 18 May 2008. He was not injured.

the North Vietnamese and the Chinese. Furthermore, it was not isolated from the general population by its ethnicity as the Chinese MNLA had been, so its potential support base was not limited. In addition, in Malaya the British had a strong rapport with the Malays after joining with them to fight the Japanese. However, the French had earned the disdain of the Vietnamese, who transferred this to the Americans, preventing the development of mutual trust.

The Americans therefore had the odds heavily stacked against them at the outset of the conflict, but in addition they made some very grave errors. In the early 1960s, they viewed Vietnam as a conventional conflict against a structured enemy in which they thought their superior firepower would prevail. As such, they failed to implement a policy to win over the population, to develop an effective intelligence network, or to emphasise the individual soldiering skills and integrity in the treatment of locals that are required in a low-intensity conflict. With such deep flaws in their strategy, the result was inevitable, but US determination, set against the threat of an ongoing Cold War, meant that the war did not end until 1975. However, the American failure caused a deep trauma to the military and America's population that took years to heal. Although the US was in a phase of maturity during this war, it can be conceptualised as a war of contraction as part of the larger WCSE cycle in that the rising ASE sought to challenge the old and declining super-empire.

AFGHANISTAN AND IRAQ

The Afghanistan War was portrayed as a defensive conflict following the 9/11 attacks, but in essence it was an operation by the US to prevent the country from being used as a base by *al-Qa'ida*. The US considered Afghanistan a special-operations invasion in which local troops of the Northern Alliance were supported by US advisers and firepower. However, once the *Taliban* had been removed (despite errors that showed the US to be culturally unable to act effectively in an agricultural and tribal country), the US made a major mistake in not maintaining sufficient troops on the ground to stabilise the country. More importantly, it failed to seize the chance to give the nation, torn by war for some twenty years, a new way of life that other Islamic countries would come to desire and want to emulate. Thus the chance to win the hearts and minds of the Afghan people was lost.

The Second Gulf War started as a conventional battle that turned into an insurgency. In 2003, when the US had won the conventional war to invade Iraq, it was totally unprepared to encounter a population that did not embrace a US presence with open arms. The US army was configured to fight an all-out war, and for years it failed to integrate the lessons that it needed from the British army, which had a vast array of relevant experience. The main problem was that Rumsfeld failed to deploy the 450,000 soldiers that the generals thought necessary to control the country. Those soldiers that were deployed were not judicious in the application of their firepower, having an over-riding belief that the protection of their own forces was more vital than the prevention of the deaths of innocent civilians. Furthermore, the US's cultural sensitivity to their environment was shockingly lacking and it took far too long for commanders to understand the differences between the *Shi'is* and the *Sunnis* – a fundamental differentiation that was vital both in winning over the general population and in distinguishing friend from foe in a confused urban environment.

It took two years for the Americans to realise that their initial strategy was doomed, and to change it to a counter-insurgency campaign, which has been increasingly effective from 2007 onwards. However, as with Britain in 1902 with the Boer War, whatever the outcome in Iraq, this asymmetric war of contraction has done enormous damage to the world's perception of the US's invulnerability. Indeed, one of the unintended consequences has been that the *Shi'is* have gained power, giving Iran extended influence over southern Iraq.

While the situation in Iraq is improving, the conflict in Afghanistan remains unresolved, once more through the inability of the US to adhere to the principles of counter-insurgency. Foremost in the list of errors is a failure to convince the US's population of the need to stay until the *Taliban* is defeated. The easing of pressures in Iraq may release more troops for deployment in Afghanistan. However, if a more passive stance is taken, this conflict will rumble on without resolution. This could have the unintended consequence of radicalising Pakistan with the grave possibility of nuclear conflict with India. Finally, having taken NATO out of its mandated region (Europe) where it was so successful in the Cold War, there is a risk that a failure in Afghanistan will have major ramifications on NATO's structure and reputation as a deterrent force.

ENOLA
GAY

The Age of Atomic Weapons

LEFT Colonel Paul Tibbets Jr, pilot of the B-29 bomber *Enola Gay*, which dropped the atomic bomb on Hiroshima on 6 August 1945, waves from the cockpit before takeoff. The B-29 Superfortress bomber was named after Tibbets' mother. The bomb, which was given the codename 'Little Boy', caused extensive destruction and heralded in the nuclear age.

The atomic weapon that was dropped on the occupants of Hiroshima was equivalent to some 12,000 tonnes of explosive. The city's population suffered from an instantaneous release of thermal radiation that lasted for only seconds but scorched all that it touched for miles around. Then, as the fireball of super-heated air rose above the city, the blast wave swept all before it within a radius of one mile, although the damage spread out much further.

In little more than a minute on 6 August 1945, the city of Hiroshima had ceased to exist, along with some 70,000 of its occupants. By the end of 1945, a similar number would be dead from their wounds, and a similar number again were to die from the effects of radiation by 1950.

Three days later, on 9 August 1945, the Americans dropped a larger and more efficient implosion bomb, which resulted in the annihilation of Nagasaki and its population.

The utilisation of nuclear weapons had changed the world. In time they would undergo developments that would lead to weapons of increased capacity and increased miniaturisation that could be deployed by a multitude of delivery systems.

For four years after Hiroshima, the US was unassailable. However, in 1949, the Soviets tested their own bomb, and from then on the sword of Damocles was poised over the head of humanity. At first, nuclear bombs had to be carried by a bomber, travelling long distances to reach its intercontinental targets and being vulnerable to interception. However, as the space race accelerated, the military offshoot was the creation of the intercontinental missile, which took only thirty minutes to travel from the USSR to the US and would be impossible to intercept. With this evolution, the world moved to within a trigger's breadth of Armageddon, as each side had to be ready to launch its missiles if the other attacked, giving them little over twenty minutes to respond. The Cold War tension was further exacerbated when tactical weapons with shorter flight times were located within Europe. In response, Russia tried to station missiles in Cuba, almost precipitating a nuclear exchange in the crisis that followed.

Meanwhile, parallel development of the submarine-launched missile to an effective operational system calmed tensions, because these submarines could hide silently and undetectably in the oceans, waiting to retaliate if their country was attacked.

The three independent weapon-delivery systems provided by bomber, missile and submarine make up what is known as the strategic triad, ensuring that MAD is credible – that one side could not destroy the other without also being destroyed. It is this nuclear stalemate that prevented the Cold War from becoming a hot war. While the US, France, Russia and China all operate a strategic triad, Britain relies primarily on its submarines. Of the five countries, the US and Russia have the largest arsenals. In 1985, there were estimated to be some 65,000 active warheads in the world, a number that had decreased to 20,000 by 2002 and may well have decreased by 30 percent by 2010. Many of the decommissioned weapons were dismantled, although they remain stockpiled in their respective countries.

As a measure of absolute power, the nuclear weapons club is unsurpassed and it is only the US, Britain, France, Russia and China that are members of the permanent UN Security Council, a legacy of the Cold War that continues today. Indeed, the ability of a nation to make nuclear war is the highest forum of resolution in the world, given that as a conflict escalates through various stages, unless a country is able to create a stalemate situation at this level, it will lose the war. As such, a nuclear arsenal grants its owner less reliance on conventional forces if it only desires to maintain the sovereignty of its territory and does not want to project power. This in turn has economic benefits in terms of a lower defence budget.

RIGHT **The ruins of Hiroshima after the atomic bomb blast. The bombings led, in part, to post-war Japan adopting Three Non-Nuclear Principles, forbidding the nation from nuclear armament. The role of the bombings in Japan's surrender and the US's ethical justification for them, as well as their strategic importance, is still debated.**

However, despite nuclear weapons having enormous power, they also have limitations. The US's possession of a nuclear force did not prevent or stop the wars in Korea, Vietnam or Iraq, which have all placed a severe drain on its economy and population. These conflicts took place below the threshold at which the US considered that it could use nuclear weapons. Even with lower-yield weapons, the stigma of global opinion against their use is so great that they can only be used for self-protection rather than for expansion or power projection. This remains the domain of conventional forces, and particularly sea and air power.

For many, the lessons of the Cold War have been consigned to the history books, but, just like the Second Industrial Revolution that now drives Asian growth, and is changing the world, a nuclear arms race is taking place in the emerging nations. Just as in the Cold War, the race to build nuclear weapons is being run in parallel with a space race to develop effective delivery systems and enhance national prestige. Of major concern are the risks associated with this phase of nuclear development and deployment before a full strategic triad has been deployed that replicates the MAD balance.

HOW HAVE WE AVOIDED ARMAGEDDON SO FAR?

Since 1945, there have been 2,000 nuclear weapons tests (see Figure 45), so the question must be asked, 'How have we survived?' The answer is that the Cold War was a war of contraction fought between two factions of the WCSE with shared imperial character, collective values, memories and experiences. It was these factors that ultimately prevented the war from becoming apocalyptic. It is of considerable importance to understand the five essential components of this shared experience because other nations that plan to develop a nuclear capability may not embrace them. The five are described below.

A LEGACY OF DESTRUCTIVE CONFLICT

The legacy of the blood-letting of the two peak wars fought over home territory has had a profound effect on inhibiting the aggressive application of nuclear weapons in times of extreme international tension. Both sides in the Cold War were aware of the massive loss of life during those wars, and the legacy was a very deep-seated reservation across the leadership of NATO and the USSR to risk such an event again. Thus, at critical moments when tensions were high, the memories of the realities of war were foremost in the minds of the leaders of both sides and acted as a vital deterrent.

The Increasing Power of Atomic Weapons (Kilotonnes) *(fig. 45)*

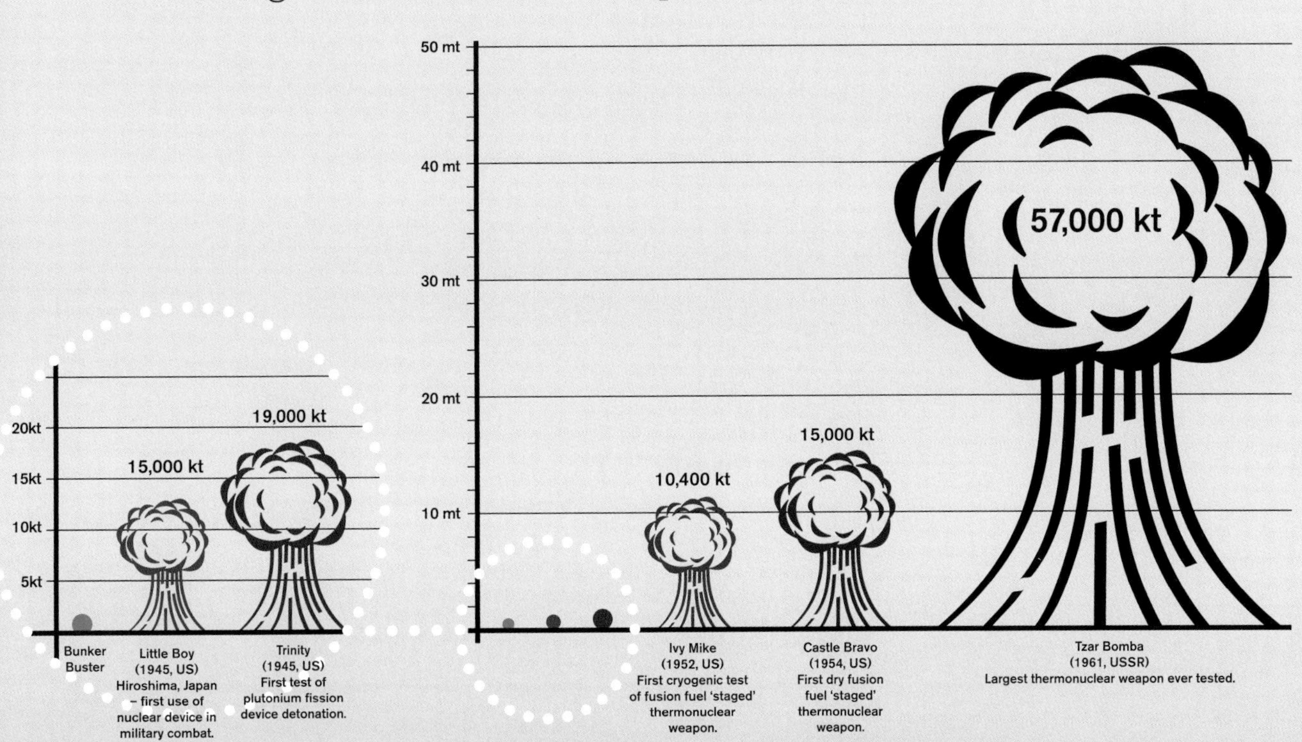

STABLE POLITICAL SYSTEMS

The political systems on both sides of the Cold War were large and well structured, and the leadership – even in the USSR – could not initiate conflict without involving a number of other politicians. Thus these systems were, in effect, greater than any one leader, preventing one fundamental error of judgement from initiating a war.

GAME THEORY

The Cold War was played out according to John Nash's Game Theory – namely there was a policy of increasing each system's equity stake, but not at the risk of losing everything. In effect, the political systems responded logically, aiming to increase their global influence but not at the risk of losing their primary equity. Thus the game was played not between the centres but at the periphery of each player's sphere of influence, through a series of proxy wars in which one side hoped to weaken the other. Essential for this model to work was that both sides had the same rule base, foremost of which was that they valued their own lives.

THE CREATION OF MAD

As nuclear technology improved, the strategic triad came of age, significantly reducing the chances of a pre-emptive strike by an aggressor who could not guarantee the absolute destruction of its enemy without facing equal retribution. In effect, MAD was nuclear stalemate that brought with it stability and, as such, the world became a safer place, despite a massive increase in the number of deployed warheads.

PREVENTION OF ACCIDENTAL WAR

Last but not least, there were enhanced systems for preventing an accidental missile launch as well as security for the plethora of warheads stored across the world. At the end of the Cold War, with the collapse of the USSR, the risks of small warheads falling into the wrong hands was very high and the US employed some very aggressive and effective measures to secure any loose weapons. This is particularly important in the era of modern terrorism, as a nuclear weapon would be a terrifying prospect in such hands. However, unlike normal explosives, which are generic, atomic weapons house radioactive metals that derive from a nuclear reactor, each of which has a unique signature through which any nuclear detonation can be traced back to its reactor source and nation. The threat of retaliation is accordingly a major limitation on anonymous state-sponsored nuclear attacks under the guise of terrorism.

THE RISK OF NUCLEAR PROLIFERATION

The restraint of the key players sharing a common heritage meant the world survived the Cold War. But in the countries outside the Security Council, many of the proliferating nations do not share the same history or values. Particularly alarming is the idea of nuclear weapons in the hands of fundamentalist Islamic states where some extremist elements place a higher value on the afterlife than on life itself, suicide bombing being one manifestation of this. In essence, this destroys the safety net of Nash's Game Theory.

Nuclear Weapon States *(fig. 46)*

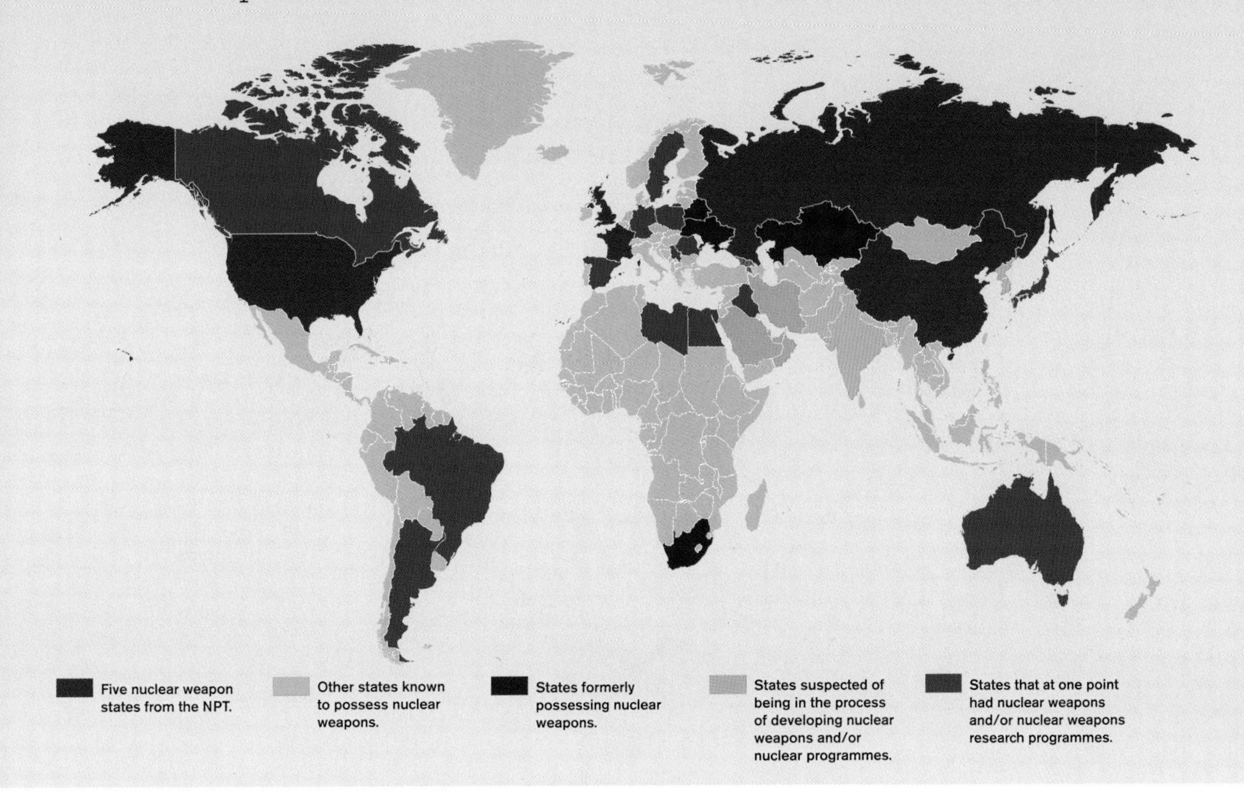

It would be irresponsible of the older nations to allow nuclear weapons to be developed by countries that have a high probability, at some stage, of using them, and killing many hundreds of thousands of people in the process. Thus, when a country succeeds in becoming a nuclear power in the face of opposition from the US, it has the same effect as a pilot war, without a shot being fired, in that it acts to demonstrate a decline in the ability of the dominant empire to effectively project power. North Korea was a case in point, diminishing US power in the Korean Peninsula, and Iran is likely to succeed in the Middle East unless the US demonstrates a higher degree of intention.

History has shown repeatedly that once a technological advance is made in one country, it is only a matter of time before it spreads. This has been the case with American nuclear technology since 1945. The Russians accelerated their own programme, testing their weapon by 1949. By 1952, Britain followed suit using the data gained from its collaboration in the Manhattan Project, and in 1960 the French, using their own research, did likewise. Then, with help from the Russians and after years of effort, China became the fifth and final member of the Security Council. It is important to note that as the Chinese approached their first test, the US seriously considered an all-out nuclear attack in order to prevent Chinese success. Fortunately, cooler heads prevailed, but it is important to understand that the issues around Iran's path to proliferation have been played out before, although China and Iran carry very different risks to world peace.

'Axis of Evil': Proliferation of Nuclear Technology and Missiles *(fig. 47)*

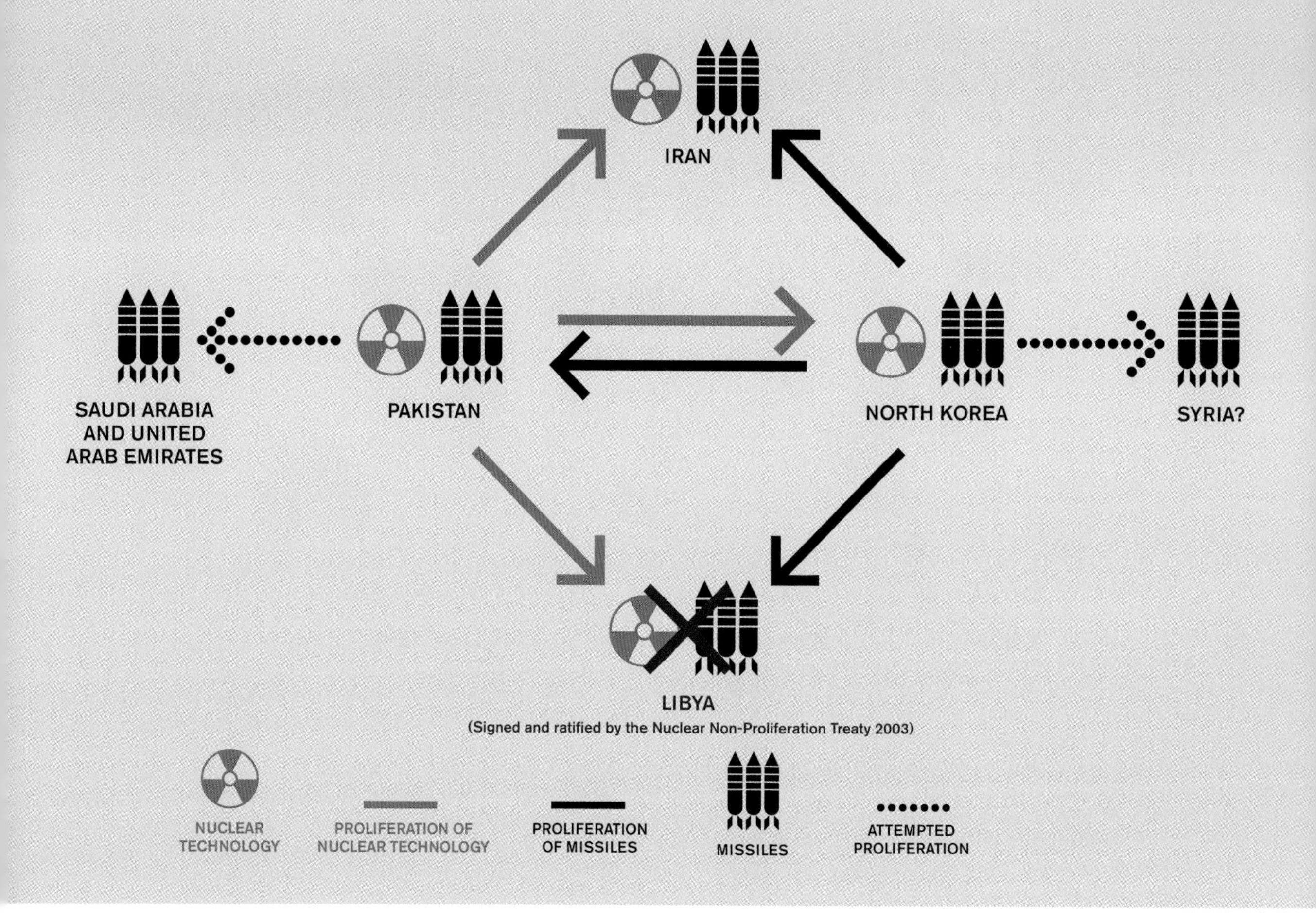

At this stage, the five members of the nuclear club, concerned that others might want to join them, implemented the Nuclear Non-Proliferation Treaty (NPT) on 5 March 1970. This was designed to limit the transfer of civilian nuclear technology into a military application. However, India, which had not signed the NPT, tested its first A-bomb in 1974. They did this by clandestinely diverting civilian technology provided by the Canadians into a military programme, in the very way that the NPT was designed to prevent. India and Pakistan were involved in a multi-decade regional rivalry, and so Pakistan became determined to develop its own weapons. It achieved this with the support of the Pakistan's intelligence services (Inter-Services Intelligence: ISI). The A. Q. Khan network stole the plans for a Dutch centrifuge facilitating the technology to refine uranium. The network went on to gather other related technologies and spread them to other aspiring nations. For Pakistan, this copy-and-assimilate phase resulted in its first nuclear test in 1998. It is significant that the expansion of Chinese regional power was at the heart of the spread of nuclear technology into the Islamic world. It was Pakistan's ownership of nuclear technology that accelerated the proliferation process, and many of the current looming nuclear crises can be traced back to this point.

Israel had been besieged by the Arab world from all sides from the day of its founding, and had fought a series of wars that at times looked as if they were going to end in the complete destruction the country. The region was also a proxy for the Cold War, so the fall of Israel would have amounted to the unacceptable loss of US control over the vital oil supply lines. Thus, for both Israel and the US, the solution was for Israel to become a nuclear power. One has to suspect that Israel achieved this with secret US support, but the upshot was that although it never officially declared its ownership, creating what it termed strategic ambiguity, by the late 1970s, the presence of Israel's nuclear weapons brought the conventional wars of the Middle East to a rapid close.

The next country to develop nuclear capability was North Korea, which surprised the world in 2003 by pulling out of the NPT after the US removed energy assistance when it gained evidence that North Korea had a uranium-enrichment programme. It has since built its arsenal to an estimated ten warheads. In a dramatic stroke, one of Bush's declared members of the 'axis of evil' (see Figure 47) had gained nuclear weapons and poked American power in the region in the eye. However, what appeared on the outside to be one country's trump in joining the nuclear weapons club was far more sinister for US interests. First and foremost, North Korea is 100 percent dependent on Chinese energy supplies. Accordingly, if China had not considered that North Korea's nuclear weapons were in China's interests, it would have literally turned the lights out. Then, cooperation between North Korea and Pakistan has resulted in both countries having rocket and nuclear technology, so each could, in effect, build a complete set of deployable nuclear weapons.

North Korea's success constricted American power in the peninsula in line with its regional strategic objectives. However, for China, there was one unintended complication: the awakening of Japanese concerns over North Korean power. This thinly veiled underlying concern about the rise of Chinese regional military power has generated a revival of the Japanese military complex. Thus, the Six-Party Talks (which aimed to find a peaceful resolution to the security concerns as a result of the North Korean nuclear weapons programme) were, for the Chinese, a balance in maintaining the North Korean thorn in the US's side while reducing its bellicose attitude for fear of further encouraging the Japanese hardliners.

In an ongoing sinister development, North Korea and Pakistan have had significant roles to play in helping both Iran and Syria in their attempts to proliferate. Both of these aspiring nations were prevented from developing nuclear weapons with bold and well-executed air strikes by the Israeli Air Force, which destroyed the reactors of each country. This should serve as a warning to Iran, which is determined to gain nuclear capability.

However, all is not proliferation: it is worth noting that some nations have peacefully given up their nuclear weapons (South Africa in the early 1990s, and Belarus, Kazakhstan and Ukraine all by 1996).

Clearly, the containment of nuclear proliferation is key to the future security of the human race, and as such we will return to this subject in FUTURE.

Chapter Nine

Disease and Empire

Man's greatest war has been fought against disease. We have engaged in protracted battles against bacteria, viruses and other disease vectors, such as fleas, lice and mosquitoes. The history of this war can be traced to the successful expansion of the human population over the past two millennia, and through the cycles of empire (see Figure 48).

Disease and the cycles of empire are inextricably linked: the rate of death by disease compared with the birth rate gives us the demographic curve of an empire. Consequently, epidemics during the ascent to empire phase, a period of rapid population growth, have a lesser effect than an epidemic that strikes a declining empire, where a falling population cannot compensate for the ravages of disease. Modern-day Russia is an example of a population, already in decline, in which the ambient disease process in this case of Human Immunodeficiency Virus (HIV)/tuberculosis (TB) has played a significant role.

RIGHT **The cinchona tree is the only known natural source of quinine. Its medicinal properties were discovered by the Quechua Indians of Peru and Bolivia, and later the Jesuits brought it back to Europe. It first appeared in therapeutics in the seventeenth century.**

BELOW RIGHT **Sir Alexander Fleming (1881–1955) was most famous for his discovery of the antibiotic powers of penicillin in 1928, for which he shared the 1945 Nobel Prize for Medicine with the two chemists who perfected a method of producing the drug.**

A Short History of Disease

The emergence of disease has gone hand in hand with man's social evolution. The shift from tribes of hunter-gatherers to settled agricultural societies led to the integration of domestic and farm animals into human communities, with a concomitant merging of the disease pool. There are a number of diseases able to jump the species barrier – avian flu, swine flu and AIDS being the most notorious present-day examples.

Over 4,000 years ago, the population concentrations of the riverine civilisations created the ideal conditions for human-to-human disease transmission, giving life to new strains of bacteria and viruses. In addition, parasites integrated their life cycle with that of humans so that, for example, there was an increase in incidence of water-borne schistosomiasis (bilharzia) in the irrigated fields of the Nile as well as common intestinal parasites (e.g. tapeworms). As urbanisation concentrated the human population to unprecedented densities, human-to-human transmission of disease accelerated. Towns and cities brought with them major challenges in social hygiene that took many centuries to understand. Diseases were also exported along trade routes with catastrophic consequences. The decimation of the Native American population of what is now the US and Canada and the Aboriginal population of Australia by smallpox and measles are but two examples. As the world moves towards a globally integrated

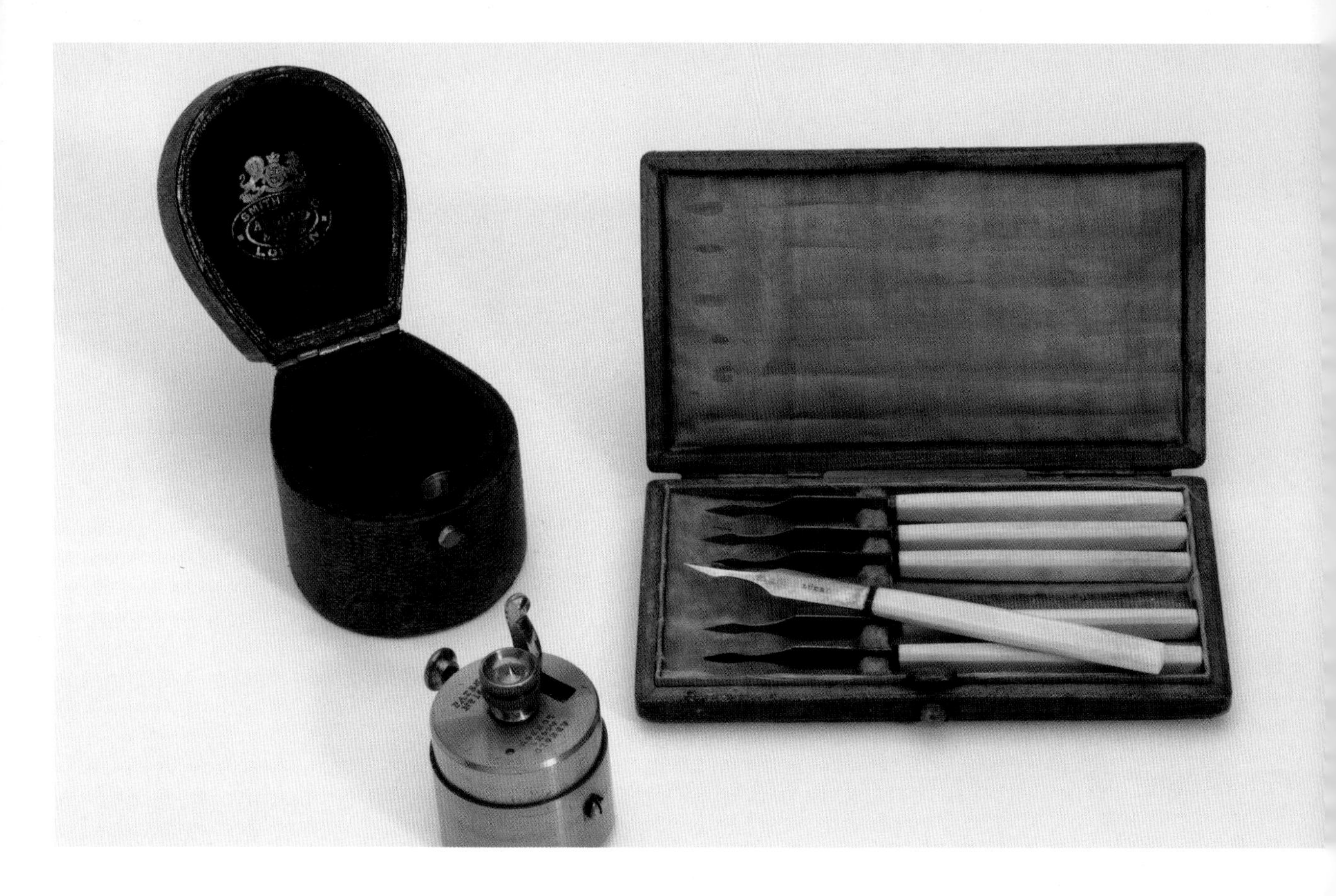

population, ironically, it is disease reservoirs in the developing world that could now threaten the world's wider population, if not properly contained.

There are four distinct types of disease: bacterial, parasitic, viral and lifestyle. In the past 200 years, medical science has made significant progress in the war against bacteria and parasites, which has enabled the acceleration of population growth. Foremost in the process were three key pharmaceutical discoveries: quinine, the smallpox vaccine and penicillin. However, our initial successes are now being challenged as diseases adapt and fight back, resulting in many new drug-resistant disease strains.

ABOVE A smallpox vaccination kit from the nineteenth-century, containing small blades used to pierce the patient's skin. Edward Jenner (1749–1823), a country doctor, developed the inoculation against the killer disease.

RIGHT Representation of the double-helix structure of a segment of DNA molecule.

The living conditions of the most socially disadvantaged (underfed, overcrowded, unsanitary) provide the ideal conditions for the spread of disease. Stress or shock to the social system can cause a local outbreak to overflow into the wider population, creating epidemics. There are degrees to this process, but it is clear that a society is only as strong as its weakest link, which should be sufficient incentive for governments to endeavour to eradicate both poverty and the diseases it shelters.

Viruses continue to represent a major threat to society through epidemics, so they are the focus of considerable modern scientific research. The unravelling of the DNA code is a significant breakthrough that promises to help in our ongoing struggle against the genetic aspect of disease.

The Disease Cycle of Empire *(fig. 48)*

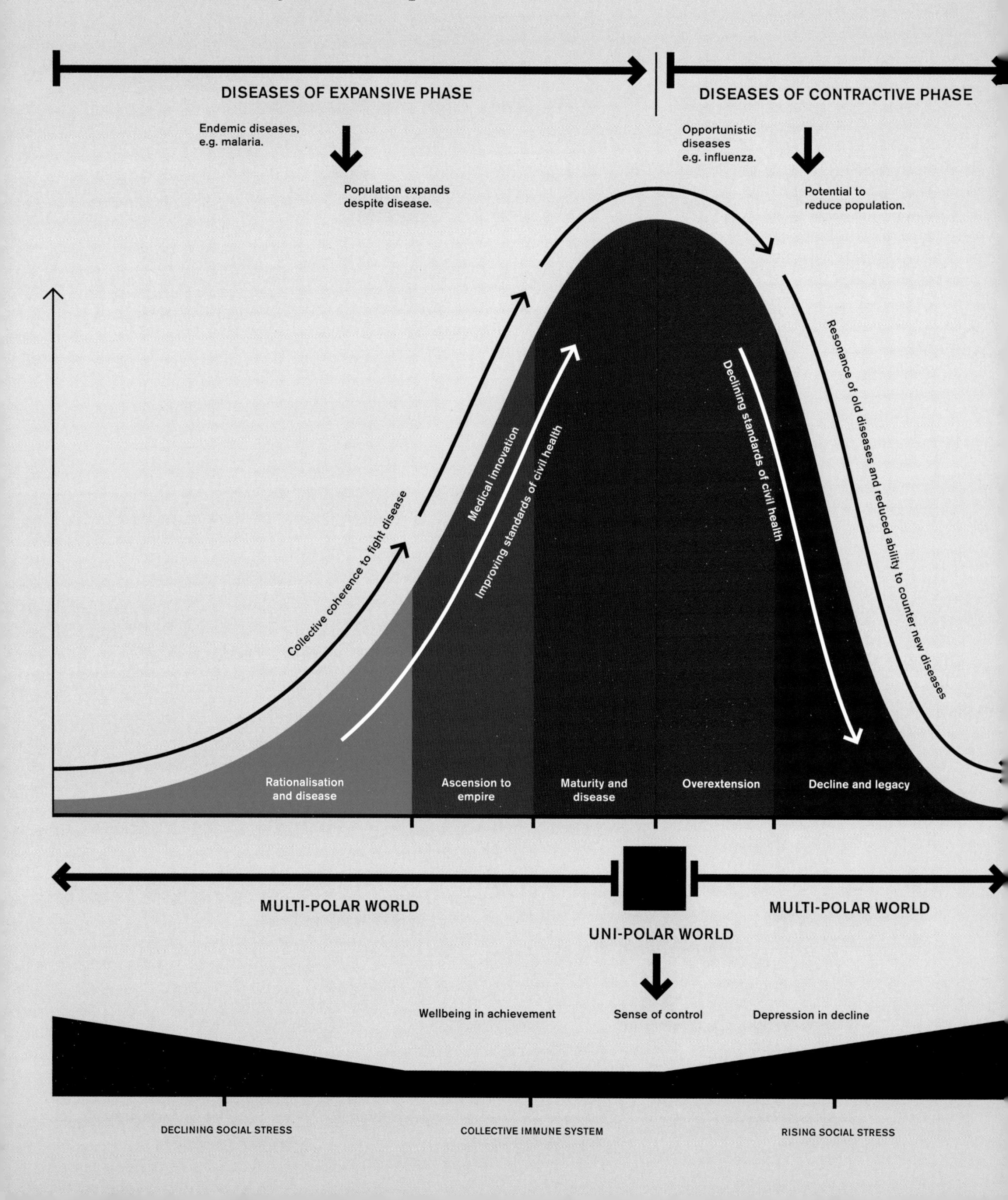

ENDEMIC AND OPPORTUNISTIC DISEASES

When looking at the cycle of population growth and demise, we can classify diseases and their effects on an empire as being either endemic or opportunistic (see Figure 49). Endemic diseases, such as malaria and typhus, have historically acted continually on the population to limit or reduce its growth. Advances in medical knowledge and improvements in urban health (at least in the WCSE) have had the effect of rolling them back. However, opportunistic diseases, such as bubonic plague, smallpox, cholera and influenza, take advantage of the population density of urbanisation coupled with social stresses, such as famine, conflict and reduced economic coherence, and these have not proved so easy to counter.

On a collective level, there is likely to be improved longevity and resilience in the population of a growing empire as it gains more control over its environment and achieves greater economic power. Conversely, there is likely to be a weakened collective immune system in periods of heightened social stress during the decline of an empire. If, as has been argued, the cycle of an empire is akin to that of an individual who suffers the diseases of youth, middle age and old age, then an empire also has to engage with diseases of its ascendancy and expansion, and its decline and contraction.

DISEASES OF EXPANSION OF EMPIRE

Historically, the coherence of organisation that accompanies the growth of an empire, with improved utilities and methods of disease control, has mitigated the effects of disease on a population. This may be further enhanced by the strengthening of the collective immune system as individuals gain a sense of control and pride in their achievements that feeds into a positive cycle. However, the strength of a collective immune system varies with advances and setbacks in the society of which it is part. In periods of stress, epidemics take hold, starting with localised outbreaks that then gain a foothold in the general population. In addition to periodic outbreaks of opportunistic diseases, there are continual struggles throughout the empire cycle against endemic diseases, such as malaria and typhus.

DISEASES OF CONTRACTION OF EMPIRE

Conversely, as an empire goes into decline, and its social and economic structure disintegrates, the collective immune system of the population is weakened by the stress of holding together a failing system, coupled with decaying public utilities. This allows disease vectors to take hold. Increased social stress facilitates an acceleration of the process of decline (on both an individual and a societal level), allowing disease to take hold and ravage the population. The decline of an empire is a not a linear process; instead it displays periods of apparent stability followed by accelerated deterioration. It is the latter period that is most likely to be associated with collective immune-system suppression directly linked to stress, with consequential epidemics.

THE PLAGUE OF JUSTINIAN AND THE DECLINE OF THE BYZANTINE EMPIRE

A key example of a disease of contraction of empire was the Plague of Justinian. This started in AD 541 at the peak of the Byzantine Empire's power, and it lasted for over a century. Named after Emperor Justinian, who caught and survived the disease, it was to change the fate of his empire. Although at the start of the outbreak the Byzantine Empire seemed intact, it had in fact been put under massive social stress by the power shift following the fall of Rome and the massive military campaigns aimed at securing the beleaguered empire's borders against the Goths and Vandals. It was not surprising then that the population of Constantinople succumbed to the plague and that, at its height, 5,000 to 10,000 people a day perished.

The plague, which is believed to have originated in Egypt, followed the grain trade routes north to Constantinople then spread across the whole of the Eastern Roman Empire. During the height of its devastation in AD 541–542, it resulted in the death of one-quarter of the Mediterranean population.

The plague impacted on the very future of the Byzantine Empire. Just before it struck, Justinian was about to embark on a campaign to restore most of the original Roman Empire, including Rome itself. However, not only was this venture aborted but, under the burden of a significantly reduced population, the empire continued to be prey to the expansion of the Goths. Indeed, so devastating was the effect of the disease on the population of the Byzantine Empire that it is not merely speculative to suggest that, even a century later, it played a significant role in the successful expansion of the Islamic Empire of the first caliphate into its territory.

RODENT EPIDEMICS

Man has become the ultimate predator on Earth, vulnerable only to disease. As a result, a key question for the future is whether medical advances will keep pace with the war against disease or whether new forms of, at least initially, drug- or control-resistant disease will reduce the human population. There are a number of animal populations that exemplify the cycle of growth of population and disease, unmodified by medical intervention, and that can inform us about what this process might be like.

Foremost among these are the mouse plagues of Australia. Every few years, a number of factors conspire to create the perfect conditions for an explosion in the mouse population. First, there must be above-average rainfall to produce a plentiful food supply with an overlap of successful Summer and Winter crops.

Once these conditions exist, the prolific breeding capability of the mouse is triggered. Mice can breed at any time of the year, and can lactate while pregnant, decreasing the time between litters. Combined with a gestation period of eighteen to twenty-one days, litter sizes of four to eight and a sexual maturity of only five to six weeks, this enables the mouse population to grow at a staggering rate over very short time frames.

Each mouse consumes some 2–3 g of food per day, so that 200 mice eat as much as a sheep (200 kg consumption per year). Up to 1,400 mice per hectare have been recorded, and the huge numbers that appear at the time of these plagues can devastate a farmer's crops, festooning their barns and homes with a moving carpet of mice.

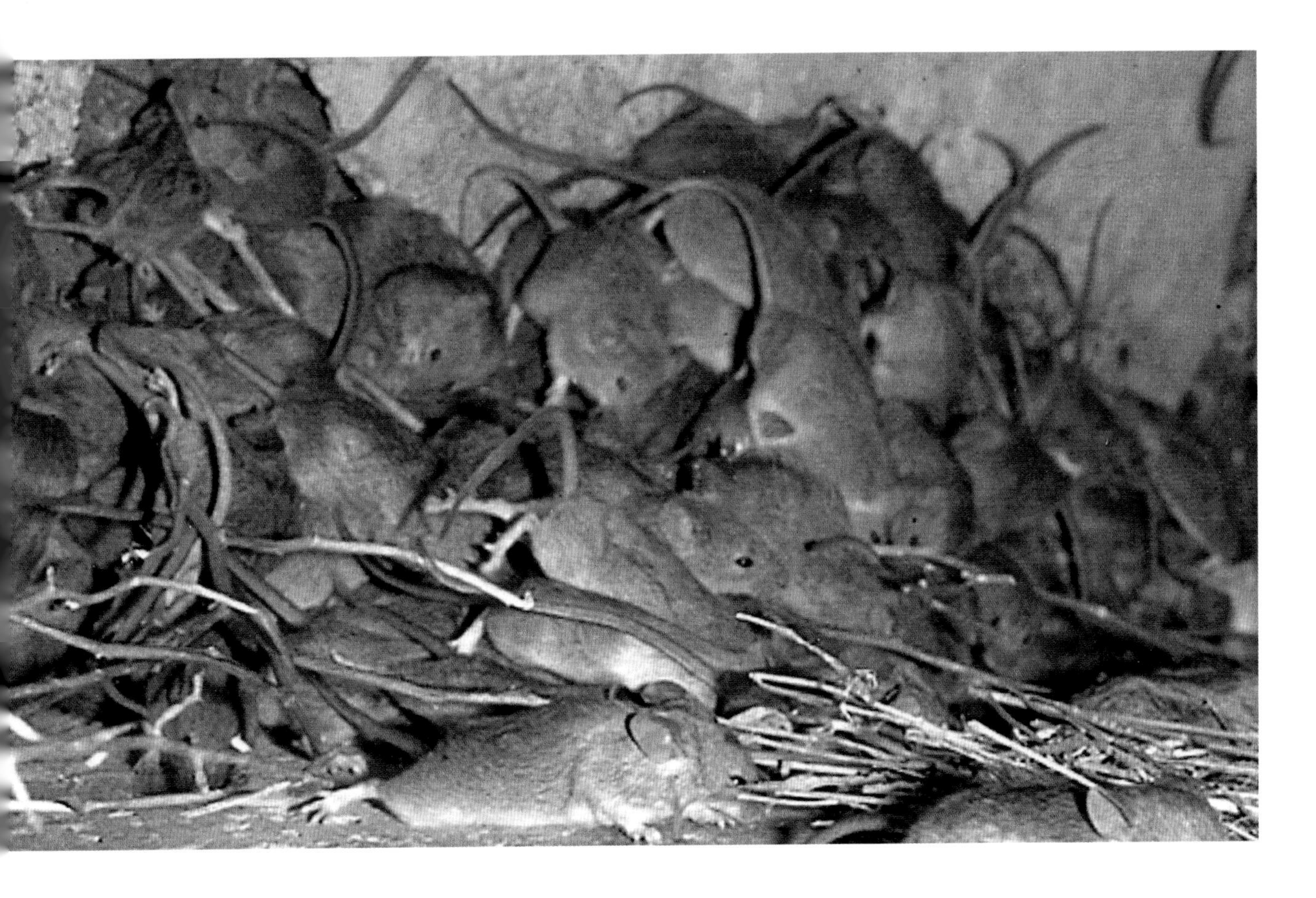

ABOVE Mice swarm during an Australian epidemic. These rodents live for up to two years. Females can start reproducing at six to eight weeks old and can have between five and ten young in each litter (gestation is three weeks). One female can thus produce more than a hundred offspring a year.

The scale of these epidemics is astonishing. For example, one morning a farmer who had baited with poison had to sweep 28,000 dead mice off his veranda, and one town recorded 544 tonnes of poisoned mice in five months.

To all intents and purposes, the mice are contained in a closed system, and their population growth is naturally limited by their source of food running out. Consequently, most plagues end in July as food becomes scarce and the weather cold, increasing the level of social stress. These factors, combined with the high population density of the mice, facilitate the rapid spread of disease. As the competition for food increases, the behaviour of the mice changes and they become more aggressive. Fights break out, resulting in mortalities and vicious wounds, which are easily infected, thereby further accelerating the spread of disease. Within two days to two weeks, the population crashes. These plagues are periodic, ebbing and flowing, constrained in their population cycle, as has been said, by existing in a relatively closed system.

Another example of the cycle of growth and disease is the Norwegian lemming – a rodent no more than 8 cm long that inhabits the Arctic tundra. The lemming population also periodically goes through rapid expansions. However, they have evolved a strategy to escape the limitations of their closed system: when the population density reaches an unsustainable level, they embark on a mass migration. Their prodigious swimming ability means that they fearlessly take to the oceans in search of new territory despite the majority perishing in the process. The staggering size of these lemming migrations is illustrated by the anecdote of a pre-war steamer in a Norwegian fjord, ploughing through a shoal of lemmings for a quarter of an hour.

The lessons of these rodent plagues for mankind are clear. First, there is a series of factors – climatic and social – that allow a population to grow to the sustainable limits of its environment. This expansion takes place despite any endemic disease processes. The population reaches its peak at the point at which the resources to sustain it run out. Social stress and hunger quickly set in, and disease soon takes its toll as the collective immune system declines, and so the destruction of the population unfolds at an alarming rate.

Historically, man has dealt with such environmental stresses by migration. Indeed, this has been the demographic driver for ascension to empire. This strategy has worked in the past because there have been new areas with lower populations to move into. However, with increasing demographics, this option will no longer be viable without

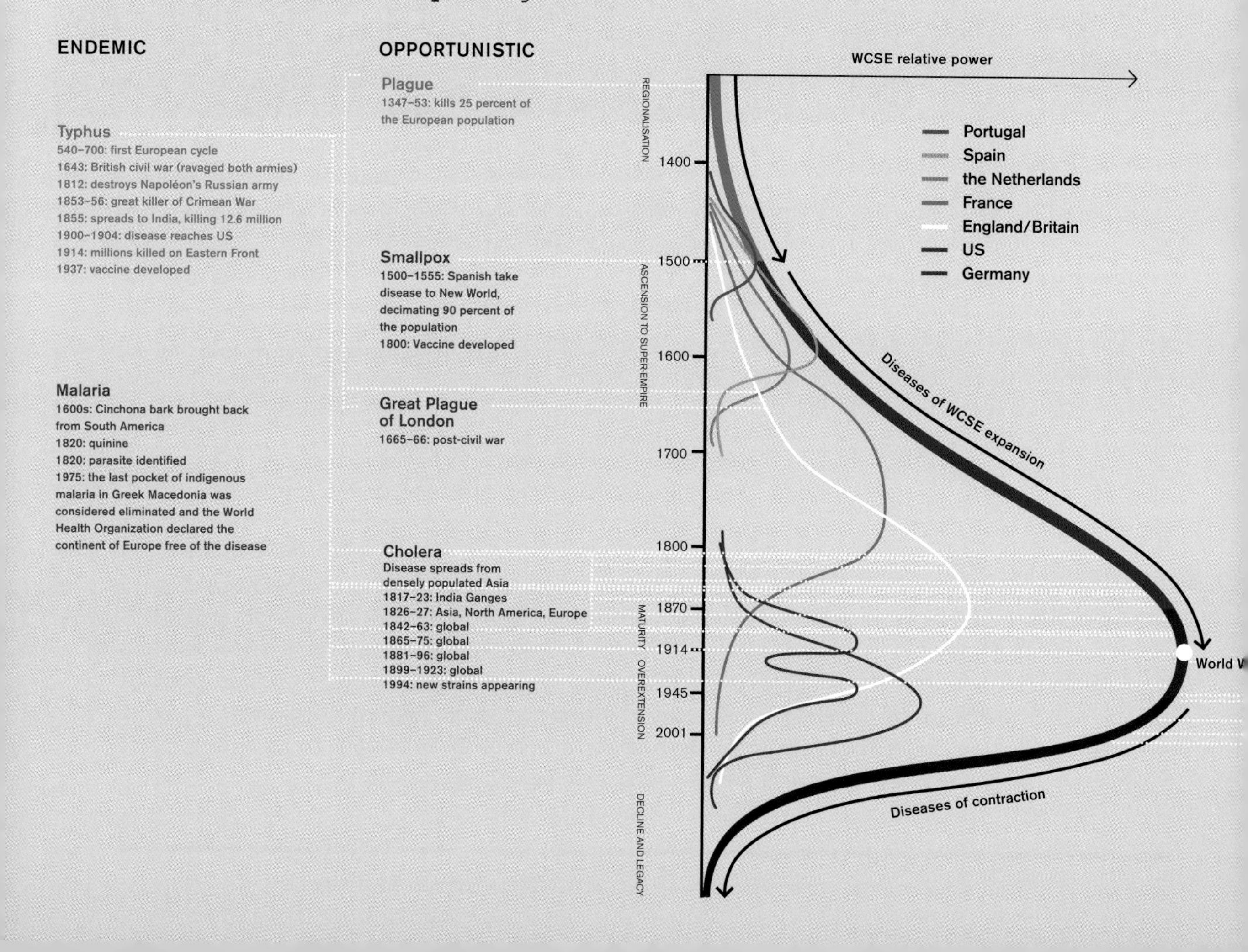

provoking conflict. Thus the closed-system fate of the mice rather than the expansive strategy of the lemmings is likely to be the destiny for huge swathes of the population of the Earth faced with depleting resources, climatic change and disease.

Disease Cycles of the WCSE

The WCSE provides an excellent example of the trajectory of population and disease through the cycle of empire. To understand the larger patterns at work within the cycle, this chapter is restricted to the major diseases that have had a significant effect on the population of the WCSE, broken down into diseases of the expansive stage of empire, diseases of its maturity and those of its contraction. These three stages further divide into endemic and opportunistic diseases.

THIS PAGE The diagram shows the empire cycles of the WCSE and the ASE. For the WCSE, it relates the endemic diseases and the milestones that pushed back the effects of both typhus and malaria. For the opportunistic diseases, it relates these to the phase of empire. The plague and smallpox were diseases of regionalisation and early ascensions to empire, taking place as populations grew and urban population densities increased, but commensurately without the later investment in public health infrastucture, which only increased the population's susceptibility to both. Of note is that within the WCSE, each country became vulnerable depending on its stage in the cycle. The Great Plague of London took place at the end of the regional phase of Britain's evolution.

Cholera was the WCSE's disease of maturity, because its imperial reach took it into the disease reservoirs of other empires, and its trading systems then imported it back into the empires' heartlands. Meanwhile, viral infection is the regional disease of the new ASE, caused by the high population densities of Asia in particular, where animals and humans live in close proximity so that viruses can cross the species boundary. Once created, these viruses can then travel round the world, where the most vulnerable populations will be those of the WCSE, due to the social pressure of economic and geopolitical decline. HIV, having been limited by technology in the developed world, is now more a disease of regionalisation, particularly for Africa. However, in combination with other infections, such as TB, it may spawn new drug-resistant infections that become globally prevalent.

HIV
1981: first discovered in Africa
2000+: AIDS, the regional disease of Africa

Influenza
1918–19: Spanish influenza kills 50 million globally
1957–58: Asian influenza (originating in China) kills 2 million
1968–69: viral influenza kills 1 million
1996: HSN1 bird flu detected
2003: SARS

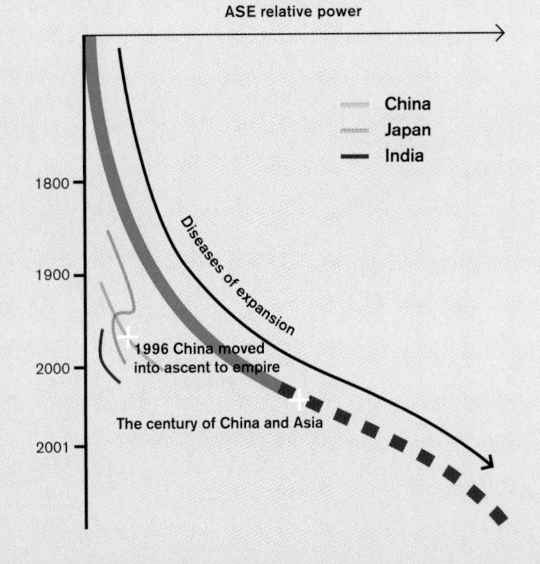

MALARIA

There are sixty species of mosquito that are capable of carrying the four species of protozoa parasite that infect humans. The most deadly is *Plasmodium falciparum*, which can cause 'cerebral malaria'. While the other three protozoan parasites can only breed inside young red blood cells (just 10 percent of the total in the human body) thereby restricting the impact on the host, this one can multiply in any of them. This leads to blood clots, which then kill the victim via a heart attack or stroke. Children are particularly vulnerable to the effects of malaria. In early twentieth-century England, a third of all infants in the marshy areas of Kent, Essex and the fenlands died of malaria before reaching their first birthday.

Malaria Cases, 2008 *(fig. 50)*

Data source: UNAIDS.

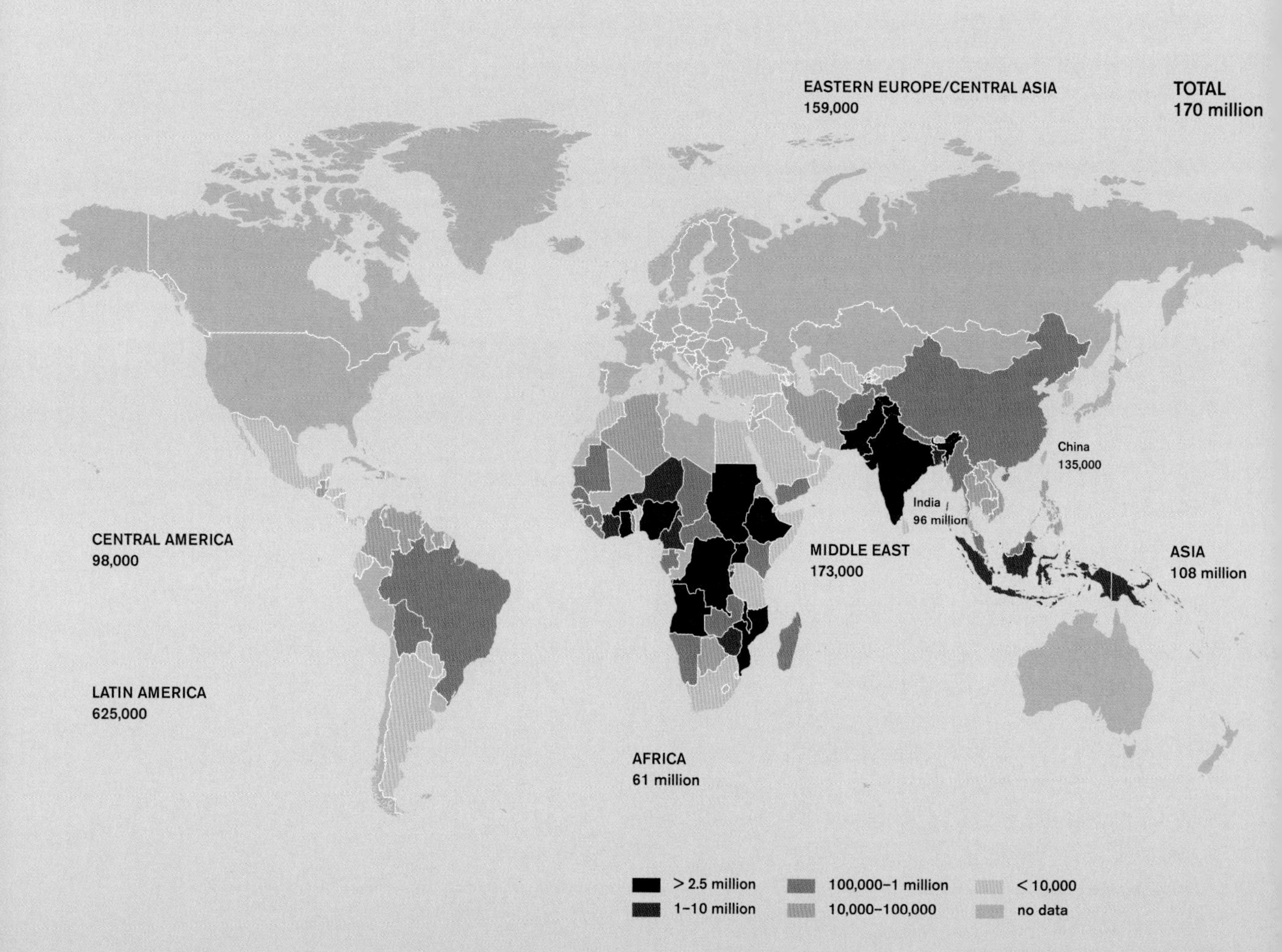

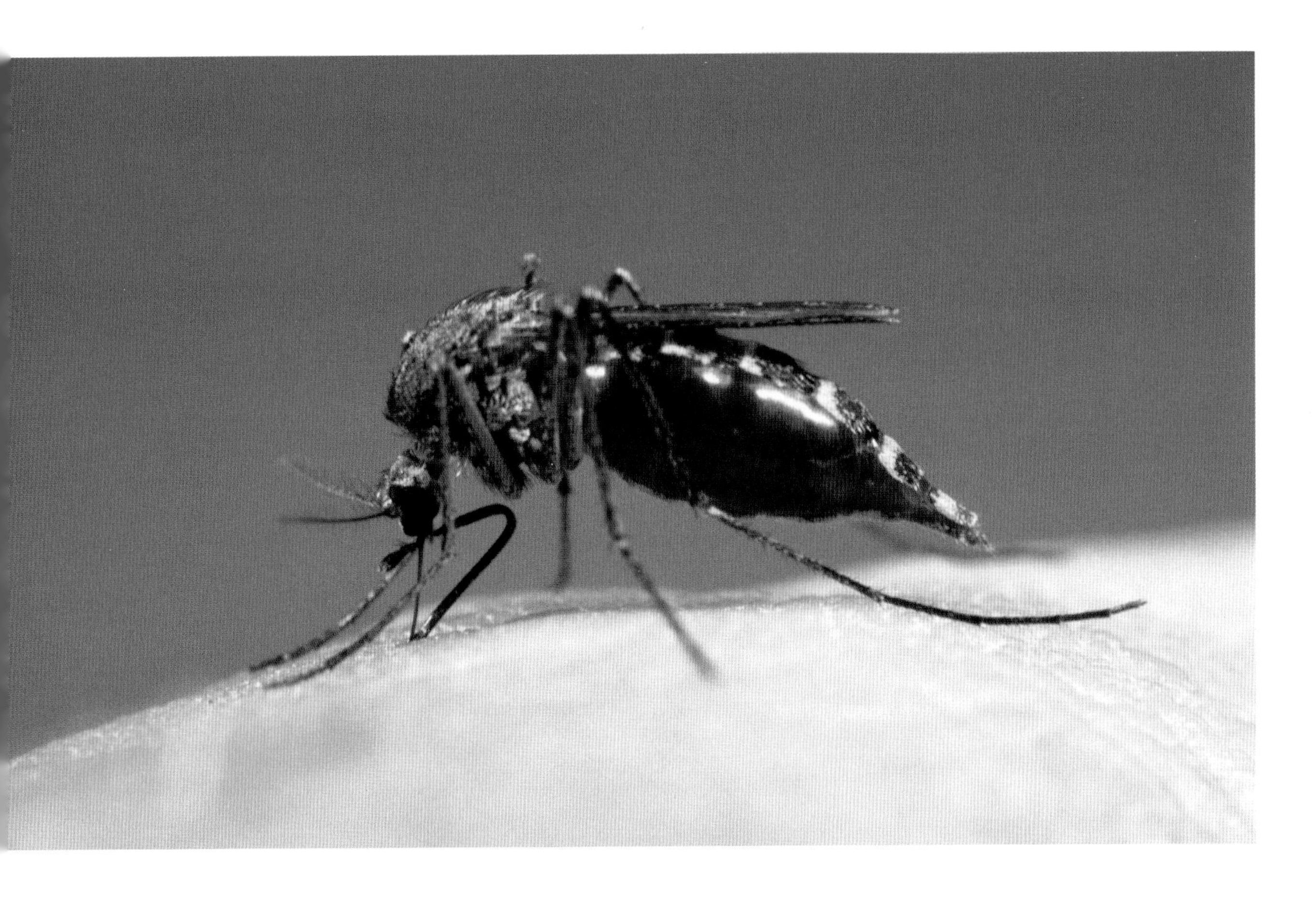

ABOVE **Mosquito (*Anopheles*) feeding on human blood. Their mouthparts are adapted for piercing the skin of plants and animals. While both males and females typically feed on nectar and plant juices, the female needs to obtain nutrients from a 'blood meal' before she can produce eggs.**

The malaria caused by the other three types of parasite – *P. vivax*, *P. ovale* and *P. malariae* – results in a repetitive debilitating disease process that blights whole communities in the tropics to this day.

The spread of malaria globally is inextricably linked to the empire cycles of the WCSE. It is believed that *P. vivax* was taken to the Americas from Europe in 1492, while *P. falciparum* spread with the slave trade to southern Europe and North America. Over the centuries, many curative responses to malaria were attempted, but it was not until the early 1600s that the bark of the cinchona tree was found to alleviate the effects of the disease. Quinine was isolated as the active ingredient of the bark in 1820, being embodied in the iconic gin and tonic of the British Empire. The protection afforded to British colonials from malaria cannot be underestimated.

Quinine facilitated the expansion of the fractal empires of the WCSE into hitherto inaccessible regions of the world. A prime example is the building of the Panama Canal, which was not just an extraordinary feat of engineering but also a milestone in the war against the mosquito – and the survival of the labour force that built it.

Urbanisation has proved to effectively counter the mosquito. The growth of a modern industrialised society, in combination with the spraying of mosquito-infested areas and the development of new anti-malarial drugs, has turned Europe into a malaria-free zone. However, the disease is still prevalent in the poorer, developing parts of the world (see Figure 50), where it currently affects some 300–500 million people and is by far the biggest cause of mortality (some 1–3 million deaths per year), mostly in infants. Indeed, the combination of malaria and HIV has proved to be a more deadly killer in Africa than the individual diseases. The development of drug-resistant strains of malaria poses a serious threat to our collective future health.

The mosquito, of which there are estimated to be a staggering 170,000 for every human, is also the vector for viral yellow fever and dengue fever. The banning of DDT, which was previously used to prevent the breeding cycle of the mosquito, has sadly proved to be an environmental own-goal.

TYPHUS

A relative of the much older human head louse, the typhus louse has lived symbiotically with man for something like 190,000 years. It lays its eggs in the warm clothes of its host. The typhus-infected faeces of the lice then enter the human body through abrasions in the skin, or other body secretions, causing illness after seven to fourteen days. The lice will only move from one victim to the next on the death of the host when they lose the warmth of the living. Symptoms of the disease are fever, headache, aches, bright red spots that look like bites on the skin and, most importantly, a stupefied and dullish demeanour caused by the build-up of the toxins that ultimately lead to death in 10–40 percent of infected cases.

Typhus was probably responsible for the plague in the Peloponnesian War of 430 BC that devastated Athens, but the first confirmed outbreak was among European soldiers who had passed through Cyprus before they laid siege to Moorish Granada in 1429. This epidemic killed 17,000 of their number.

Typhus thrives in an environment of social stress, with the accompanying poor hygiene, overcrowding, cold, hunger and, most importantly, unwashed bodies and clothes. Prior to the development of chemical controls for lice, the privations of long military campaigns provided the perfect conditions for the disease, and so typhus was responsible for decimating armies. Of the 600,000 soldiers who set off in Napoléon's 1812 advance into Russia, only 90,000 reached Moscow. Of those, only 30,000 returned to France and only 1,000 were able to fight again. It was not distance, or the vicious Winter, but typhus that destroyed Napoléon's dream of defeating the Russians.

The surviving soldiers infected the population of Europe on their return, and thousands died in the ensuing epidemic. Consistent with the empire cycle, this epidemic struck Napoleonic France just as its empire was slipping from maturity into precipitous decline (it had just lost the war against England).

Even as late as World War One, some 3 million Russian soldiers succumbed predominantly to typhus in the combat zone. This huge decimation of the army, largely made up of ordinary conscripts, assisted in catalysing the Communist Revolution.

Typhus was also rampant in prisons up until the twentieth century. There were periods when as many as four times as many inmates died of the disease as were subjected to hanging. It was not until 1909 that lice were understood to be the vector

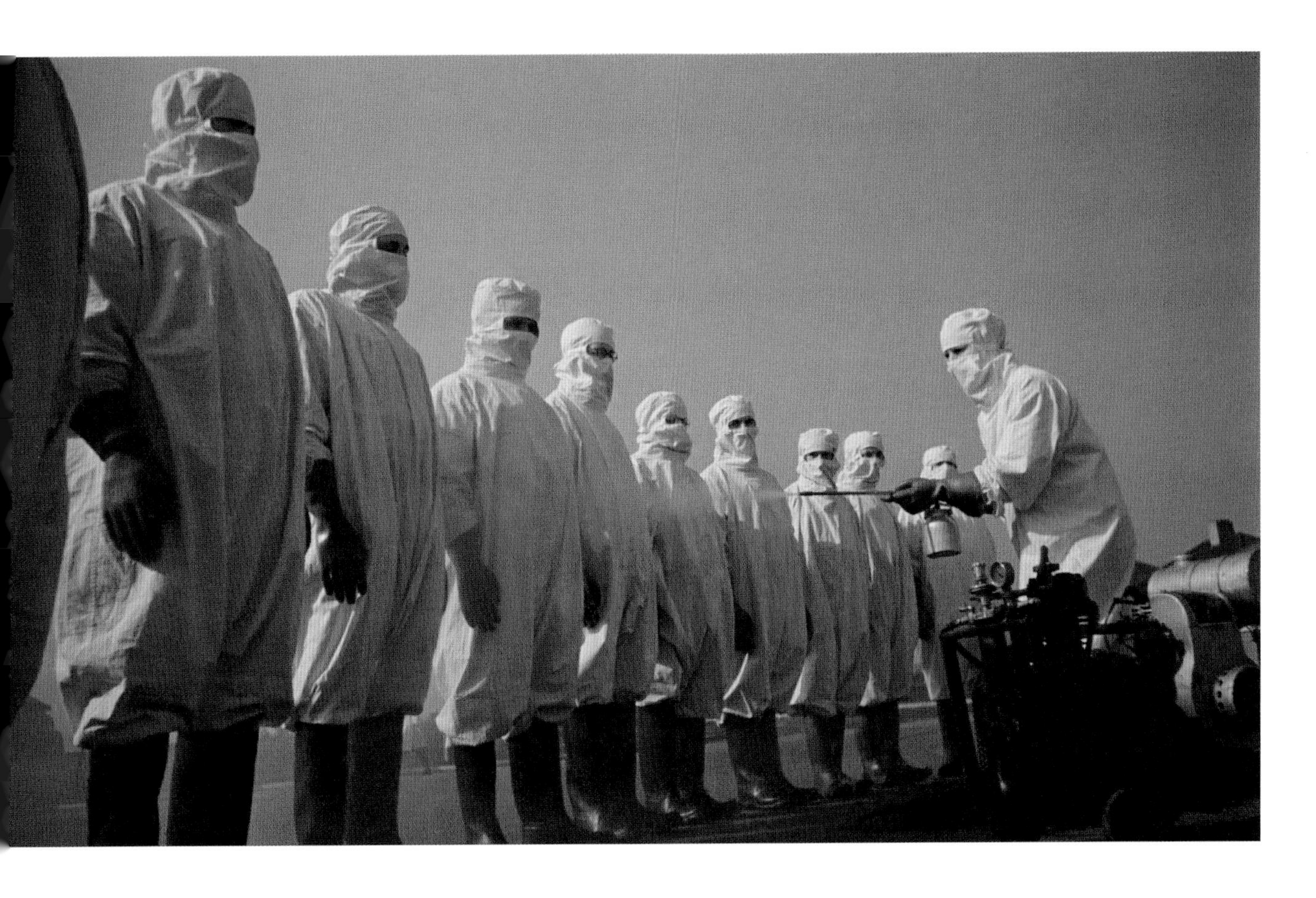

ABOVE A line of port medical officers in May 1942 being hosed down after possible contact with typhus during their work to monitor the disease at British ports. Epidemic typhus is considered a potential bioterrorism agent and was tested as such in the former USSR during the 1930s.

responsible for the disease's transmission, and only in 1937 was an effective vaccine produced. However, even during World War Two, typhus was rampant in Nazi concentration camps, as well as on the Eastern Front.

OPPORTUNISTIC DISEASES OF EXPANSION OF EMPIRE

In addition to the endemic diseases of expansion of empire, there are the opportunistic diseases and plagues that proliferate when key environmental conditions are optimal. As the WCSE grew in wealth and power, the effects of vectors, such as fleas carrying bubonic plague, typhus-infected lice and contaminated water, were reduced by higher standards of civil health, utilities and medical advances. With the exception of smallpox, for which various preventative vaccinations were used from the early 1700s, most diseases were a consequence of humans living at greater population densities than ever before, where infection could be transmitted freely from vector to human, and then from human to human.

During the growth of the WCSE, there were three main epidemic cycles, with each phase representing a stage in the super-empire's development. First, there were the diseases of regionalisation, in the form of fourteenth- to seventeenth-century plagues and outbreaks of smallpox. Second, as the WCSE expanded, these diseases were exported: smallpox to the Americas and Australia, where it devastated the indigenous populations who had no natural immunity, and plague to India in the mid-nineteenth century. The third and final phase unfolded as the WCSE expanded into all areas of the globe and world trade connected the different disease reservoirs. Thus, cholera, which is thought to have originated in the Ganges River, was brought back to the heart of the WCSE via trade routes and then spread around the world, creating the nineteenth-century cholera epidemics.

BELOW **The angel of death presides over London during the Great Plague of 1664–66, holding an hourglass in one hand and a spear in the other. This illustration was published in the 26 June 1665 edition of the *Intelligencer*. With no understanding of the disease, including its prevention or cure, Londoners were powerless to prevent its spread.**

BUBONIC PLAGUE

The Bubonic Plague, also known as the Black Death, was the predominant disease of the regional phase of empire of the WCSE.

Carried by the rat flea, its host thrived in the increasingly populous urban centres with their poor standards of hygiene. Like the Australian mouse, the rat had cycles of population growth and, if these coincided with phases of high plague infection, the rats would die very quickly. The rat fleas, seeking a new food source, would then jump to other hosts, including the human population. If the disease reached the lymphatic system, the victim had a 60 percent chance of death; if it reached the lungs, this increased to 90 percent, along with the risk of infecting others because this form was directly transmissible from human to human. Once the plague reached the bone marrow, in the septicaemia version, death was inevitable.

Mortality in London, 1665–66 *(fig. 51)*

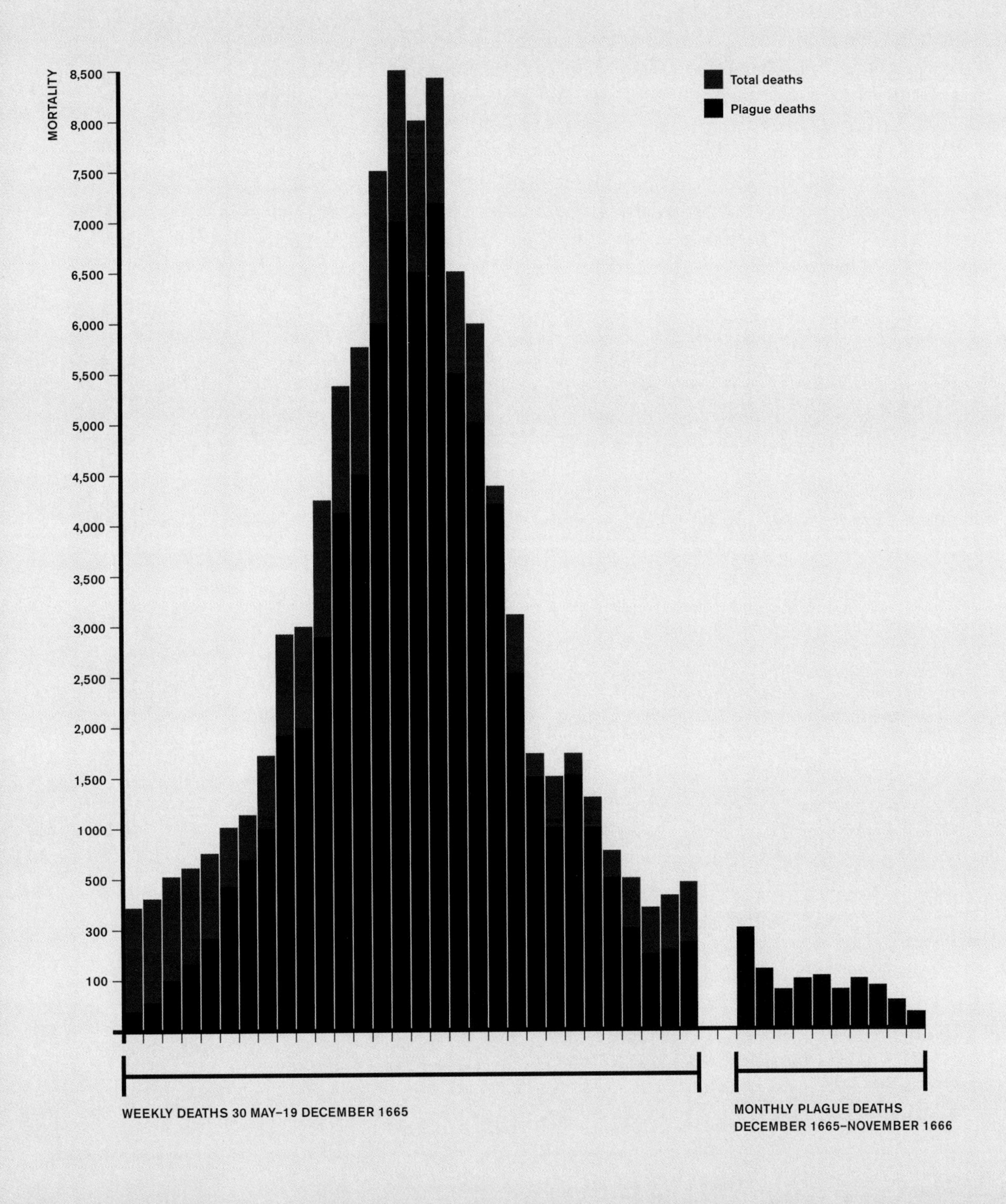

It is estimated that over the course of recorded history, some 200 million people have died of plague. After the Plague of Justinian, the disease appears to have lain dormant for 800 years, until it reappeared in the human population of the Gobi Desert in China. The vibrancy of the Chinese Empire facilitated its export along its trading routes to the west, and by 1347 it had once more reached Constantinople, where it ravaged the population. From there, it spread across Europe, killing thousands daily between 1347 and 1353, and becoming known as the Black Death. During this time it is believed to have resulted in the deaths of 30 percent of the total population of Europe. This prodigious death rate slowed the regional phase of the WCSE by at least a hundred years.

The plague continued across Europe until the 1700s, with more than a hundred individual outbreaks, including the Italian plague of 1629–31, the 1647–52 plague of Seville, the great 1679 plague of Vienna, the plague of Marseilles in 1720–22 and that of Moscow in 1771, but fortunately none of them turned into a second European pandemic. Most notable of these was the Great Plague of London in 1665–66, which struck England in its final stages of regionalisation, killing some 38,000 Londoners (see Figure 51).

The effects on the psychology of the population were sobering, as people attempted to explain the outbreaks. This resulted in the persecution of scapegoats, such as the Jewish communities, lepers and foreigners, who were blamed for the disease.

By the beginning of the nineteenth century, Europe had developed beyond its regional stage, and the WCSE's wealth had improved general living conditions to the point that the plague had all but vanished.

However, it would later reappear in a third major cycle in 1885 as the Asian Plague, killing 12.6 million in that region. From there, it spread to the US, where it appeared opportunistically in small clusters, such as after the San Francisco Earthquake in 1907–1909. At this point, the US had just passed into ascension to empire. These outbreaks occurred at much the same point in the empire cycle as the Great Plague of London had occurred in the cycle of the British Empire.

SMALLPOX

During the period when plague dominated Europe, smallpox was also prevalent, as it had been for millennia. Indeed, there are records of the Chinese inoculating people by taking the dried scabs from the disease and pounding them into dust, which was inserted into an incision in the skin. The disease is transmitted via airborne droplets from the lungs or through contact with bodily secretions, and it is extremely virulent, taking only twelve days to incubate and begin its invasion of the whole body. The external symptoms are the tell-tale skin macules, which cause terrible disfigurement.

Smallpox originated in Egypt some 3,000 years ago, but during the second millennium AD it became more virulent, and by the 1500s it accounted for 10–15 percent of all deaths in Europe. It was an epidemic of cities, in which 80 percent of children under ten were infected, with 25–40 percent mortality rates. The surviving population was largely immune to the disease. However, when the Spanish exported it to their growing empire in the Americas, it had a devastating effect on the local indigenous population (who had no immunity), killing as many as 90 percent of its victims.

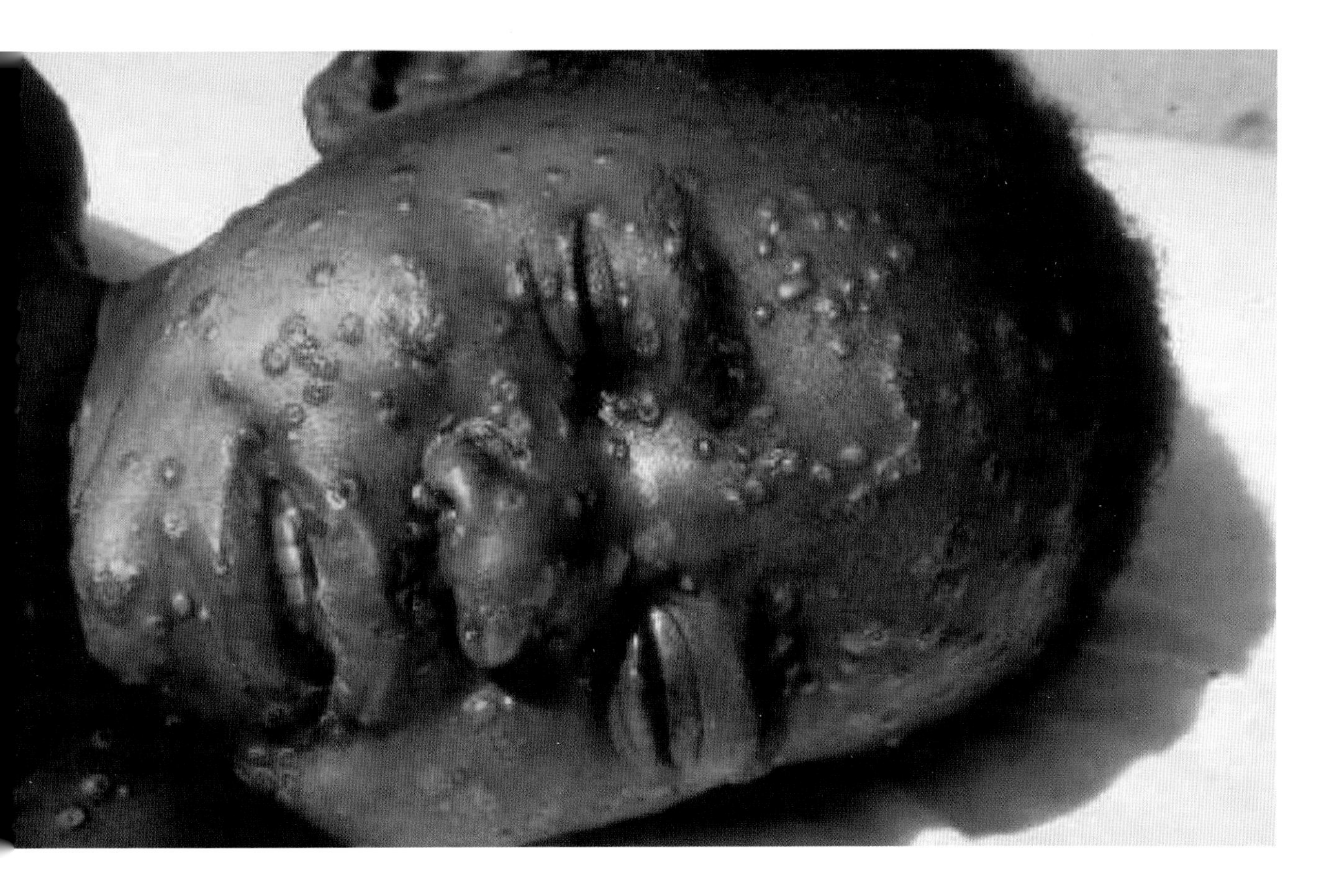

ABOVE A young boy with face lesions that are characteristic of smallpox. This is another potential bioterrorism agent. Routine vaccination against the disease came to an end in 1972.

However, in the context of empire cycles, the indigenous empires of the Americas were already in a state of maturity and decline, making them ripe for the incoming smallpox virus to ravage their populations. The Aztecs, for example, had founded their empire in the twelfth century, some 400 years before the arrival of Cortés in 1521. On the basis of the 500-year empire cycle that has been posited, they were in overextension/decline at the time of invasion. Meanwhile, the Incas on the west coast of South America, having founded their civilisation in 1200, expanded to empire in 1442, and then in 1533 suffered a massive peak civil war that greatly weakened their empire. This allowed the Spanish to launch their successful conquest, and the smallpox that they spread to the Inca people had devastating consequences.

Thus both the Aztecs and the Incas were in the latter stages of their respective empires when struck with smallpox. It is not surprising then that the disease hastened the end of these ancient civilisations.

It was not until 1796 that Edward Jenner developed a cowpox vaccine that proved effective against smallpox. The global inoculation process that followed was extremely effective. In 1801 in England, 100,000 people were inoculated; the French followed with 1.7 million inoculations in 1811, and the Russians with 2 million in 1814. However, despite widespread inoculation programmes, it is estimated that smallpox continued to kill 300–500 million people during the nineteenth century. Indeed, such was the virulence of this disease that it was not until December 1979 that the World Health Organization (WHO) could declare that it had finally been eliminated.

DISEASES OF MATURITY OF EMPIRE

CHOLERA

During its maturity phase of empire, the WCSE continued to expand into the few undiscovered regions of the globe. Ironically, having previously exported diseases from Europe, this expansion led to contact with new disease reservoirs that were carried via trade routes back to the core of the WCSE and then outwards to the rest of the world. This process first occurred with syphilis, which was imported by the Spanish from the Americas. Then, in 1817, the British encountered cholera in India. This proceeded to spread around the globe in a series of six epidemics during the next century.

Cholera is a waterborne disease, also transmitted via food, carried by the bacterium *Vibrio cholerae*. It kills by causing acute diarrhoea and dehydration, and it has the unpleasant reputation of being one of the most rapidly fatal diseases. At the extreme end of the spectrum, a victim can die in just three hours, but usually death occurs somewhere between eighteen hours and a few days. At its height, cholera killed one in four infants before their first birthday.

Like other diseases, cholera thrives in areas of social stress, poverty and poor water hygiene. Epidemics persisted until improved living standards, coupled with advances in medical science and the recognition of the need for clean water, finally eradicated the disease from Europe. However, it is endemic in the majority of developing countries where water hygiene is still lacking (see Figure 52).

The most recent cholera epidemic took place in Zimbabwe in 2008–2009 after the country had suffered severe economic deprivations and social stress under the tyrannical rule of Robert Mugabe. The public water supply became contaminated and some 80,000 people were infected, with an estimated 3,000 deaths.

TUBERCULOSIS

Tuberculosis (TB, caused by *Mycobacterium tuberculosis*) attacks the lungs in its pulmonary form. It has long infected man, but in the nineteenth and twentieth centuries it took on a new virulence in areas of urban poverty, most probably exacerbated by the pollution of industrialisation and high population density. It was also called the White Death, as its victims suffered pallor, a bloody cough, breathlessness, pain, sweats and wasting. In the nineteenth century it was a major cause of death, but by the early twentieth century it was the greatest killer in the newly globalised world. A very high percentage of urbanised populations were exposed, with 10 percent developing the disease and 80 percent of those dying, suggesting that a staggering 8 percent of the urbanised population, over a number of decades, died from TB.

From the 1920s, vaccination, combined with a new drug-treatment regime in the 1950s and vigilance in keeping cattle herds free of bovine TB, made significant progress in containing the disease. However, despite the widespread use of the vaccine (known as the BCG after its inventors Calmette and Guérin), the disease is still with us. Indeed, it has become one of the biggest single killers in the modern world. An estimated 2 billion people or almost one-third of the world's population are infected with TB, 8–10 million are manifesting the illness and 1.5–2 million die each year (see Figure 53). The majority of these deaths are in Africa, Bangladesh, China, India, Pakistan and the Philippines. Problems are now arising with drug-resistant strains of TB, and there is a risk that the unresolved disease pool will lead to mutations that will not be protected against by the current vaccination programme.

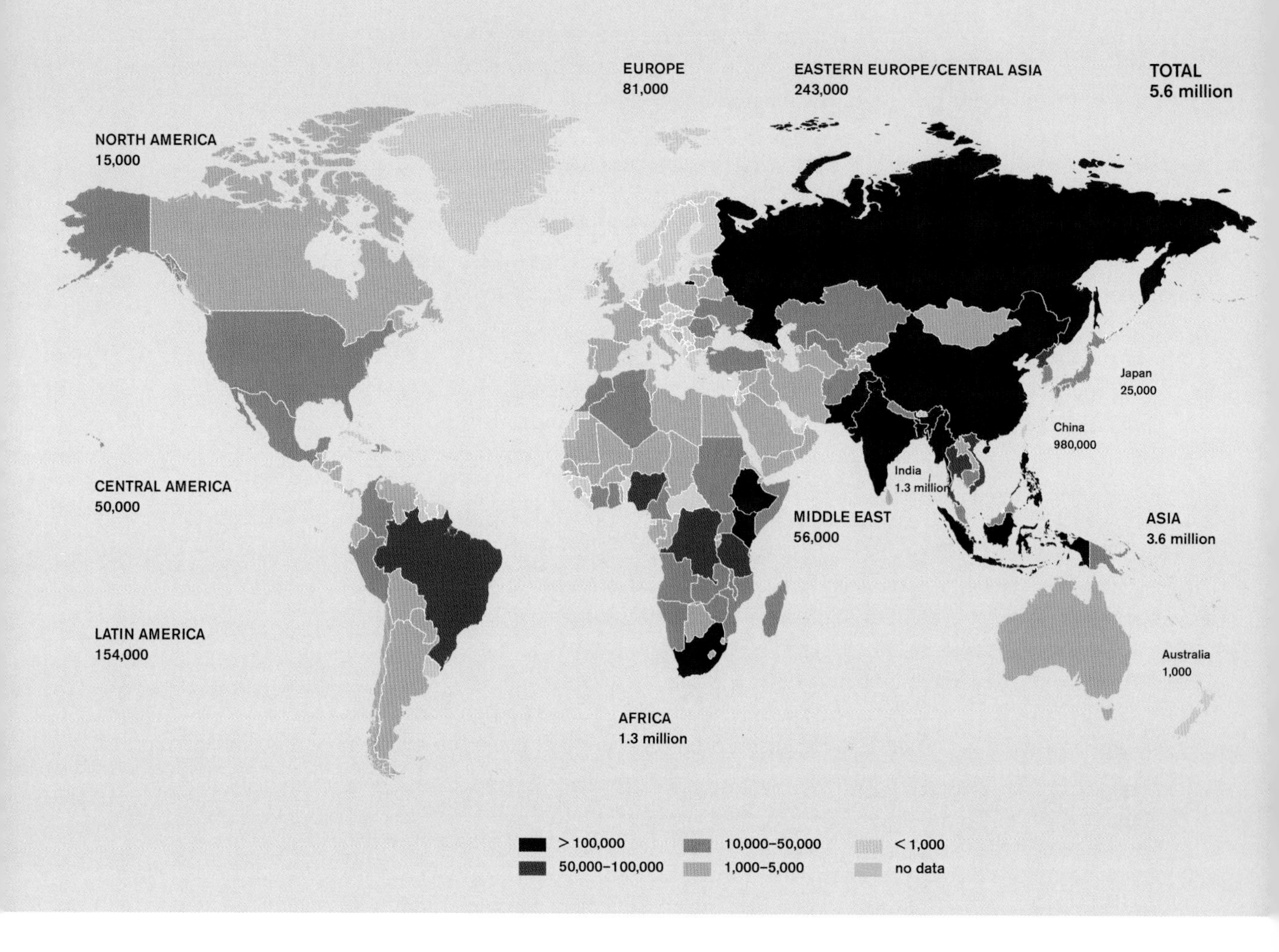

DISEASES OF CONTRACTION OF EMPIRE

OPPORTUNISTIC DISEASES OF CONTRACTION OF EMPIRE

Following the successes against bacterial and parasitic diseases, it is now viral epidemics that present the biggest threat to the WCSE in overextension and decline. The main disease reservoirs are in Asia, which is just moving out of its regional phase into ascension to empire, and so has the capacity to export its diseases through extensive trade links to more vulnerable areas of the world, such as the declining WCSE. Africa, in early regionalisation, is also a threat, having been the crucible for both HIV and ebola.

VIRAL INFLUENZAS

After the trauma of the peak civil war of 1914–18, the WCSE was ravaged by influenza. The epidemic is believed to have started in Flanders in May 1918 and it went on to kill 50 percent of its combat-exhausted victims. The news of the outbreak was suppressed by wartime censors for obvious military reasons, but it was reported in the neutral Spanish press, gaining the name Spanish flu.

From its epicentre and following the lines of Allied troop movements, a second wave of the epidemic unfolded that reached Boston in the US, Brittany in France, and Sierra Leone by August 1918. By November that year it had become a global pandemic. The disease killed up to 20 percent of its victims within forty-eight hours; their lungs filled with fluid and blood oozed from their orifices. Victims even dropped dead while apparently active. By Spring 1920, the disease disappeared as mysteriously as it had arrived, by which time extensive quarantine measures had been introduced, including the banning of public gatherings, physical contact and spitting in public. During its eighteen month rampage, Spanish flu killed at least 50 million and possibly as many as 100 million people worldwide.

Modern-day studies of exhumed victims of this epidemic indicate that the strain was H1N1, an avian flu that had adapted to humans, and the forerunner of the strain that has recently reappeared.

Consistent with Asia being the most likely new disease reservoir, the Asian flu H2N2 killed 2 million people in China in 1957. An outbreak of H3N2 flu followed in Hong Kong in 1968–69, killing 1 million people. However, both outbreaks were contained by Asia's then relative isolation, being in regionalisation, and effective control mechanisms.

BELOW A man sprays the top of a bus with an anti-flu virus in London, in March 1920, during the epidemic that followed World War One. Public health measures such as this were powerless in preventing its widespread transmission.

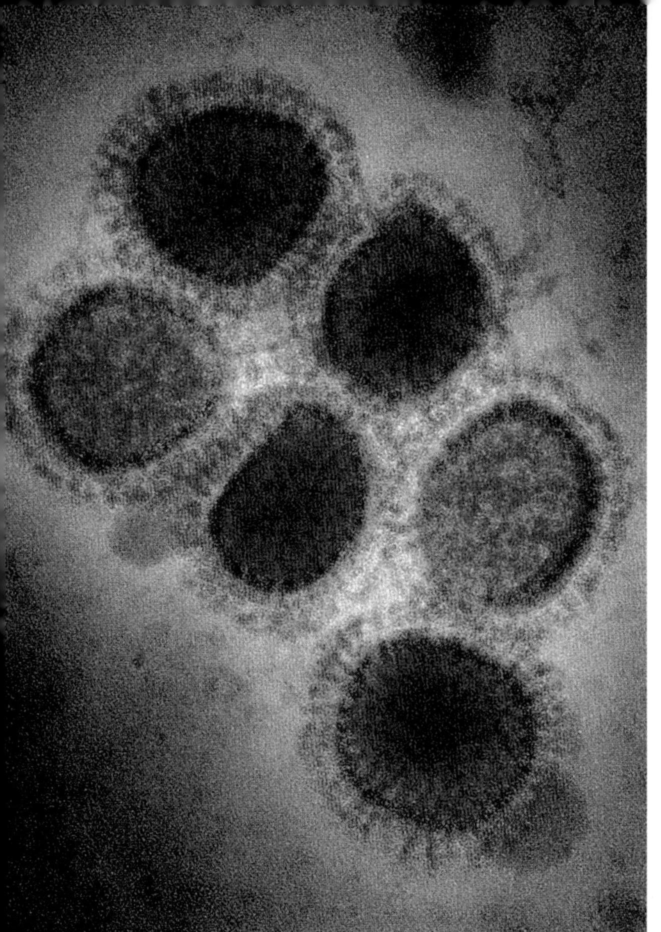

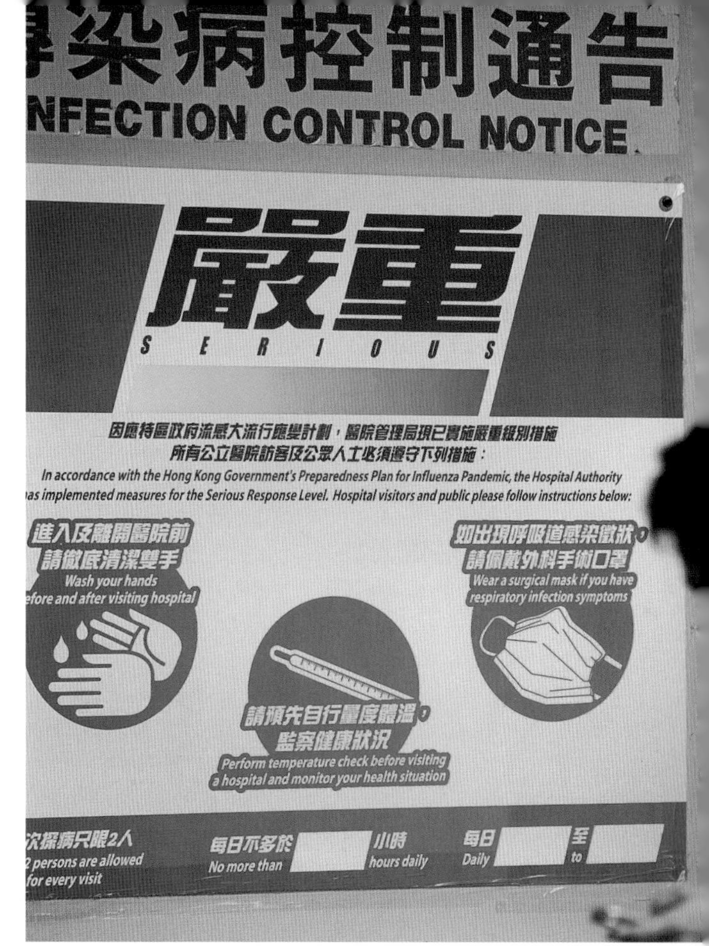

However, in 1996, H5N1 (avian flu) was identified in Guangdong, China, and by 1997 it had reached as far as Hong Kong, where it killed six people. Rapid and exemplary action by the authorities contained what could well have become a global epidemic.

Avian flu lives in the intestines of birds and is initially transmitted by contact with their droppings (following which human-to-human transmission can occur). Taking this infection route, avian flu jumped to the local wild-bird population, from which, carried along migratory paths, it passed to the European bird population. To date, 315 H5N1 cases have been recorded in the human population, of which a sobering 191 victims died.

More recently, in mid-February 2003 and again in China, a mysterious respiratory infection appeared, killing five people. The Chinese authorities covered up the crisis, so it was not until 15 March 2003 that the WHO declared a travel emergency to limit the spread of the disease called Severe Acute Respiratory Syndrome (SARS). In reality, the disease was probably five months old by the time it was recognised in China, as its first victims had gone undetected by government agencies. The pathology of the transmission of SARS is a salutary reminder of how quickly a new disease could spread across the world. By July 2003, SARS had vanished as quickly as it had appeared,

ABOVE LEFT Transmission electron microscope images of the newly identified H1N1 influenza virus captured in the CDC Influenza Laboratory. The swine flu virus was first isolated from a pig in 1930.

ABOVE RIGHT Precautionary signs posted permanently at a Hong Kong hospital in April 2009. The city was at the forefront of the SARS epidemic in 2003 and was on alert for signs of bird flu.

but in its short lifespan it had spread across the globe via airborne droplets exhaled by its carriers, infecting 8,000 people and killing 700 of its victims.

Most recently, we have had the threat of worldwide epidemics from avian flu (2008) and swine flu (2009), both originating from emerging nations. Although causing public alarm – swine flu in particular was very effective in its ability to transmit itself – neither resulted in widespread epidemics, although the latter strain prompted the mass-production and stockpiling of Tamiflu, the antiviral drug.

This record makes it likely that Asia will act as a disease reservoir for future serious epidemics in the waning WCSE. Despite the anticipated economic hardship of the next decade, it is imperative that national health organisations and the WHO remain vigilant and are provided with sufficient resources to combat a potential future outbreak. Combating flu is difficult for modern medicine because a vaccine has to be created for each and every strain, although generic anti-viral treatment is able to slow the spread of a virus to allow time for a suitable vaccine to be produced. Whatever the viral source, the next pandemic will be a great challenge for the Western nations with their declining power, resource bases and corresponding social stress.

HIV AND AIDS

HIV depresses the immune system to a point where an opportunistic disease kills the victim. Medical advances have reduced its impact in developed nations, but it is still ravaging the poorer regions of the world.

This virus was first discovered in 1980, although it is thought to have originated in western equatorial Africa some time between 1930 and 1950. The HIV-1 virus is closely related to the harmless Simian Immune Deficiency virus of chimpanzees, which may have crossed the species barrier to hunters via cuts or consumption. The virus has a very long incubation period and a nebulous symptomatology, which helped to delay its recognition for decades. The disease is a retrovirus that is transmitted sexually from human to human, or via contaminated needles or blood products. HIV breaks down the immune system, which then opens the door to Acquired Immune Deficiency Syndrome (AIDS). As such, it can be viewed as a disease of the decline of the WCSE, associated with a reduced collective immune system.

Like syphilis before it, HIV initially suffered from social stigma due to its modes of transmission (e.g. homosexual sex). The virus had incubated slowly and finally unleashed its devastating effects in an HIV pandemic that has since killed millions. Ironically, this coincided with the eradication of smallpox and the belief that science would in time win the war against disease. Because AIDS at first carried the stigma of affecting only high-risk groups, denial was a common response and politicians were slow to mobilise funding to counter the threat. The trigger for change was its leap into the heterosexual community and the risk of the infection of blood banks, which could affect the wider population. Isolating the victim was not an option due to its long incubation period. Indeed, with the exception of the first wave of syphilis in the fifteenth and sixteenth centuries, doctors had no prior disease that could be used to model and combat its spread.

In Africa, the disease spread from its source in the west along the trucking trade routes, soon to be known as the AIDS highways, and from there to all corners of Africa and to all sections of its society. In Uganda it was given the named 'slim disease' because its victims wasted away and died. The pattern was matched in the US, as cases jumped from zero in 1980 to 7,699 in 1984, of which 3,665 died, with a similar pattern of mortality rates emerging in Europe. Its expansion continued, so that by the beginning of the twenty-first century the pandemic was causing 3 million deaths a year.

The first scientific breakthrough came in 1987 with AZT (azidothymidine), the first anti-viral drug. This was extremely expensive and only slowed the effects of the virus rather than curing the disease. A decade later, another treatment regime, Highly Active Anti Retroviral therapy (HART), was developed – an anti-viral cocktail that allowed HIV suffers to return to normal life, even though they still carried the virus. Consequently, death rates dropped significantly across the Western world. However,

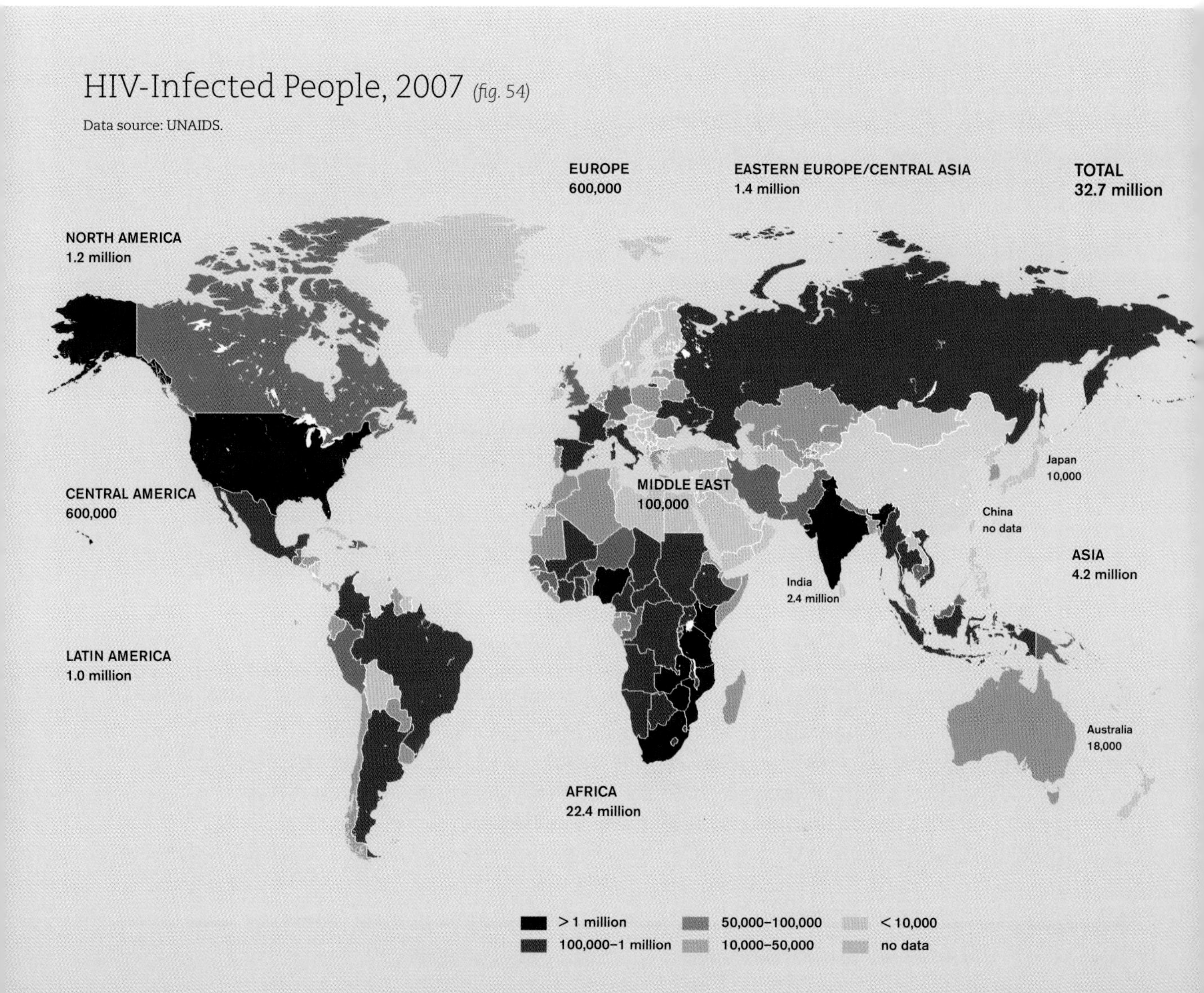

HIV-Infected People, 2007 *(fig. 54)*

Data source: UNAIDS.

ABOVE Indian HIV-AIDS activists dressed in skeleton costumes participate in an awareness rally on 1 December 2008 to mark World AIDS Day in Agartala, the capital of India's northeastern state of Tripura. The National AIDS Control Organisation of India estimates that around 2.3 million people were living with HIV in India in 2007.

the expense and complications of the treatment regime have meant that the HIV virus has continued unabated in poorer areas of the world, such as Africa.

The statistics are shocking: some 65 million infected with HIV; 25 million already dead; 3 million dying each year, half of whom are children; and 6,000 people infected every day (see Figure 54). The lowered immune systems of the victims have offered TB a new lease on life, compounding the death rates of HIV sufferers.

ENDEMIC DISEASES OF CONTRACTION OF EMPIRE

The endemic diseases of the contraction phase of the WCSE are so-called lifestyle diseases, such as cancer and diabetes, rather than the endemic diseases of the expansionary phase, such as malaria and typhus. However, as we have seen, there are signs that bacterial diseases, like TB, are resurging in new, more virulent strains. This demonstrates again that the declining phase of a system provides fertile ground for new and old diseases. In addition, there is a growing body of evidence that some diseases, like peptic ulcers and some cancers, may be associated with precursor viruses and bacteria, adding a new layer of complexity.

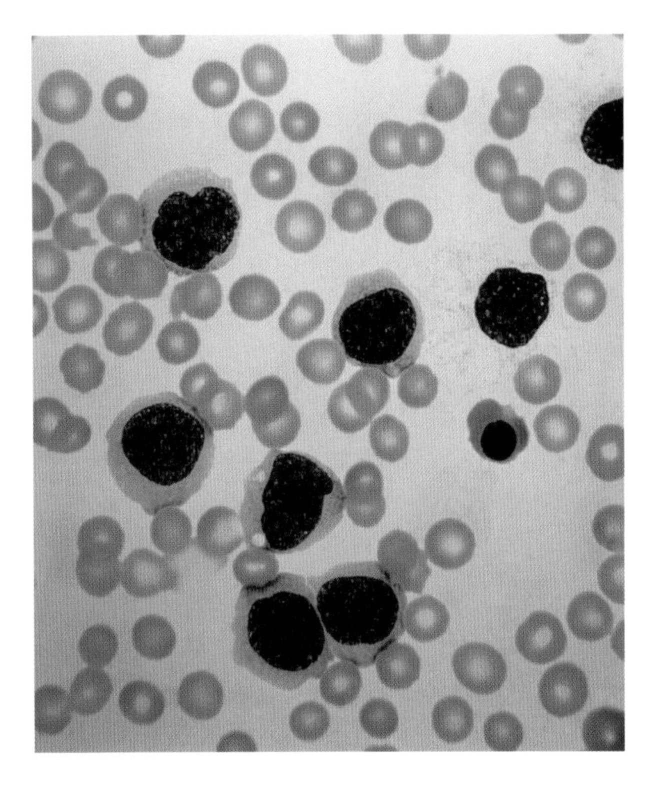

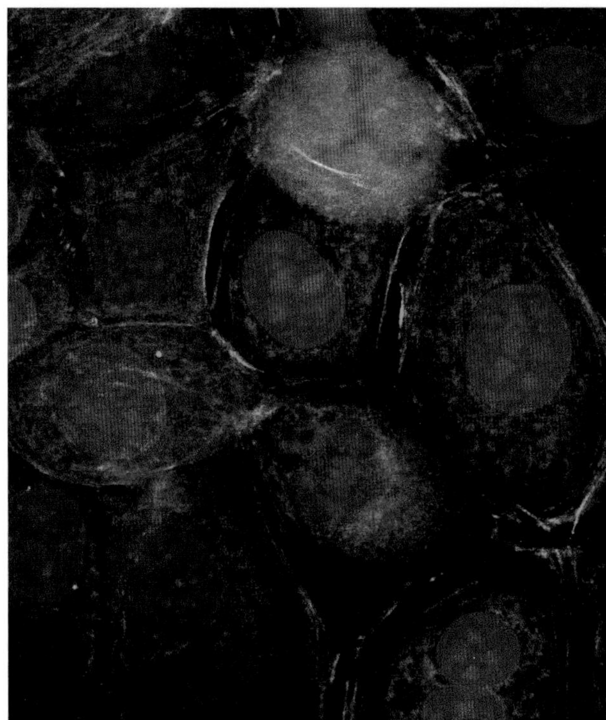

ABOVE LEFT **Immature white blood cells resulting from congenital leukaemia. This is a cancer-like disease of these blood cells.**

ABOVE RIGHT **Cancer cells from human breast. This a common site for benign and malignant tumours.**

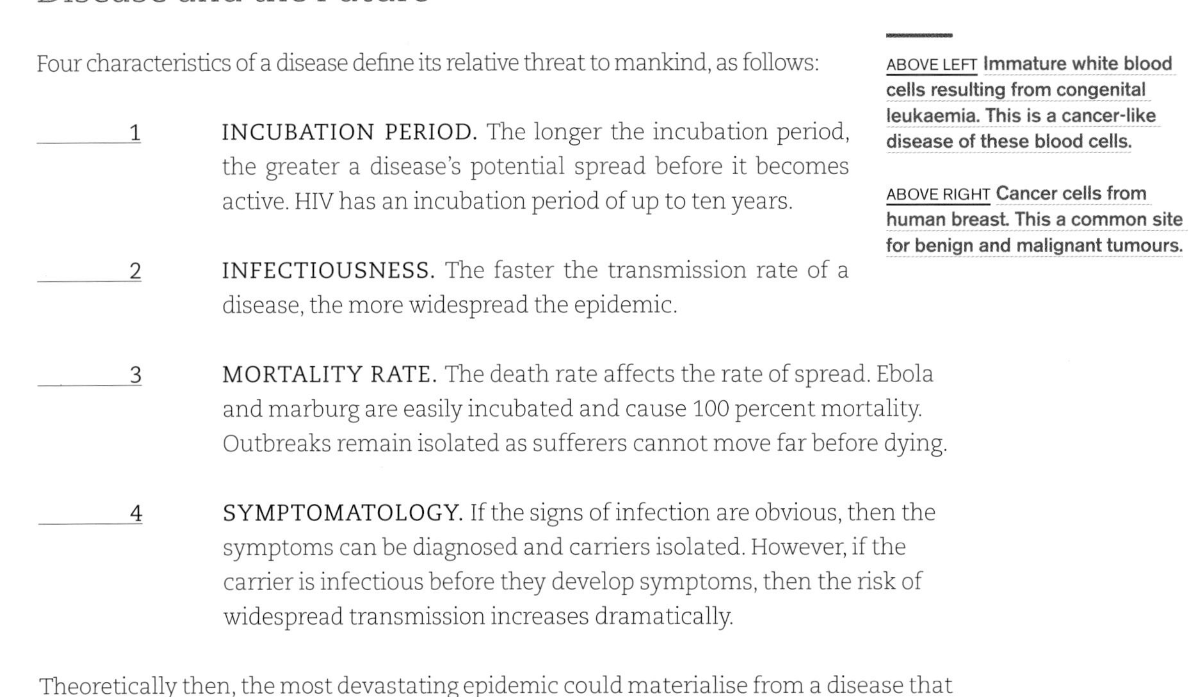

Disease and the Future

Four characteristics of a disease define its relative threat to mankind, as follows:

1. INCUBATION PERIOD. The longer the incubation period, the greater a disease's potential spread before it becomes active. HIV has an incubation period of up to ten years.

2. INFECTIOUSNESS. The faster the transmission rate of a disease, the more widespread the epidemic.

3. MORTALITY RATE. The death rate affects the rate of spread. Ebola and marburg are easily incubated and cause 100 percent mortality. Outbreaks remain isolated as sufferers cannot move far before dying.

4. SYMPTOMATOLOGY. If the signs of infection are obvious, then the symptoms can be diagnosed and carriers isolated. However, if the carrier is infectious before they develop symptoms, then the risk of widespread transmission increases dramatically.

Theoretically then, the most devastating epidemic could materialise from a disease that incubates slowly, shows no symptoms while the host is infectious, and results in a high mortality rate. This would be able to spread across the globe before the authorities could recognise the threat and then quarantine the infected areas.

As we have seen, the emerging nations with their high population densities and relatively low public health standards are potential disease reservoirs. China and India are the foremost candidates, and they have historical precedent. However, Latin America, particularly Brazil and Mexico, have massive cities that are potential incubators for disease and could well provide the source of the next global pandemic. Indeed, swine flu, which threatened a world epidemic in 2009, emanated from Mexico. A disease reservoir on the doorstep of the US, currently facing the aftermath of the credit crisis with its associated social stress, is a sobering reality, yet consistent with epidemics gaining traction in the declining phase of the life cycle of an empire.

At present, diseases such as HIV/AIDS appear to be contained – in the WCSE at least. However, we will need to be vigilant about the development of more virulent versions that, in tandem with TB and malaria, which prey on the reduced immune systems of victims, may result in a more deadly epidemic than any we have yet seen.

If our thesis of the cycle of empires and disease is correct, it is in the interests of the developed world to invest in medical research to fight against diseases prevalent in the emerging nations, because with globalisation any new epidemic will inevitably and

BELOW **A farmer in Mexicaltzingo, Mexico, seeks to protect himself from swine flu after the outbreak in 2009. The feared worldwide pandemic did not materialise, although the World Health Organisation raised its alert to an unprecedented level four.**

quickly reach their own shores. Indeed, the WCSE may well suffer more than the expanding empires of Asia due to the social stresses of the west to east power shift.

More broadly, as the expanding population of the world competes for limited resources during the coming decades, causing ever higher levels of social stress, the ground may well be set for a global epidemic of proportions akin to the Black Death. This would rebalance the human population within the Earth's ecosystem. Indeed, at times, man's war against disease seems to be an ever-accelerating race against the Earth's desire to rein in its own disease – humankind – in the quest to maintain its equilibrium.

RIGHT **Members of a 'Haz-Mat' response team at the US post office in West Trenton, New Jersey, on 25 October 2001. The US post office was closed after two letters containing anthrax were traced back to the facility. Following the 9/11 attacks the previous month, fear of bioterrorism was strong.**

A discussion about disease would not be complete without a brief word about biological weapons. As if our struggle against naturally occurring disease were not challenging enough, our downfall could be at our own hands should any of the biogenetically modified biological weapons that have been developed ever escape the laboratory through human negligence or terrorism. It is uncertain whether the containment of such a catastrophe would even be possible.

However, even with modern medicine and DNA analysis, constant vigilance is required to contain the power of the naturally occurring bacterium and virus. Perhaps an understanding of the nature of disease within the cycle of empire may provide the context for understanding the nature of disease itself. Indeed, we have only viewed individual diseases within the context of the empire cycle, but if we look at the sum effect of diseases against the cycle, we have to conclude that the coherence of thought and energy that are required to build an empire also push back the boundaries of disease and its effects on the population; conversely, as coherence declines, the disease process regains the ground it had lost. Despite our best attempts at scientific advances, disease will constantly find ways to mutate to counter new cures, ensuring that man's war against disease will be ongoing and protracted.

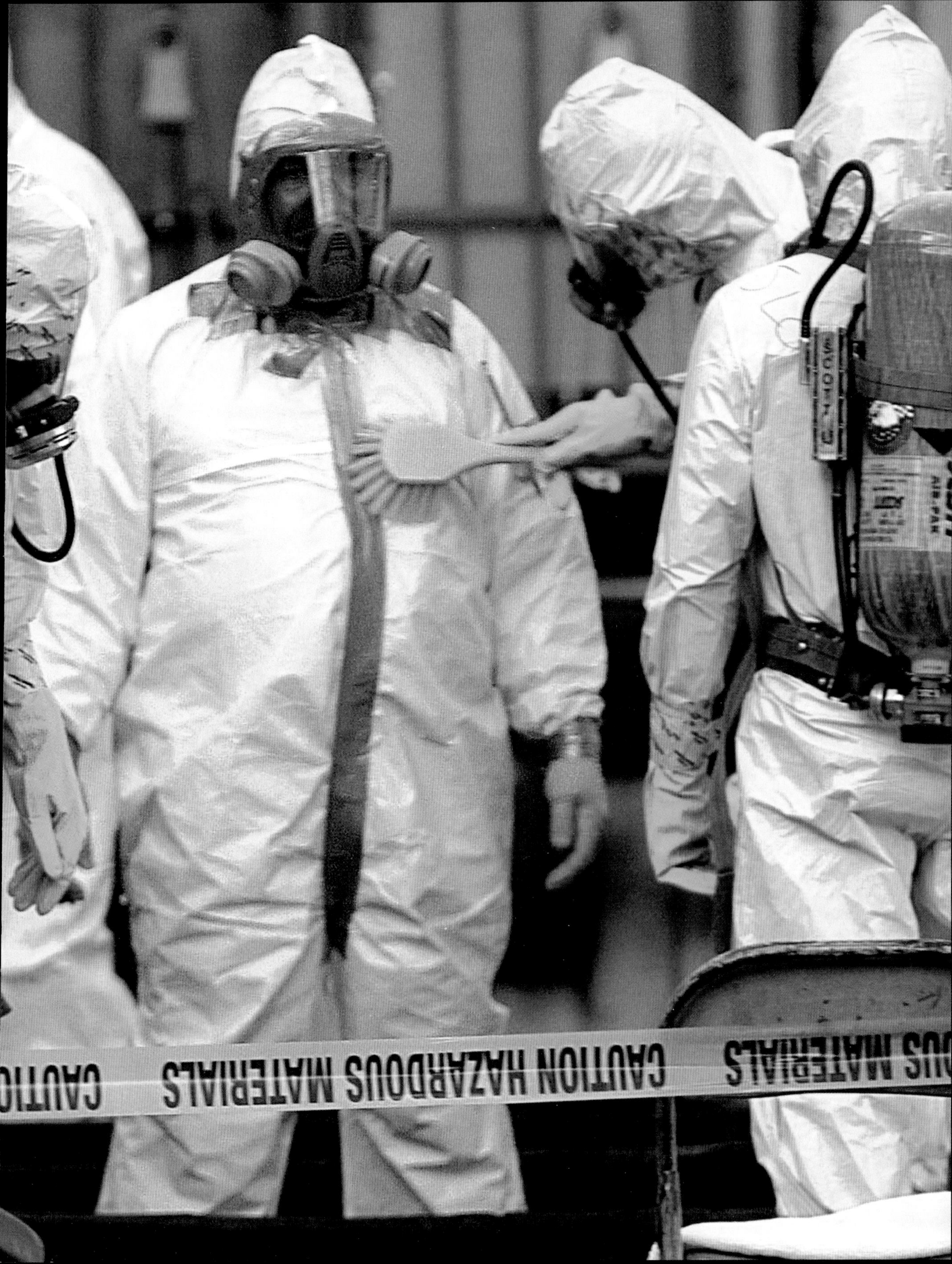
CAUTION HAZARDOUS MATERIALS

Chapter Ten

Climate Change

Climate Change in Context

'Climate change' is now a common phrase, yet few people understand the changes to the Earth's biosphere or their historical context. Indeed, there are still those who refute its very occurrence despite significant and incontrovertible evidence to the contrary.

It has become possible to make longitudinal assessments of the Earth's past temperatures from the study of ice cores up to 400,000 years old that contain trapped air bubbles dating from when the ice was formed. The study of these air bubbles, particularly of the heavier isotopes of hydrogen and oxygen, can reveal average temperatures and levels of environmental carbon dioxide (CO_2) and methane. This information allows us to place today's climate in a historical context.

The Earth's temperature has varied within a 13°C range during a time period of 90,000–110, 000 years, the low points marking the ice ages and the high points the warm phases. It is noticeable that the cooling phases were gradual, with average declines of 1°C over 10,000–12,000 years, while the phases of warming were much sharper, with each jump of 10–13°C taking roughly 10,000 years.

The last ice age ended some 18,000 years ago, and over the next 6,000 years the temperature rose by 8°C at a rate of 1°C every 750 years, to reach current levels. Temperatures were relatively stable for the next 10,000 years, varying only in a range of plus or minus 1°C. It was this warm and stable environment that provided the optimum conditions for the growth of civilisation.

Since 1900, when reliable temperature records were begun, land temperatures have risen by 0.9°C and those of the oceans by 0.5°C. The seas are in effect the heat reservoir of the planet. Slower to respond to atmospheric temperature changes than the land, they act as a stabilising element for the temperature of the biosphere. However, once they begin to warm, momentum will hold those temperature changes for some time, delaying their overall cooling.

To put the current average 0.9°C increase into perspective, this represents 1°C every 140 years – which is five times as fast as the rate of 1°C per 750 years that followed the last ice age. Current climatic modelling suggests that a temperature rise of more than 2°C would create the potential for a climatic shift of major proportions. Taking a linear view, if we extrapolate the current rate of change into the future, we should expect to reach the temperature risk point by 2180. However, the Earth's atmosphere is not a linear system, and all of the evidence suggests that it is now undergoing an accelerating rate of change. Indeed, eleven of the last twelve years rank as the warmest since 1850. It is therefore highly likely that the catastrophic effects of climate change will become increasingly evident in the much shorter term.

Climate Change and the Earth's Past

The concept of climate change has become one of the hottest political topics of the first decade of the twenty-first century. However, before we examine the evidence and potential consequences, it is worth stepping back to gain a bigger perspective.

The universe is around 6 billion years old and the Earth approximately 4 billion years old. During its long life, the Earth has cooled from a hot, molten mass to a lush planet of which 75 percent of the surface is water.

Some 550 million years ago, after 200 million years of glaciation, the Earth was icebound. Then a giant volcanic eruption suddenly raised CO_2 levels to 12 percent of atmospheric gases – 350 times current levels. This brought an end to the Precambrian period and led to the Phanerozoic period of warmth, which gave rise to simple multicellular life, followed by more complex plants and animals. Atmospheric CO_2 levels were ten times today's concentrations, dropping to four to six times in the Mesozoic era and rising to ten to fifteen times at the start of the Devonian period. This latter era, some 400 million years ago, witnessed the extensive spread of plants, which reduced CO_2 concentrations significantly and stabilised the atmosphere.

For millions of years, the Earth has evolved and hosted numerous ecosystems, each with its own adaptations to the prevailing climatic conditions. So when we talk emotively about 'saving the planet', we reveal the egocentricity of our species. Earth will survive, even after we have finished destroying the ecosystem that gave us life. Indeed, we may simply be fulfilling a role similar to that of the asteroids that acted as an instrument of change in prompting the next phase of the Earth's history. Viewed from this perspective, talk of 'saving the planet' really translates into 'saving ourselves'.

During the 1990s, the initial observation of scientists – that the Earth was undergoing a climatic shift – were met with general scepticism. This was founded upon the premise that there have always been fluctuations in global weather patterns, and we simply did not have sufficient historical context to understand the current temperature variations. For reasons we will consider later, opposition to the concept of climate change has been consistent and relentless. Thankfully, there has been a mass of scientific work, foremost of which was the analysis of the 400,000-year-old ice cores mentioned earlier, which have given scientists a clear understanding of historical levels of atmospheric greenhouse gases. Furthermore, the relationship between the levels of particular atmospheric gases (including CO_2) and global temperatures has now been clearly established (see Figures 55 and 58). Extrapolating from this relationship, it is clear that the increased human population and associated industrialisation are causing an increase in atmospheric levels of greenhouse gases and, hence, global warming.

Analysis of CO_2 levels within the last 110,000 years clearly shows a natural

Correlation Between Global Temperature and CO_2 Levels *(fig. 55)*

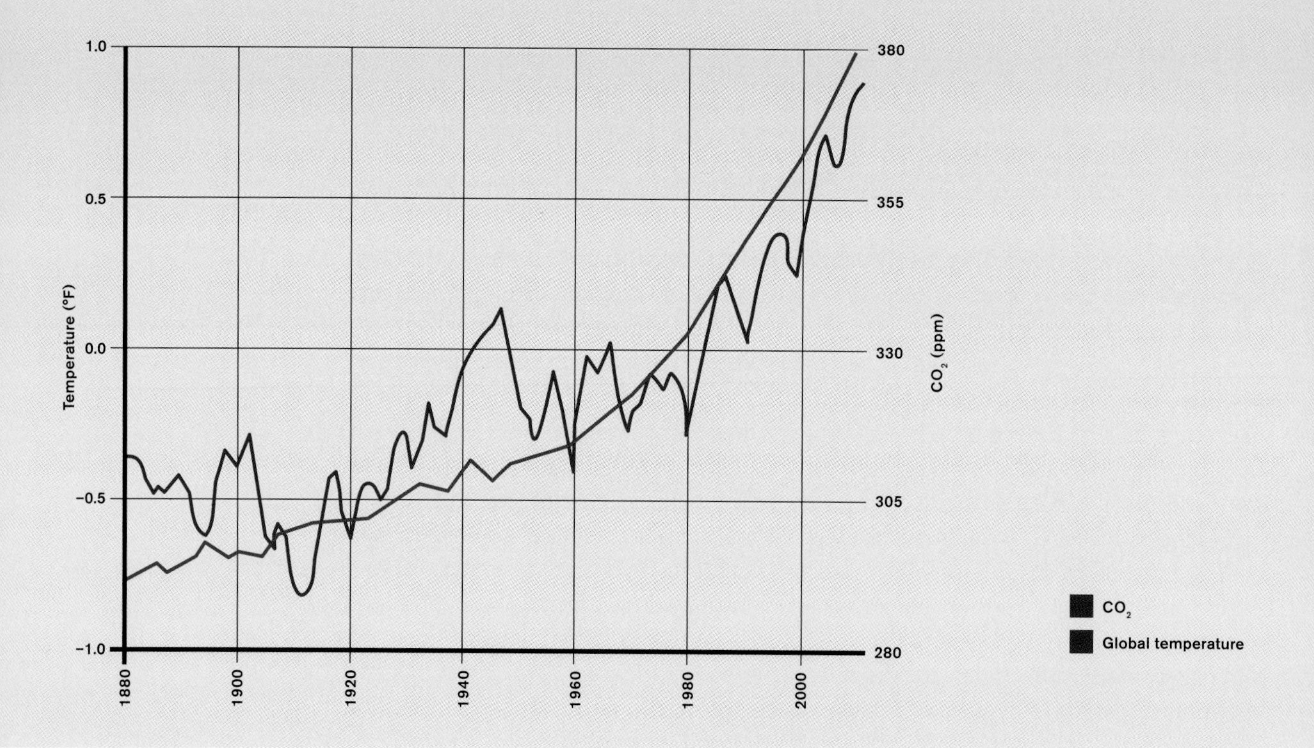

variation of between 180 and 300 ppmv (parts per million by volume) until the middle of the twentieth century. The lower levels of CO_2 correlate to the ice-age cycles, the last of which reached its peak some 18,000 years ago. Since then, the planet has been warming, resulting in an environment conducive to the expansion of the human population.

However, despite our relentless expansion, by the 1800s the world's economies were still largely agrarian and only 3 percent of the population lived in cities. During this period, the average temperature of the planet was 14°C. At the end of the first decade of the twenty-first century, following the First Industrial Revolution and the start of the second, the average temperature of the planet is 15°C.

As the Second Industrial Revolution unfolds, a staggering 3.2 billion people (50 percent of the Earth's population) now live in cities. By 2030, this is projected to reach 5 billion (60 percent). By 2020, the world is expected to have nearly 500 cities with more than 1 million citizens, with the majority having more than 5 million and the ten largest containing more than 20 million. Given the weight of evidence that urbanisation and industrialisation are both responsible for the dramatic increase in CO_2 levels that are changing our climate, we can expect that trend not simply to continue but to accelerate.

Industrialisation compounds our influence on the environment by many multiples. The climatic changes wrought by the developed world as a result of the First Industrial Revolution have, to a large extent, been absorbed by the Earth's ecosystem.

However, as the emerging empires of the East seek to emulate the lifestyle of the WCSE, the additional carbon outputs from their rapid industrialisation is tipping the balance of the Earth's closed ecosystem. For all practical proposes, we have already passed the point beyond which the biosphere of our planet has been destabilised. The danger now is that this process of change has its own momentum that will only cease when it finds a new equilibrium. Undoubtedly, that equilibrium will result in a planetary ecosystem very different from that of the last 800,000 years.

Our overpopulation of the Earth is alarming at present but is expected to reach a staggering 9.2 billion by 2050. Without having the option of space travel, we are more like the territorially constrained mice of Australia than the lemmings that at least have the option of escaping their immediate overpopulated environment. As such, we now face the consequences of our success and will have to cope with the severe climatic changes that we have induced. This will add massively to the social stress of our overpopulated planet. The message is clear: climate change and population expansion are inextricably linked, and the solution, if we choose to genuinely attempt to work towards one, will involve some radical changes to the way we live.

Climate Change and the Cycles of Empire

Climate change has not occurred at some arbitrary point but at a specific phase in our history – as the First Industrial Revolution segues into its successor. As we have seen, the US – the last offspring of the WCSE – peaked in 1990 and has since been in decline. Thus it is safe to suggest that the response of the developed nations of the WCSE to the problem of climate change has been shaped by the psychology typical of decline and overextension, in which it has become an ever-increasing challenge to maintain the high standard of living that the developed nations have come to expect and demand of their politicians. This being the case, the diversion of public resources to counter climate change, and the reduction in lifestyle and choice that would be required on an individual level, are both likely to meet with substantial resistance.

This resistance is compounded by a society set in its ways and unable to adapt effectively to new challenges, the interest of a declining empire being the maintenance of the *status quo*. In addition, the decline in creativity of a waning empire further limits its range of responses, new problems being met with old solutions. Furthermore, empires in decline face immediate threats from rising empires, and these take precedence over future, less tangible dangers.

Thus the US, invested in consumption and heavily reliant on oil, has been slow to respond to the threat of climate change. Its response has been partially mitigated by that of Europe. Being in legacy and reformation, and therefore not as overstretched as the US, it has been one of the most responsive territories to the dangers of climate change and the protocols of Kyoto and Copenhagen.

The populations of the WCSE are now in decline, consistent with its decline in empire. In contrast, the emerging nations in regionalisation, such as Africa, India, Latin

The Effect of Increased Global Industrialisation as a Function of the Addition of the ASE *(fig. 56)*

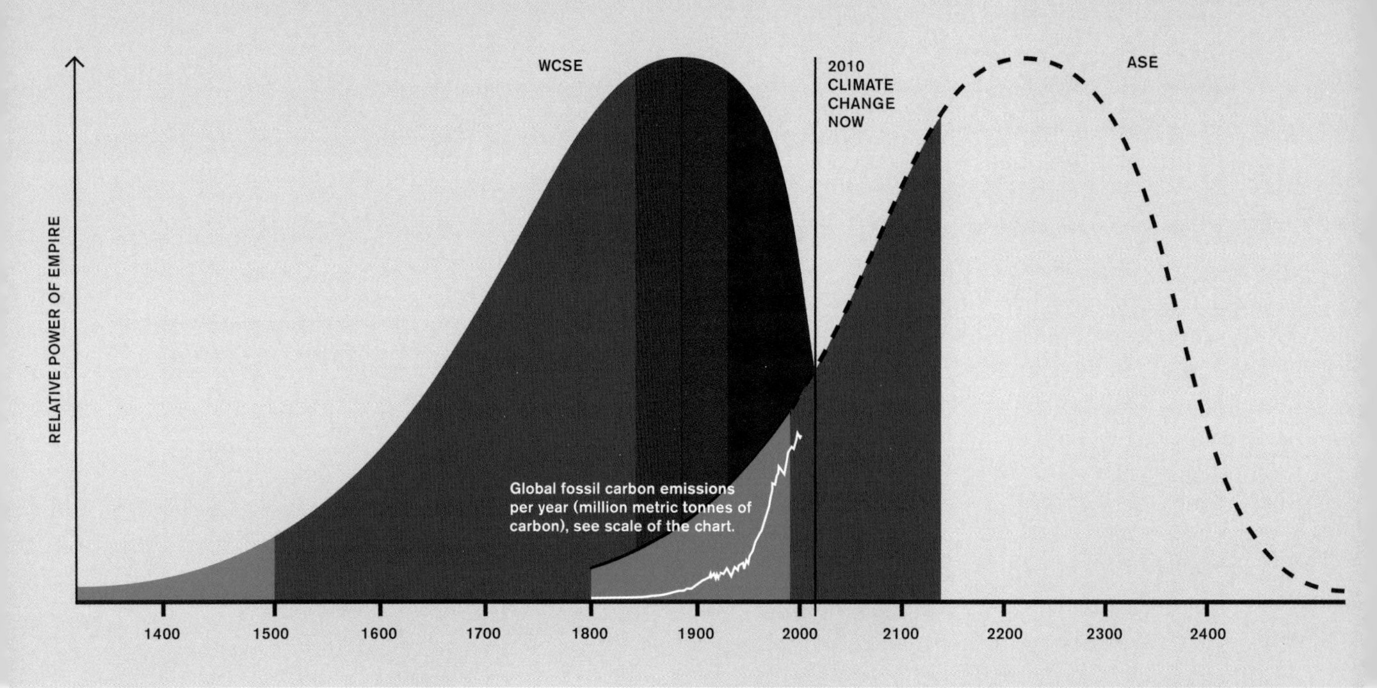

America and the Middle East, and those in ascension to empire, such as China, are expanding demographically. As has been posited, these nations have a single goal: to expand and emulate the standard of living of the developed world. As yet, the poverty of the majority of the emerging nations means that the day-to-day challenges of existence override the concerns of climate change. However, it is these nations, currently building new modern infrastructures, that could theoretically be capable of integrating green technologies into their growing economies if the cost differential were compensated for by the developed nations. Without such support, the emerging nations are unlikely to opt for cleaner energy, justifying their CO_2 emissions by accusing the developed world of creating the problem in the first place. The net effect is a great deal of talk but no action of the magnitude that is urgently needed to address the current changes to the climate.

BOTTOM LEFT **Smoke billows from several chimneys at a factory in Jilin, in north China's Jilin Province, on 29 November 2009. The EU said on 30 November 2009 that to avoid cataclysmic climate change requires Chinese leadership, as the two sides wrapped up a summit with China defending its efforts against global warming.**

Understanding the Science

It may be helpful at this point to elucidate some of the science behind climate change.

GREENHOUSE GASES

The Earth's atmosphere is a gaseous layer that absorbs thermal radiation from the sun and re-emits it. It is vital in controlling the temperature of the surface of the planet. Naturally, in such a complex system, there are many variables, such as the emission patterns of the sun and the water vapour content of the atmosphere. However, an essential component is the so-called 'greenhouse gases', which are able to trap heat within the atmosphere, preventing it from being radiated into space. Without the presence of natural greenhouse gases, the Earth would be 20°C colder than current temperatures. Thus, if all other conditions are stable, the net effect of these gases is to regulate the temperature of the planet.

The greenhouse gases – water vapour, CO_2, methane, ozone and nitrous oxide – occur naturally, but since the mid-twentieth century the cumulative addition of man's industrial outputs has shifted the delicate balance. The gases created by human sources are known as anthropogenic, and it is this ever-growing component of the atmosphere that is responsible for the acceleration of global warming.

Water vapour accounts for the largest percentage (65–85 percent) of greenhouse gases and, fortunately, human activity does not significantly affect its concentration. It is CO_2 that is the main determinant of the greenhouse effect. Prior to the First Industrial Revolution, natural sources contributed more than twenty times the amount created by man and were countered by natural capture, such as weathering rocks, the growth of plankton and the photosynthesis of plants. Our biosphere was in perfect balance.

Since the 1800s, this balance has been destroyed. We now produce 50 billion tonnes of CO_2 per year, which increases by 2–3 percent a year. Pollution by CO_2 is now 30 percent higher than at any of the previous peaks during the last 800,000 years.

US Greenhouse Gases Emissions Flow Chart *(fig. 57)*

Emissions data come from the Inventory of US Greenhouse Gas Emissions and Sinks: 1990–2003, US Environmental Protection Agency (EPA; using the CRF document). Allocations from 'Electricity & Heat' and 'Industry' to end uses are World Resources Institute (WRI) estimates based on energy use data from the International Energy Agency (IEA, 2005). All data are for 2003. All calculations are based on CO_2 equivalents, using 100 year global warming potentials from the IPCC (1996), based on total US emissions of 6,978 $MtCO_2$ equivalent.

Emissions from fuels in international bunkers are included under 'Transportation'. Emissions from solvents are included under 'Industrial Processes'. Emissions and sinks from land use change and forestry (LUCF), which account for a sink of 821.6 $MtCO_2$ (metric tonnes of CO_2) equivalent, and flows of less than 0.1 percent of total emissions, are not shown.

For detailed descriptions of sector and end user/activity definitions, see Navigating the Numbers: Greenhouse Gas Data and International Climate Policy (WRI, 2005).

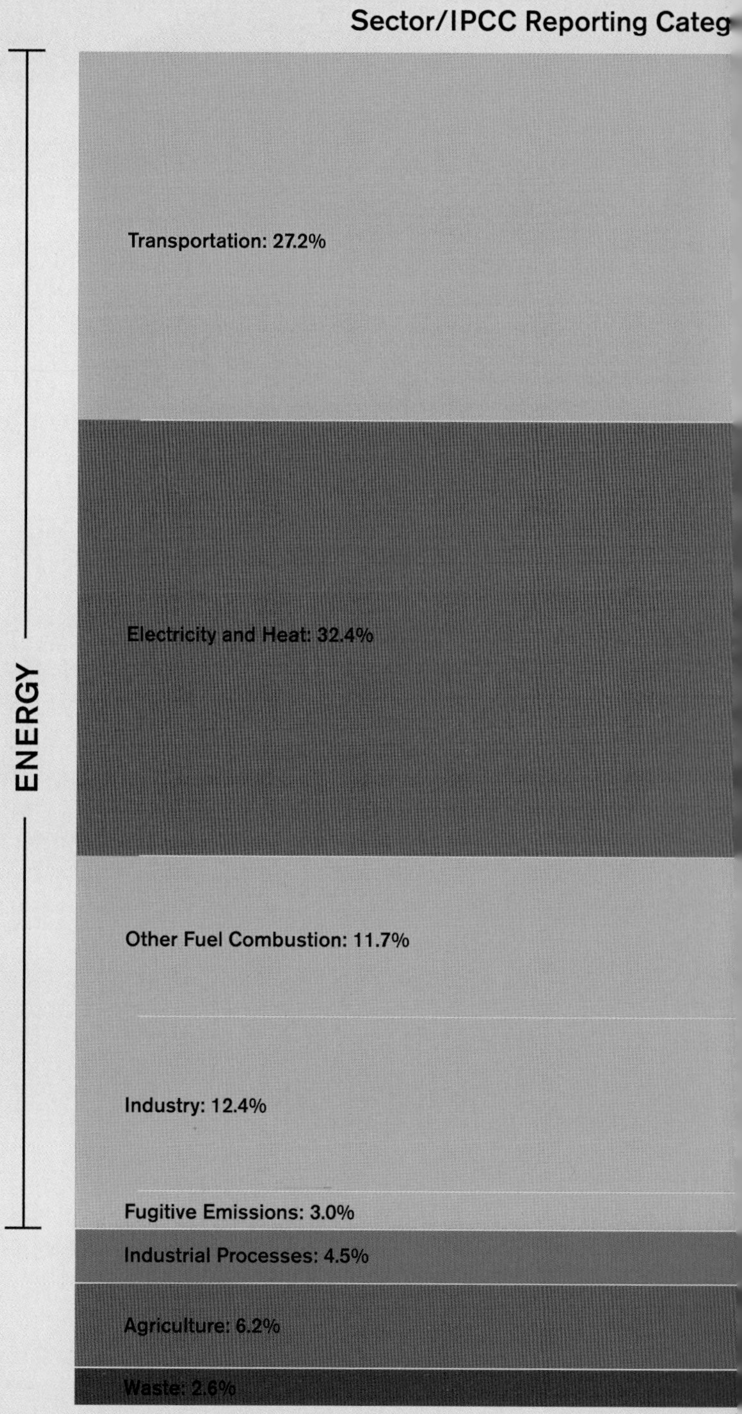

Sector/IPCC Reporting Categ

ENERGY

Transportation: 27.2%

Electricity and Heat: 32.4%

Other Fuel Combustion: 11.7%

Industry: 12.4%

Fugitive Emissions: 3.0%

Industrial Processes: 4.5%

Agriculture: 6.2%

Waste: 2.6%

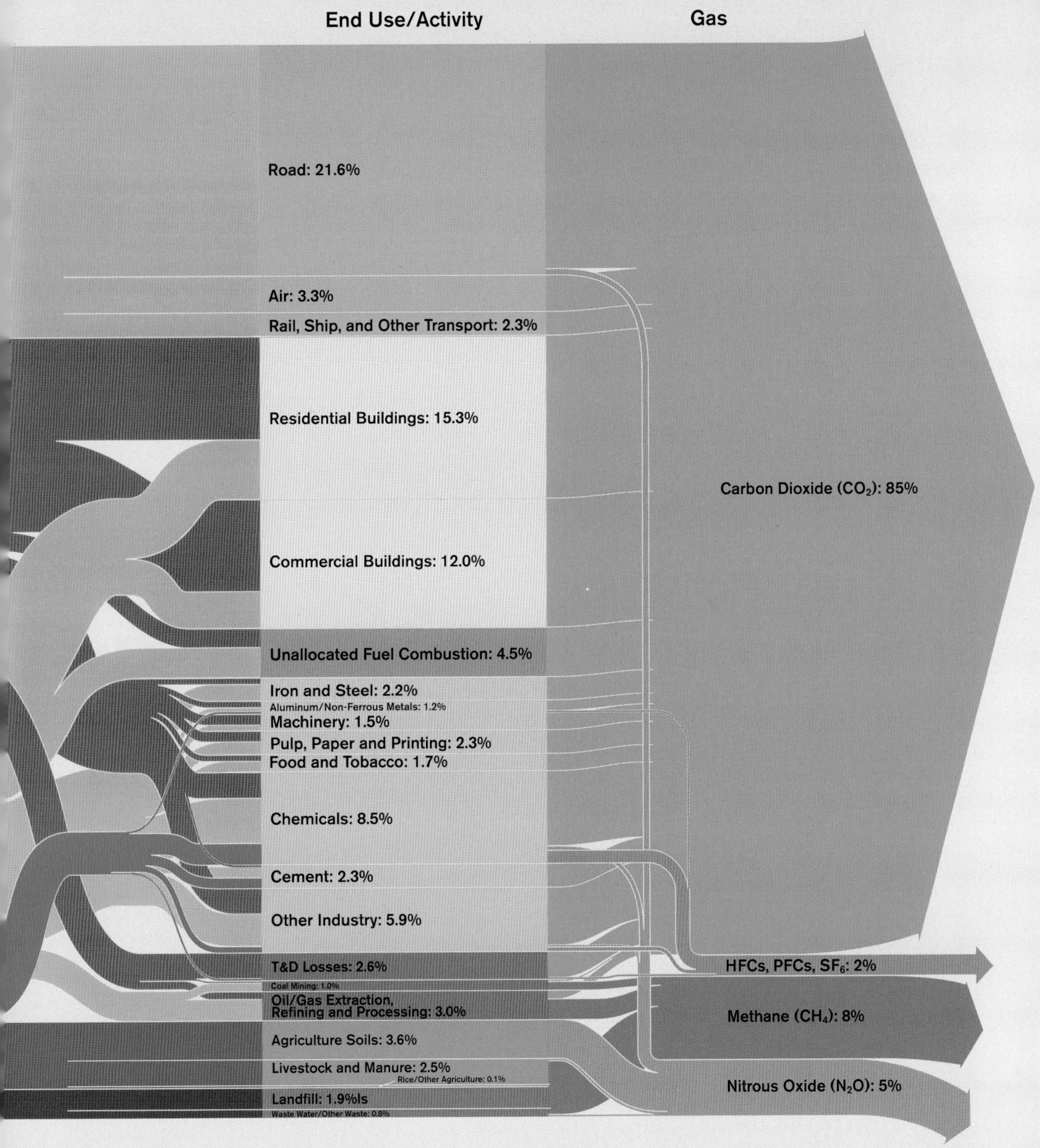

End Use/Activity

Gas

Road: 21.6%

Air: 3.3%

Rail, Ship, and Other Transport: 2.3%

Residential Buildings: 15.3%

Commercial Buildings: 12.0%

Unallocated Fuel Combustion: 4.5%

Iron and Steel: 2.2%

Aluminum/Non-Ferrous Metals: 1.2%

Machinery: 1.5%

Pulp, Paper and Printing: 2.3%

Food and Tobacco: 1.7%

Chemicals: 8.5%

Cement: 2.3%

Other Industry: 5.9%

T&D Losses: 2.6%

Coal Mining: 1.0%

Oil/Gas Extraction,
Refining and Processing: 3.0%

Agriculture Soils: 3.6%

Livestock and Manure: 2.5%

Rice/Other Agriculture: 0.1%

Landfill: 1.9%ls

Waste Water/Other Waste: 0.8%

Carbon Dioxide (CO_2): 85%

HFCs, PFCs, SF_6: 2%

Methane (CH_4): 8%

Nitrous Oxide (N_2O): 5%

CO_2 is a by-product of the burning of fossil fuels and the worst polluter is coal. The foremost offender is therefore China, with its rapidly growing (one a week) tally of coal-fired power stations. Between 2000 and 2010, China alone will have increased its CO_2 emissions by a staggering 600 million tonnes.

Another major contributor to CO_2 levels is deforestation. The burning of rainforests not only releases CO_2 but also destroys the planet's vital carbon sinks. Methane is twenty times as destructive as a greenhouse gas than is CO_2. The single greatest source of methane (some 40 percent) is livestock. A cow produces 100–200 litres of methane per week – equivalent to that emitted by an average 4×4 vehicle travelling 33 miles. With some 3.3 billion ruminants worldwide, this is a considerable problem, and one that will only become worse as more people aspire to a meat-rich diet.

The next major producer of methane (some 29 percent) is the energy sector. Again, China's coal-powered stations are the worst offenders. Other sources are waste disposal (18 percent), emissions from land-fill sites, paddy farming of rice and the melting of the permafrosts of Siberia (itself an effect of the warming of the planet).

RIGHT The effects of deforestation in Central Mexico. The removal of large numbers of trees without sufficient reforestation has resulted in damage to habitats, biodiversity loss and aridity. These in turn have an adverse impact on biosequestration of atmospheric CO_2. Deforested regions typically suffer significant adverse soil erosion and very frequently degrade into wasteland.

BELOW RIGHT Icebergs in Jökulsárlón Lake, the largest of a number of glacial lakes in Iceland. It is situated at the southern end of the Vatnajökull Glacier and is filled with icebergs that are calving off the Breidamerkurjökull Glacier.

Other contributors to the greenhouse effect are the chlorofluorocarbons (CFCs) used to cool refrigerators, halo-carbon-gas fire extinguishers and fertilisers, which lead to higher nitrous oxide concentrations. Although these gas emissions are relatively small by volume, they are many times more powerful as greenhouse gases than is methane.

Given that the Earth has maintained its equilibrium for so long, it is reasonable to ask how long it would take for the planet to rebalance itself if we stopped all emissions now. The answer is sobering: projections suggest that after 200 years, 70–80 percent of the additional CO_2 would have been reabsorbed by the oceans, but it would take tens of thousands of years for the remainder to reach a new equilibrium at about 20 percent higher than pre-industrial levels. This is a very short time for the planet but too long for man. The painful reality is that the changes we are now making to the atmosphere will be ones that we will have to live with for millennia to come.

THE DEMISE OF ICE

THE DISAPPEARING ARCTIC

The most visible manifestation of climate change is the diminishing Arctic ice cap. Historically, this has expanded in Winter and contracted in Summer. Over the past few decades, it has expanded to a maximum of 5.8 million sq. km and contracted to 2.7 million sq. km. However, in Summer 2008, it contracted to just 1.6 million sq. km. This represents a 43 percent contraction since 1979, when accurate satellite records began. The shrinkage began almost imperceptibly but is now accelerating, so by 2013 the Arctic ice cap is expected to have disappeared. This will result in the opening of both the Western and Eastern Northwest Passages for the first time in 130,000 years.

The rate of decline of the Arctic has exceeded all scientific expectations, and it has been accelerated by two key factors. First, when solar radiation is reflected off the ice but absorbed by the darker water, this warms and melts the ice. The less ice there is,

ABOVE A polar bear (*Ursus maritimus*) on pack ice. Global warming is the most significant threat to the animal because the melting of its sea ice habitat reduces its ability to find sufficient food. The polar bear was listed as a threatened species under the Endangered Species Act by the US Department of the Interior in 2008.

the greater the effect. Second, the reduction in the thickness of the ice sheet, which over two decades has decreased by 40 percent, makes it more prone to breaking up under stress. Once individual pieces of ice have broken away, their volume to surface area ratio increases, causing them to melt faster than when still part of a sheet. These factors are in addition to the general warming of the waters around the Arctic, which consistently erode the ice in Summer.

THE ARCTIC AND NEW OPPORTUNITIES

When the Arctic Ocean opens to shipping, it will cut some 4,000 miles off the journey from Europe to Japan and 5,000 miles off that from Europe to the west coast of the US. It will also bypass the existing size restrictions for shipping travelling through the Panama Canal. The strategic choke points of the Suez Canal, the Cape of Good Hope and the Panama Canal will no longer be of significance.

The Arctic Ocean is estimated to contain 25 percent of the world's remaining hydrocarbons, although they are considered to be some of the most difficult to extract. At present, there are three known fields containing over 10 billion barrels of reserves, and a fourth containing 6 billion barrels. This does not include any of the areas that have been permanently covered by ice and were hitherto inaccessible to seismic surveys.

Temperature and CO_2 Records for the Past 800,000 Years *(fig. 58)*

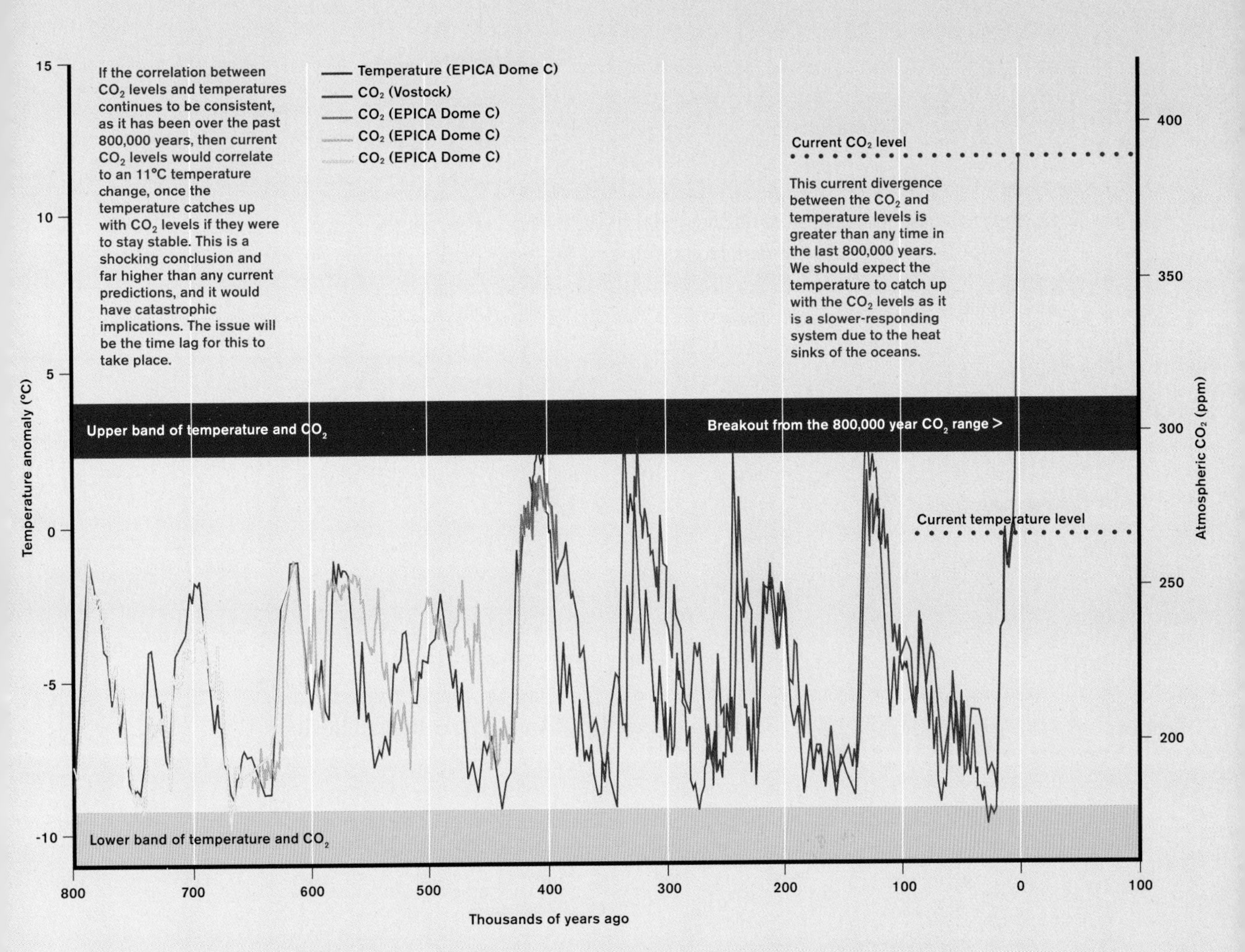

CO_2 Levels and Projected Temperature and Climate Effects

Level (ppm)	Observations
280	Pre-industrial level; temperature 0°C and sea level of 0 m
280–300 ppm	Equates to 1.7–2.7°C and possible sea level rise of 4–6 m
350	Level suggested by climate scientist James Hansen for a safe climate future
380	2008 level, 2.7–3.7°C global temperature rise possible, 15–35 m possible sea level rise possible
450	Level targeted for international negotiations in 2008 but considered dangerous due to risk of exceeding 2–3°C global temperature rise
550	Considered very dangerous due to likely global temperature rise of 4°C or higher
650	Level predicted for 2050 based on current carbon emissions. Considered extremely dangerous. 5.7°C or higher temperature rise possible. Possible sea level of 75 m.

Notes to Rate of CO_2 Level Increase

Scientists at the Mauna Loa Observatory in Hawaii say that CO_2 levels in the atmosphere now stand at 387 parts per million (ppm), up by almost 40 percent since the First Industrial Revolution and the highest for at least the last 650,000 years. The figures, published by the US National Oceanic and Atmospheric Administration on its website, also confirm that CO_2, the chief greenhouse gas, is accumulating in the atmosphere faster than expected. The annual mean growth rate for 2007 was 2.14 ppm – the fourth year in the past six to see an annual rise greater than 2.0 ppm. From 1970 to 2000, the concentration rose by about 1.5 ppm each year, but since 2000 the annual rise has leapt to an average of 2.1 ppm.

As the ice melts, the nations surrounding the Arctic are working to maximise their access to the newly available resources. However, they are restricted by the United Nations Convention on the Law of the Sea (UNCLOS), which permits exploration only to a 200-mile offshore limit, although this can be extended to 350 miles if the continental shelf extends that far. As the Arctic Ocean is relatively shallow, this will benefit Russia in particular, which will be one of the few nations to profit directly from global warming. As the potential beneficiary of a vast resource of undersea natural oil and gas, Russia is already seeking to extend its continental shelf by staking a claim on the Lomonosov Ridge, whose area is 1.2 million sq. km (the size of Western Europe). The retreat of the ice will provide Russia with the strategic access to the world's oceans that was denied it during the Cold War, coupled with the ability to control the shorter sea route from the west to the east – the Northeast Passage.

RIGHT Calving Icebergs on Dawes Glacier in Alaska. Iceberg calving is a form of ice ablation or disruption and involves the sudden release and breaking away of a large mass of ice from a glacier, iceberg, ice front, ice shelf or crevasse.

The US, Canada, Norway and Denmark are also in a position to benefit from the exploitation of the Arctic. This is sure to result in the friction of power politics, with both regional resources and the control of the Northwest Passage at stake. Inevitably, these nations are currently involved in spirited negotiations with their neighbours to try to clarify their mutual borders.

THE ANTARCTIC

The Antarctic ice cap is the largest of the ice masses on our planet, covering approximately 14 million sq. km and containing 30 million cubic metres of ice – equivalent to around 61 percent of the world's freshwater. One of the major problems of long-term climate change, of course, is that as this ice melts, it will cause sea levels to rise.

To date, the Antarctic has been protected by the cold waters of the circumpolar current, which create a barrier against the warming of the planet. However, this has only slowed the changes, not stopped them. Since 1950, the Antarctic Peninsula has warmed by 3°C. This is higher than any other part of the world and has caused accelerating damage to the continent's surrounding floating ice fields, which are now breaking away from the landmass. Although the eastern ice shelf is currently stable, the western shelf is starting to lose mass as it warms up. During the past fifty years, ten massive ice shelves have broken away from the land, the most recent being the 10,000-year-old, 3,400 sq. km Wilkins shelf. Structures like this can crumble very rapidly, with areas of up to 700 sq. km collapsing in as little as seven days. In time, the sea ice will all but disappear and then the land ice will be fully exposed to potentially the greatest melt in the Earth's recent history.

GREENLAND

The Greenland ice cap and its peripheral glaciers represent the world's second-largest reservoir of ice – some 8 percent of the world's 30 million cubic kilometres. It covers an area of 2.85 sq. km to a maximum depth of 3 km. Greenland, like the Arctic, is particularly vulnerable to climatic warming. Indeed, it is already showing signs of shrinking and thinning at its lower elevations.

ABOVE A receding glacier in Sonamarg, about 86 km (53 miles) northeast of Srinagar, India. Due to global warming, Himalayan glaciers in Kashmir are melting fast, causing the levels of regional watercourses to rise by two-thirds. The glaciers are headwaters for Asia's nine largest rivers and are crucial for the 1.3 billion people of the south Asian region.

Recent estimates put the current rate of shrinkage at 239 cubic kilometres per month – well above the 195 cubic kilometres average measured between 2003 and 2005. This is taking place on the smaller surrounding glaciers and along the edges of the main ice cap. However, like all other heat-loss processes, the speed of the melting is related to the volume to surface area ratio, so it will only accelerate with time. The majority of current scientific projections using linear mechanics talk about a time frame of a century for the melting of the entire ice cap. However, like every other area of climate change, the actual timeframe may well be only decades.

The melting of the Greenland ice cap will cause sea-level rises of 7 m. This is sufficient to submerge every coastal city in the world. In addition, the cold, fresh meltwater will reduce the circulation flow of the Atlantic currents, which could in turn create further climatic shifts in the northern Atlantic. It is thus no exaggeration to say that the imminent melting of the Greenland ice cap represents the greatest single threat to our world as we know it.

THE WORLD'S GLACIERS

Given the distance of the Arctic, Antarctic and Greenland from the centres of human population, their fate is viewed by many with a sense of detachment. However, the melting of the glaciers, which contain only 4 percent of the world's ice mass, will have a more direct and immediate impact on a large percentage of the world's population.

Glaciers naturally expand and contract with the seasons, so melting is a natural part of their life cycle. However, they are retreating at a rapid rate. Data from a global sample of thirty mountain glaciers clearly show that the melting process has been accelerating since 1990 and further accelerating since 2005. This is due partly to the increase in volume to surface area ratio as the glaciers shrink, and partly to global warming. As the glaciers melt, their surface area decreases, reducing their ability to reflect solar radiation, which is instead absorbed as heat by the Earth.

These huge, slow-moving masses of ice are not just architects of the landscape but a vital source of water during the Summer months, essential to the survival of millions of people. Seasonal meltwater feeds streams that become rivers, providing water and irrigation for the valleys. Initially, the melting of the glaciers has been manifested as flash flooding, but in the medium term it will lead to increasing water shortages across the regions whose rivers they feed. This will usher in a new phase of water insecurity and agricultural crisis. New dams to capture and store rain will be required, placing huge cost burdens on the world's economies. Failure to build such dams will ultimately prompt mass migrations.

THE ALPS

Current predictions are that before the beginning of the next century, 75 percent of the alpine glaciers will have disappeared. However, it may be much sooner – a case of only one or two decades rather than ten – that this occurs.

THE ANDES

In countries like Bolivia, small glaciers, like Chacaltaya, 18,000 years old, have lost 80 percent of their area during the last twenty years and within a few years will have vanished completely. At one time, scientists estimated that it would be 2030 before this glacier disappeared, but this estimate has been revised to 2015. These changes are echoed by fifteen other glaciers in the area, five of which have disappeared already. In the next decade, perhaps sooner, the area will lose its water supply and will go into a water deficit that will have to be replaced by new dams. It is one of the injustices of climate change that nations like Bolivia that have not contributed to the phenomenon are the ones most suffering from its consequences.

THE HIMALAYAS

The glaciers of the Himalayan region contain one of the largest stores of freshwater outside the polar regions, supplying no less than seven Asian countries, and feeding the Ganges, Indus, Brahmaputra, Mekong, Thanlwin, Yangtze and Yellow Rivers.

Thermohaline Circulation *(fig. 59)*

The part of the large-scale ocean circulation that is driven by global density gradients, created by surface heat and freshwater flux.

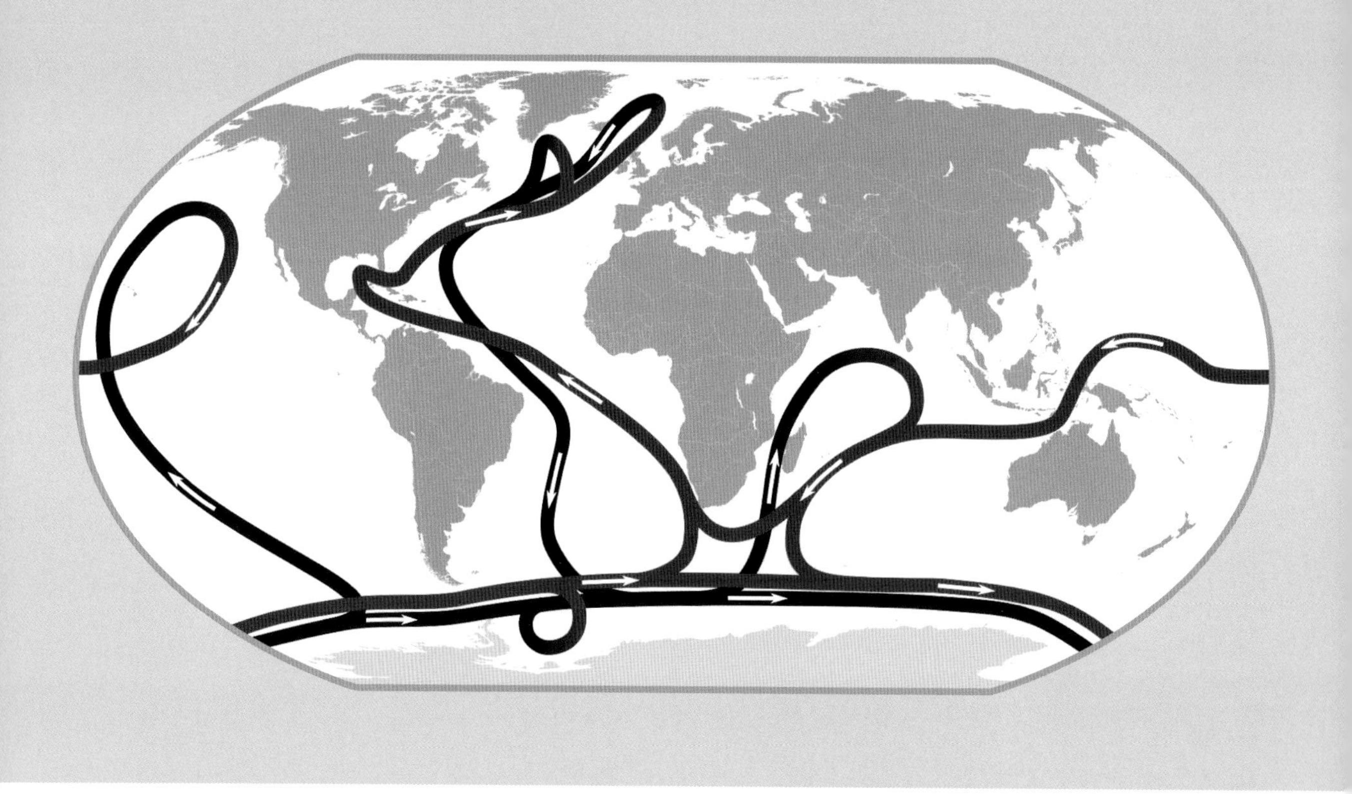

There are hundreds of millions of people in both India and China dependent on Himalayan water. The source of the Ganges in India is fed by tributaries from Nepal. Before it reaches the plains of Bangladesh, it supplies some 450 million people. The key question is whether the depletion of their rivers (predicted by 2030) caused by the effects of climate change on the Himalayan glaciers provides the reality check that India and China need to self-regulate. It is to be hoped that it is, because the water deficit that these countries will inevitably experience raises the very real risk of resource conflict.

THE CHANGING OCEANS

THE OCEAN CONVEYOR BELT

At the peak of the last warm cycle some 125,000 years ago, sea levels were 6–7 m higher than they are today. Around 20,000 years ago, at the end of the last ice age, they were about 120 m lower than they are today. It was only about 3,000 years ago that sea levels stabilised to the levels that we consider normal.

Over the past fifteen years, sea levels have risen by 3.4 mm per year and current accepted projections are for rises of between 0.6 and 1.0 m by 2080. However, this, like many current climate-change predictions that have already been discussed, is both linear and conservative, and the real increase is likely to be far greater.

Historic Changes to Sea Levels *(fig. 60)*

Rate of sea-level rise (centimetres per century) vs year (from 18000 BC to AD 2000); derived from graph by Lambeck cited in IPCC's Climate Change 2001.

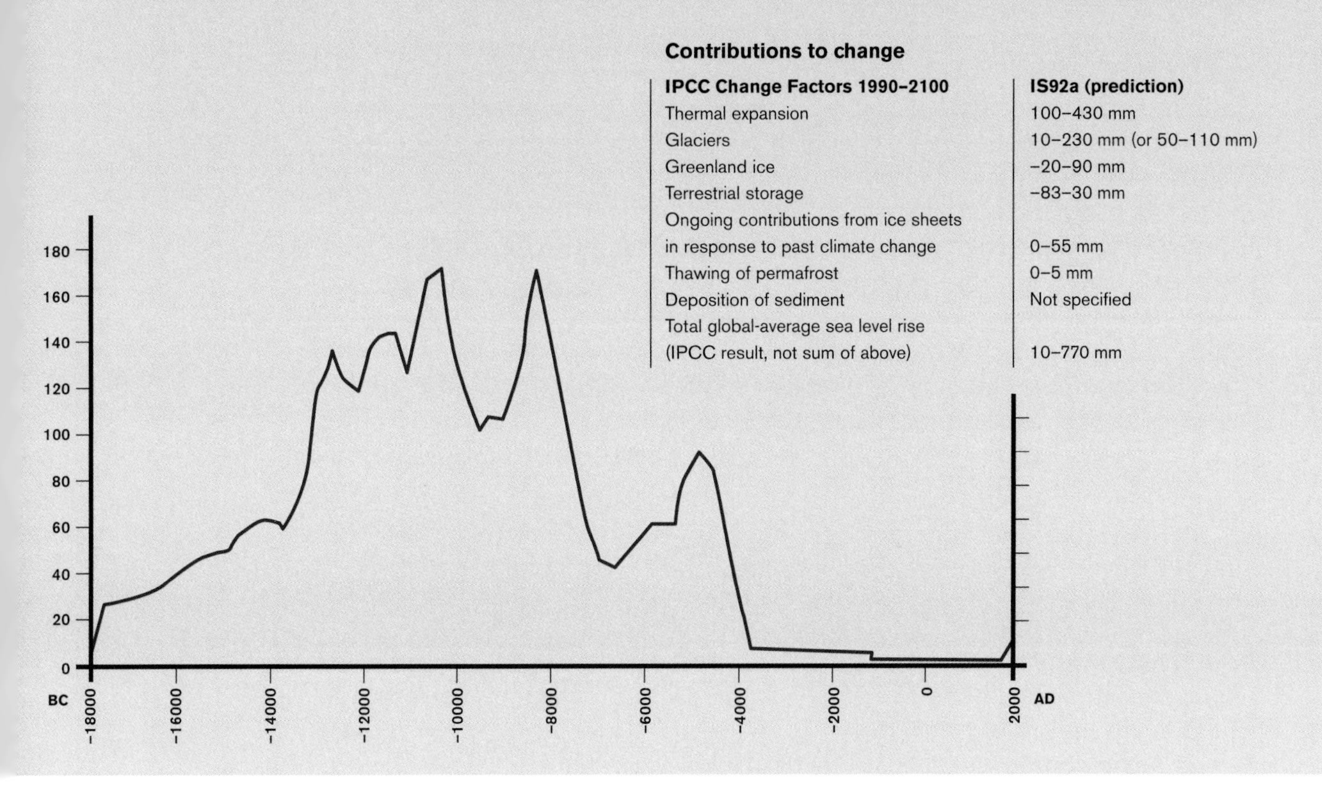

Contributions to change

IPCC Change Factors 1990–2100	IS92a (prediction)
Thermal expansion	100–430 mm
Glaciers	10–230 mm (or 50–110 mm)
Greenland ice	–20–90 mm
Terrestrial storage	–83–30 mm
Ongoing contributions from ice sheets in response to past climate change	0–55 mm
Thawing of permafrost	0–5 mm
Deposition of sediment	Not specified
Total global-average sea level rise (IPCC result, not sum of above)	10–770 mm

As meltwater from the glaciers in Greenland, the Himalayas and the Antarctic reaches the oceans, it will add to their volume and cause sea levels to rise. Due to their magnitude, the oceans are slow to change temperature and the melting ice will initially cool them. However, as the rate of ice melt decreases, the oceans' average temperature will begin to rise and their waters expand, causing further elevations in sea levels.

The two most vulnerable ice sheets – the Greenland ice cap and the West Antarctic ice shelf – could alone add 6–7 m to current sea levels. Such a rise would inundate most coastal cities in the world, but long before that point it would lead to the loss of several small island countries, such as Tuvalu and the Maldives. Given that 60 percent of the world's population lives within 75 miles of a coastline, the prospect of rapid sea-level rise has catastrophic social and economic implications for the modern world. It has been estimated that a 1 m rise in sea levels (far less than any current predictions) would directly affect some 600 million people globally.

To place the magnitude of what faces us in context, it took 18,000 years to melt 60 percent of the ice of the last ice age. However, in the 200 years since the beginning of the First Industrial Revolution, the planet has warmed at such a rate that we now face the melting of all of the grounded ice caps in only a few generations. This would result in sea levels rising by some 65–75 m, thereby significantly altering the face of the Earth and all life that it sustains. (Figure 60 shows a summary of sea-level changes to date.)

The oceans play a major role in the distribution of heat from the equator to the poles through current circulation. The Coriolis effect caused by the spinning of the Earth generates currents that flow clockwise in the Northern Hemisphere and anticlockwise in the Southern Hemisphere. Thermohaline circulation (see Figure 59) drives deeper currents as a result of the global density gradient created by surface heating and freshwater fluxes.

Wind-driven, warm, low-density surface currents travel northwards from the equator to the poles. In the process, they pass their heat to the atmosphere. Once they have cooled, they sink in the northern latitudes as denser water created by the salt concentration, as a result of the effects of evaporation. In addition, the formation of ice concentrates salt in the polar regions, creating a highly concentrated brine that works its way down to the bottom of the ice packs and into the sea. There it sinks, adding to the density of the cold, deep water masses in both polar latitudes.

It is this cold-water reservoir that becomes the return current to the equatorial latitudes, as it moves along the ocean floor of the deep abyssal plains, completing the current loop. This deep-water, salty, current flow is very slow moving, taking some 1,600 years to travel from the Arctic to the Pacific. The total system of surface and deep-sea currents is known as the Oceanic Conveyor Belt. It mixes the world's oceans into one great system that has a major impact on the planet's climate.

The forces provided by thermohalinic circulation drive a complex set of currents across the world's oceans. These were first discovered and used by the explorers of both the West and the East to traverse the world. Indeed, our success as a species has been deeply entwined with the equilibrium of the Oceanic Conveyor Belt. Yet the complex system of currents that underpin the Earth's climate are vulnerable to climate change. The two main variables are the equator to polar heat differential, and the dynamics at the oceans' sink points at the poles.

TEMPERATURE AND SALINITY

This circulation system is complex, but essentially it is powered by the heat differential between the warm equatorial area and the cold poles. However, as the ice melts, less solar radiation is reflected and the heat absorption rate of the poles increases, resulting in a narrowing of the temperature differential. This causes the power behind the oceans' currents to decrease or fail, changing the way the oceans mix.

Another driver that could be reduced is the mechanism of the downward seawater sinks at the poles. This is where the cooled water from the tropics falls to the ocean floor to create the cold-water masses that drive the return current. As the planet warms and there is reduced or no ice at the poles, one of the mechanisms for increasing the water's density at the sink points – the formation of ice – is removed. However, in compensation, the elevated atmospheric temperature increases evaporation so that by the time the water arrives at the sinks, it has already become denser. In the short term, the melting ice at the poles, particularly that from the melting Greenland glacier and Siberian rivers, will contribute cold, low-density water to the sink points. This may have the potential to stop the circulation altogether before a new equilibrium is established.

The evidence that these polar water sinks, as with all other aspects of climate change, are beyond current scientific modelling came in 2005 from submarine studies under the Arctic. These reported that of the seven to twelve expected columns of cold water, only two, with weak dynamics, were found. Later that year, the UK's National Oceanography Centre stated that, since 1992, there had been a 30 percent reduction in the warm currents that carry water north from the Gulf Stream. This ocean current has been a vital part of Britain's climate, creating warmer Winters because the North Atlantic drift current protects us from the Siberian Winters that we could otherwise expect at such latitudes.

ABOVE A catch on a fishing boat off Lofoten Islands, Norway. There has been a dramatic decrease in fish stocks in the North Sea since the 1980s due to overfishing. The introduction of non-indigenous species, industrial and agricultural pollution, trawling, construction on coastal breeding and feeding grounds, sand and gravel extraction, offshore construction and heavy shipping traffic have all contributed to the decline.

THE RISKS

As with any complex system, there will be a point in time where the conveyor belt reaches a tipping point and shuts off. Current models are not clear on where this point might be, but there is palaeo-climatic evidence that air temperatures can drop by 10 percent very suddenly under certain conditions. The shut-off stage could be sooner than we think, bringing with it unpredictable weather changes as the system goes into unstable flux. This will cause the collapse of plankton stocks and the marine food-chain in the region, and in the much longer term (over centuries) the reduction of oxygen

levels at the floor of the ocean. The oceans' currents will eventually find a new equilibrium, but before that happens we will be subject to instability that will prove catastrophic for mankind.

LEFT Aerial view as the Coast Guard conducts initial Hurricane Katrina damage assessment overflights in New Orleans, Louisiana, US, on 29 August 2005. Katrina made landfall as a category 4 storm with sustained windspeeds in excess of 135 mph.

BELOW LEFT Two men paddle their boat in high water in the Ninth Ward in New Orleans following Hurricane Katrina. Devastation was widespread across the city, with water about 12 ft deep in places.

ACIDITY AND THE MARINE ENVIRONMENT

The oceans are a complex environment and are currently home to 230,000 recorded species that have adapted to a multitude of specific marine conditions across the globe. Shockingly, it is estimated that, since 1950, the number of species has halved as a result of overfishing and habitat change. For many centuries, the seas have been a rich source of food, but, as with many other manifestations of the industrial age, our demand on this ecosystem is now outstripping its ability to regenerate. Stocks of fish like tuna, marlin, cod, halibut, skate and flounder have been reduced by 90 percent since the beginning of industrialised fishing in the 1950s.

In addition, there is the problem of acidity. Almost 50 percent of all CO_2 emissions from human sources between1800 and the present are stored in the oceans. However, the increased levels of CO_2 have rendered the water more acidic. This is good for some species but bad for the majority, which are suited to the previous pH levels that were constant for hundreds of thousands of years. Indeed, there is evidence that fish avoid breeding grounds that have a pH that is more acidic than the average. Where such areas are localised, fish can swim to an area that is more favourable. However, as the general levels of acidity increase, the risk is that breeding is inhibited and fish stocks decline even faster than at present. We now face the prospect of a looming shortage of food from one of our most important sources.

WEATHER SYSTEMS

The energy of a system increases with temperature so, as the planet warms, its weather systems will become more energetic in both frequency and intensity. This phenomenon was apparent in Hurricane Ivan in 2004, the sixth most intense hurricane on record, which caused $13 billion worth of damage on the US mainland.

As climate change accelerates, we can expect more frequent and powerful storms and hurricanes (see Figure 61) to hit the coastlines of the world with consequent destruction, loss of life and economic trauma. A more insidious danger to marine traffic is an increase in average wave height, along with numerous freak waves capable of sinking and damage large ships, making the seas a more perilous place for shipping. Conservative projections have wave heights rising by 2 percent annually in the western approaches to Europe, accompanied by stronger winds

Long before sea levels rise to the point where key population centres are submerged, the combination of exceptionally high tides and low-pressure weather systems will create periodic surges that will breach flood defences. Initially, this will cause isolated floods, but, in time, these will become more frequent. The type of storm surge from Hurricane Katrina that wreaked havoc on New Orleans in 2005 will, unfortunately, become a more frequent event.

In Europe, London and the coasts of East Anglia and Holland are particularly vulnerable to such catastrophic flooding. However, inland areas will not be immune because they are likely to experience flash flooding as a result of increasing regional rainfall rates. Crops will also be affected, as happened in the Summer of 2008, when England was so wet that the majority of the wheat harvest was lost.

RIGHT **The rainforest is burned in Sumatra, Indonesia, to clear land for oil palm plantations. Indonesia already has 6 million hectares put to such use but has plans for another 4 million by 2015.**

BELOW RIGHT **A logging operation devastates virgin rainforest in Madagascar. Commitments from various governments to increase the biofuels being sold are pushing the rise in demand for palm oil as a quick fix to reduce greenhouse-gas emissions. By 2020, 10 percent of fuel sold in the EU and 15 percent sold in China will be biofuel.**

THE CHANGING LANDSCAPE

We cohabit with 5–10 million other species. However, our effect on this shared environment is astonishingly disproportionate. Mankind's average footprint per capita is 2.2 hectares each year, yet only 1.8 hectares can be regenerated by the planet over the same timescale, meaning that each person is running at a 0.4 hectare deficit. This will only get worse as populations expand. Essentially, our success is destroying other species of life in all habitats at an alarming rate. Since 1970, land and marine species have declined by 25–30 percent.

DEFORESTATION

Trees have a significant value in areas like Africa, where wood is the primary fuel source for cooking and boiling water. The resulting deforested land is used for pasture, agriculture and human settlement, but if left unattended it often becomes a wasteland subject to desertification in temperate zones.

While deforestation can be seen as a problem of the emerging nations, it is worth remembering that this is chiefly because forests have already largely vanished from the developed world. Britain's forests, for example, were decimated in the construction of the country's shipping fleet, in the days when a Nelsonian battleship required 6,000 mature oak trees. The majority of trees that survived the ship-building programme were felled during the Industrial Revolution.

Some 13 million hectares of trees are lost each year, half being from virgin forests. Today, the worst affected areas are the poorest, with Africa's rate of deforestation double that of the global average. In Afghanistan, where war has raged for two decades, of the original 16 million sq. km of forest that existed before 1947, only half stands today.

Trees are a vital resource, as they encourage rainfall, store water, capture CO_2, and recycle and clean our supply of oxygen. The burning of the tropical forests around the equator is estimated to be responsible for almost 20 percent of all greenhouse-gas emissions when the carbon stored in the wood is released. The saying that 'the forests are the lungs of the planet' is not an exaggeration – without them, we could suffocate.

THE EXPANDING TROPICS

The zone of the tropics has expanded between 2 and 4.5° of latitudes, or some 200 miles, since 1979 as the waistband around the Earth becomes hotter at an increasing rate. This will have a direct impact on the 250 million people who live within this zone, particularly on their ability to access water and food.

ABOVE **Aerial view of a village outside Nouakchott in Mauritania, Africa, which is rapidly being absorbed into the sands of the Sahara owing to desertification.**

DESERTIFICATION

The encroachment of deserts on fertile land is a gradual, non-linear process, as the desert zones have fragile and delicately balanced ecosystems that can suddenly collapse. Climate change is impacting on this process – the Sahara Desert is shifting south at a rate of 48 km per year into the equatorial forests, and the Gobi Desert is moving south into China's heartlands, losing 3,600 sq. km of grasslands each year. Dust storms are a regular occurrence in this region, which has further devastated the country's agricultural capacity. To counter the threat, the Chinese are planting a 5,700 km long 'green wall of China', but the African nations lack such resources and will suffer the consequences in the form of famine.

The encroachment of deserts onto fertile land is also caused by the destruction of rainforests. In Madagascar's highland plateau, for example, 10 percent of the land has become desert after deforestation. But this is a worldwide phenomenon. Forests take groundwater up into their canopy where it evaporates and produces rainfall cycles. Without trees, the local climate becomes drier. Trees also reduce the rate of water run-off, so without them the rain runs straight to the oceans via rivers. Any land that is deforested, overgrazed or overcultivated has a reduced water content and increased salinity, increasing the risk of desertification. As these dry areas expand, they feed back into the climate-change process and accelerate the transformation of the Earth.

The Kyoto and Copenhagen Protocols

The Kyoto Protocol of 1997 was the first attempt by the world's governments to tackle the issues of climate change. This was to be done by the limitation of greenhouse gases created by human emissions. The agreement, which only came into force in 2005, outlined legally binding emission limits for CO_2, methane, nitrous oxide, sulphur hexafluoride, hydrofluorocarbons and peruorocarbons. At the core of the agreement was the target of collectively reducing emissions by 5.2 percent of 1990 levels. National reduction limits varied from nation to nation, with the EU expected to decrease emissions by 8 percent, the US by 7 percent and Japan by 6 percent, but allowing Australia an increase of 8 percent. When compared with the unfettered emission levels of 2010, the agreement represented a 29 percent global decrease.

However, the agendas of individual nations resulted in the creation of an accord that was not capable of properly addressing the rate of climate change. The majority of the world's nations – 181 in total – signed and ratified the agreement, with the notable exception of the world's leading industrialised power, the US. The US Senate, with protectionist motivations, decided that it should not be ratified if it harmed the US economy, and George W. Bush opposed the treaty because it did not place limits on China.

However, the agreement did not anticipate the explosive industrial growth of India and China, which at the time were not major contributors to greenhouse gases, and were viewed as emerging nations with low per-capita emission rates. However, as of 2008, China has become the largest emitter of CO_2 in the world, mainly from its power stations, while India is the third largest emitter after the US. The net effect of the Kyoto Protocol has been extremely limited. The reality is that the three largest emitters of CO_2 competing for geopolitical power have no limitations on their emissions.

The Copenhagen Conference in 2009 attracted 192 nations. Twelve years after Kyoto, the evidence for climate change had by then become overwhelming. Indeed, many nations had experienced its realities, even if some still refused to acknowledge the need for political intervention. Also, the rise of Asia was now a palpable reality in the new geopolitical world. The collective rhetoric reached new heights, with the stated goal being to defend against a temperature rise across the planet. However, the US, which was struggling to maintain its position in the world and was unable to invest in new technology without a decline in economic growth, was neutral in its commitment, as were the rising Asian powers of both China and India. The only meaningful action came from the EU, which unilaterally instigated binding measures to reduce emissions significantly by 2020. This was no coincidence, as Europe, in legacy and reformation, has no aspirations to build an empire or defend one, yet has sufficient resources and technological abilities to instigate emission-control mechanisms, foremost of which is an effective carbon-trading programme.

The resulting non-binding, voluntary agreement fell short of expectations, and the measures needed to defend the 2°C barrier. So we must assume that the temperature rise is inevitable. The world's nations have failed to work together, so they now need to find their own ways to mitigate the effects of climate change on their territories.

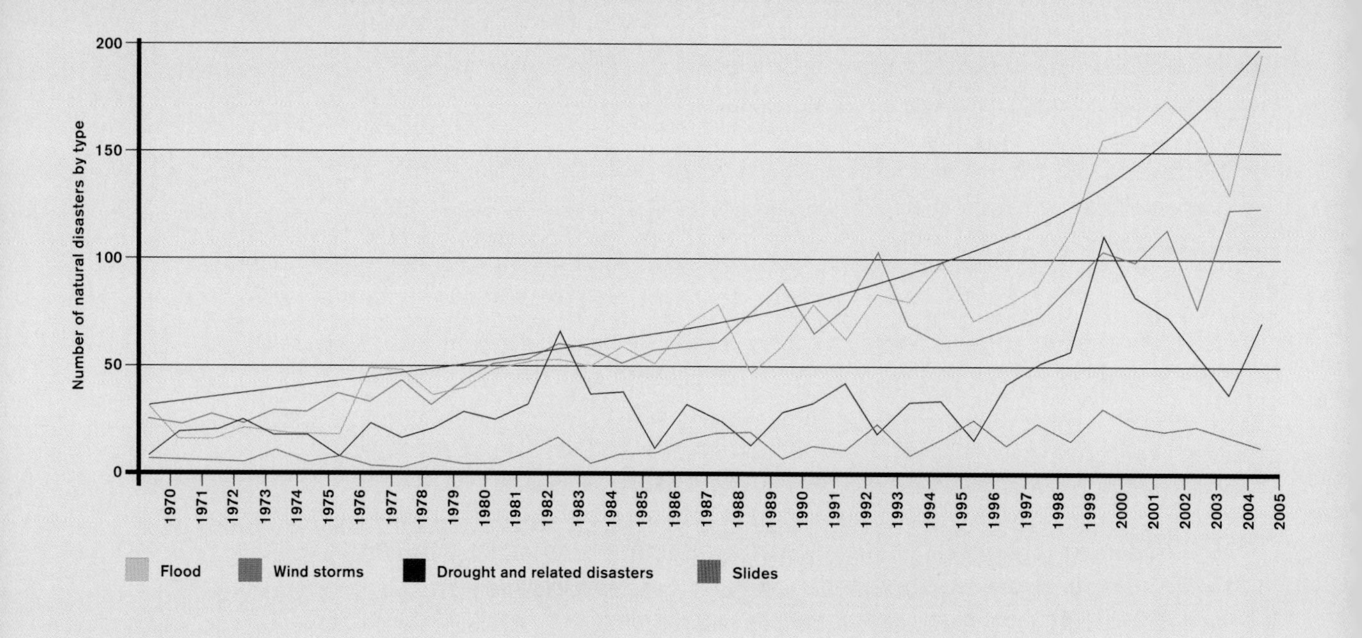

Recognition of Climate Change

The result of two decades of scientific research has given us clear and obvious evidence that climate change is a reality and, worse, that it is accelerating. So why have we been so slow to face up to the unfolding changes? There are five key reasons:

1. The message of scientists in the Western world was initially stifled by the politicians, often at the request of the hydrocarbon lobby. Even now, when changes have accelerated to the point where they are undeniable, the understanding of the magnitude of the problems and commitment to solve them still fall short. The Intergovernmental Panel on Climate Change (IPCC), set up in 1988 by the UN and the World Meteorological Organisation, has over time mounted an unanswerable case in a series of reports. The most recent, in 2007, stated categorically that the probability that climate change is the result of natural processes alone is less than 5 percent. However, the IPCC is a skeletal organisation employing only six people and reliant on the gratis contributions of thousands of scientists. It is also subject to political machinations. In 2002, it was revealed that ExxonMobil sent a memo to the Bush White House calling for the removal

of Robert Watson, the atmosphere scientist who was the IPCC's first chair. The Bush administration successfully lobbied for his removal. In addition, despite repeated assurances that it would not sponsor organisations that dispute the human source of climate change, it has now been revealed that ExxonMobil gave almost $1 million last year to that very cause.

2 The different perspectives of the world's developed and undeveloped economies – one which seeks to maintain their lifestyles and the other to achieve explosive growth, means that neither is motivated to focus on, or solve, the problem.

3 The institutions that advise governments are often slow to recognise change and unable to cope with non-linear dynamics. This is a reflection of the general inertia of the collective consciousness of all human societies, particularly those in decline (e.g. those in the West). While those in expansion are better at coping with non-linear shifts, climate change has been a very low priority for the emerging nations.

4 The rate of climate change has been matched by the rate of power shift from West to East. As noted earlier, this has been accompanied by a huge wave of industrialisation in China and the emerging world. It is perhaps because the scientists and politicians studying climate change are from developed countries that they have failed to foresee the effect that the new wave of industrialisation would have on the established world order.

5 We did not fully comprehend the impact of industrialisation on the Earth's ecosystem when the pace of change was slower and more manageable. The reality is that the changes now greatly exceed our understanding and modelling, and therefore we are struggling to keep up with accelerating environmental shifts that are creating new cycles of their own.

'Too little, too late' aptly describes the current collective global measures to prevent climate change. We must now find new ways to mitigate the effects of climate change on our civilisation as the current counter measures will not prevent or arrest the changes that are daily unfolding.

Our New Climate

Future temperature rises represent a major challenge to our civilisation. Projections of how our climate will develop are at best estimates, and, as has been repeatedly stressed, most models to date have consistently underestimated the rate of climate change. Although the speed of change may have escaped these projections, the predicted environmental effects at various average temperatures should be considered reasonably reliable. It is important to note that the average temperature is just that – an average – and that various latitudes will undergo different rates of temperature change, with the greatest occurring across the north polar latitudes, the equator and the South Pole.

Climate change is measured against the climate in 1800 at the start of the first industrial age. To understand the effects of the changes as the temperature increases, it helps to look at the effects in 1°C increments:

+1°C represents a 0.3°C rise from the current 0.7°C level, leading to an acceleration of all of the present climate-change trends: higher levels of greenhouse gases, melting ice, deforestation and desertification.

+2°C is the level that the Kyoto Protocol was designed to defend against. However, this rise is now seen to be inevitable, although with the time lag between temperature change and greenhouse-gas trends, it will probably be reached sooner than forecast. A world 2°C warmer will be blighted by equatorial deserts, although in Northern Europe increased warmth could improve agricultural outputs.

+3°C is the level at which the Alps will have lost all snow and glaciers, there will have been widespread wildlife extinctions on land and in the oceans, sea levels will be consistently higher than expected and coastal cities will suffer from storm surges.

+4°C will cause heatwaves in Northern Europe, the rainforests will be disappearing, and the melting of the Greenland ice cap, along with meltwater from the Antarctic, will cause sea levels to rise by 7–14 m, changing the geography of all landmass.

+ 5°C will lead to global crop failure, starvation and drought, accompanied by continuing catastrophic rising sea-levels.

Climate change will increasingly cause massive social stress on mankind by reducing the available productive land per capita to sustain the human population. Indeed, long before the temperature of the Earth increases by 2°C, the competition for basic food and water resources will have initiated new waves of conflict across the globe.

Cities will inevitably become much hotter. Buildings store their heat for longer than open land, and urban areas produce more heat as a by-product of energy use and population density. This means that built-up areas could be as much as 10°C hotter than their surroundings during the Summer, well up from the current 2–3°C differential. As more than 50 percent of the world's population live in cities, this heating effect will be a constant reminder of the effects of climate change for all urbanites.

Climate change will have different ramifications for the various zones of the Earth. In turn, this will affect the commitment that each nation is prepared to make in addressing and mitigating its effects.

ABOVE Sydney Harbour Bridge on 23 September 2009 during severe wind storms in the west of New South Wales that blew a dust cloud over Sydney and its surroundings.

EUROPE

Europe will see an increased risk of inland flash flooding, rising sea levels, and flooding of low-level coastal areas and population centres. Most at risk are Venice, the Dutch lowlands, London and East Anglia. There will be reduced snow and glacial coverage of mountain areas, all but destroying the Winter sports industry. There will be an increase in crop productivity in Northern Europe and a decrease in the southern Mediterranean countries. The differential effects across Europe may exacerbate economic differences, creating a two-zone continent with a richer north and a poorer south.

NORTH AMERICA

The US will experience a loss of snow and glaciers in its western mountains. While some areas may see a 10 percent increase in crop yields as a result of greater rainfall, the west coast of California will increasingly suffer from a dry, hot climate, with heatwaves and fires. It is no coincidence that this state is a leading proponent of the development of new clean technologies. In the Gulf of Mexico, the trend towards more frequent, more powerful hurricanes will prove destructive to the regional economy. In Louisiana, 1 million hectares will be lost to flooding, and low areas in Florida will be extremely vulnerable to rising seas. Indeed, many of the US's east coast cities are close to water and

are therefore likely to suffer greatly from higher sea levels. Interestingly, where the federal government has ignored the Kyoto Protocol, a number of states have voluntarily capped emissions in response to their citizens' desire to mitigate against climate change.

LEFT A fish skeleton in a dried-up riverbed in Niger, Africa. The problem of water shortage on this continent is acute. In June 2010, it was reported that a severe drought across the Eastern Sahel in West Africa was affecting 10 million people across four countries. In Niger, the worst-affected country, 7.1 million people were left hungry.

LATIN AMERICA

Tropical forests will be replaced with savannah, leading to a loss of biodiversity as thousands of species become extinct, particularly in the eastern Amazon. As a result of deforestation, the area will suffer considerable reduction in water supplies to agriculture and to its growing population. Brazil has built a large renewable-energy-based economy, which is unique in the modern world, but set against this it has failed to prevent illegal logging and burning.

AFRICA

Africa will face major climate-change challenges. Some 50 percent of the population is fed from dry-land farming, which is extremely vulnerable to reduced rainfall. Without dramatic changes towards irrigated agriculture, the continent faces a 50 percent reduction in grain production, which will inevitably lead to famine, and wars over food and water. These effects will be most prevalent in the tropical zone.

ASIA

As already discussed, the melting of glaciers in the Himalayas will initially produce flash flooding as meltwater breaks its natural dams. However, it will later dramatically reduce the availability of fresh water, leading to widespread drought that will affect nearly half of the world's population. The most conservative estimates predict severe disruption within twenty years, but like all such predictions they may prove to be underestimations.

Flooding resulting from rising sea levels and increasingly erratic and intense monsoons will impact on the many cities and population centres close to the coasts. Floods will lead to an increase in the incidence of diarrhoeal diseases, linking climate change through social stress to disease.

Desertification will also be a growing problem. This will significantly reduce the amount of available agricultural land, particularly around the Gobi Desert.

THE POLAR REGIONS

While the loss of ice will have significant effects, it will also reveal huge swathes of land in the north of Canada, the US, Russia and Greenland, which may support new settlements. This could reduce some of the friction generated by population expansion. However, it is unlikely that the nations involved will open their doors to the populations of the more crowded equatorial zones.

BELOW RIGHT A family in Nowshera, Pakistan, scramble to safety in a flood-hit area on 29 July 2010. Flash floods and building collapses brought on by heavy rains affected some 160,000 sq. km of the country. The UN estimated that more than 20 million people were suffering and left homeless.

Man's Giant Footprint

The sobering reality is that the Earth's resources are being used approximately 25 percent faster than they can be replenished. The average individual ecological 'footprint' is 2.2 hectares per year, but this figure conceals quite dramatic differences. Taking two extremes, the average US footprint is 9 hectares, while that of Afghanistan is 0.04 hectares. Globally, our personal allowance is 1.8 hectares, so many of us are operating at a significant ecological deficit, which will only become worse as the population expands and average footprints increase in industrialising nations. (For ecological footprint by region of the world, see Figure 62.)

Even if we could stop all greenhouse-gas emissions instantly, due to their long lifespan, we would still have to cope with the effects of climate change. Unfortunately, the failure of the world's governments to formulate effective measures to stop emissions makes the situation worse. For the nations of the WCSE, the damage to their economies and infrastructure will exacerbate the decline of the US and increase the period that Europe spends in legacy and reformation. For the poorest of the emerging nations, which lack the resources to adapt, there will be wars over water and food. For the richer emerging market nations, like India and Brazil, there will be economic consequences that will drag on their growth, but they can be expected to adapt and still grow.

The effects of climate change on China could well catalyse greater aggression in its efforts to expand to empire status. Indeed, as it becomes the strongest nation on Earth, there is a risk that China will choose self-interest over global considerations, taking what it needs to fuel its expansion from nations that cannot defend themselves. It has yet to be seen if China will instead choose to join the other nations of the world in finding solutions to the enormous problems that collectively face the planet before the effect on our ecosystem is totally catastrophic.

Ecological Footprint (Number of Global Hectares), 2006 *(fig. 62)*

Data source: Global Footprint Network. Australia, Brazil and others without data are included in the regional totals.

FUTURE

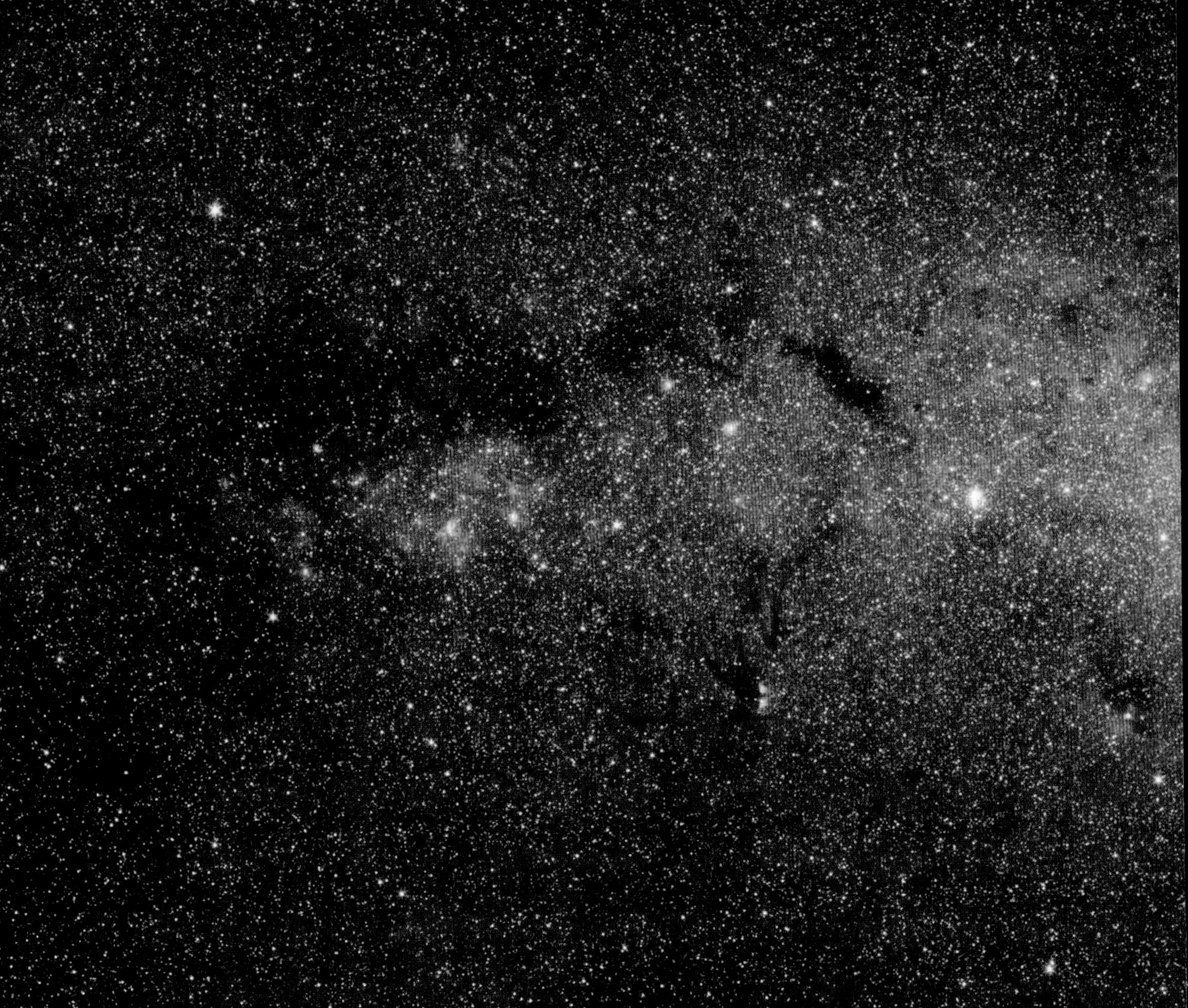

Chapter Eleven

Our Brave New Multi-Polar World?

PREVIOUS **The Spitzer Space Telescope's infrared cameras penetrate the dust that usually hides the centre of the Milky Way, revealing the stars of the crowded galactic centre region, which is some 26,000 light-years away. At that distance, this image spans approximately 900 light-years.**

At the turn of the twenty-first century, the notion that within a decade the seemingly indomitable power of the US and Western Capitalist democracy will be in decline would have been unthinkable throughout most of the world. However, the power shift from the West to the East is now only too apparent, and the indications are clear for those who choose to see them that the rate of this massive geopolitical shift will now only accelerate and its consequences exceed anything most people can perceive. Yet being able to entertain a vision of this future is not enough; we also need an insight into its mechanisms.

In this section I shall consider not only what can happen but also how it might happen. Using an understanding of the principles of the empire cycle, we might choose to navigate the rough waters of our unconscious collective behaviour in order to arrive at a new geopolitical paradigm.

Imagine, if you will, two tectonic plates of enormous scale that have, over centuries, built up tremendous forces at their interstices. They appear immobilised and inert until, without warning, all of that pent-up energy is released in less than a minute. The result, of course, would be an immensely destructive earthquake. Such tectonic stresses are analogous to geopolitical tensions, and since 2000 we have been witness to the greatest build-up of such tensions in history – not merely between empires but, for the first time, between the world's first two super-empires: the declining Western Christian Super-Empire (WCSE) and the burgeoning Asian Super Empire (ASE).

Given the six geopolitical drivers that were discussed in PRESENT – multi-polarity, commodities, polarisation, global military balance, disease and climate change – as the West-to-East transition unfolds, I believe that it is inevitable that, unless addressed, the geopolitical tensions caused by Chinese expansion and the competition for resources will bring the world to a third global war. This will happen not at some indefinable date in the future but by 2025, as the current commodity cycle reaches its peak. Indeed, in many ways, the second decade of the twenty-first century bears an uncanny resemblance to the ten years preceding World War Two. The question is, faced with such a potentially disastrous future, what steps can be taken to change the patterns of human behaviour to prevent the occurrence of another world war?

Sun Tzu and *The Art of War*

The world we live in is being changed beyond recognition by the increasingly rapid growth of China. In this process of transformation, the Chinese have an important advantage over the West. While they have a complete grasp of Western cultural values on account of the last two centuries of global dominance by the WCSE, the converse is not true: the West is trailing woefully behind in knowledge of its rival. The West, blinded by its ideal of the primacy of democracy and the Capitalist model, fails to recognise that China's culture, arguably the oldest and most sophisticated in the world, has its own primacy in terms of its appreciation and application of strategy. Of particular and chilling import for the West, given China's recent rise as a challenger on a global scale, is its historical and philosophical understanding of the art of war.

The Art of War, a book of the sixth century BC written by the great Chinese military tactician Sun Tzu, is a masterpiece of strategic thinking. It remains as valid and eminently useful today, in both its analytical and predictive capacities, as when the general himself fought his battles. Any book concerning itself with geopolitics is by definition a book on strategy. *The Art of War* is designed to encourage the most effective thought processes when considering a situation from a strategic point of view. It is invaluable in understanding thought processes in China, which, as has been said, are very different from those in the West, but, more importantly, it also offers great insight into the new geopolitics of our day. Its many maxims illustrate this point.

> You see the opportunity for victory; you don't create it.
> *The Art of War*, 4: 1.9–10

Opportunities can only be taken advantage of when they materialise; they cannot be contrived. This point underscores the importance of conserving energy until the right opportunity is recognised, then using maximum resources to ensure that it is fully exploited. In many ways, this concept runs counter to Western ways of thinking, which tend to the view that opportunities can be manufactured by sheer force of will, against the flow of events in larger scale. China's progress from regionalisation to the cusp of empire has taken place in the vacuum created by the decline of the US. Before this time, the country was unable to impact or affect the empire cycle of the US, and challenging the US too early would have been disastrous for the People's Republic. All it could do was ready itself and patiently wait for the opportunity to present itself.

> You must know the battleground. You must know the time of battle.
> You can then travel a thousand miles and still win the battle.
> *The Art of War*, 6: 6.1–3

Awareness and understanding are paramount: to be able to act effectively, we must have complete knowledge of where we find ourselves. In this respect, by fully comprehending the six geopolitical drivers that I discussed in PRESENT, we can understand the world as

it evolves in a new direction and thereby identify well in advance those points at which conflict is most likely to erupt.

> We say: know the enemy and know yourself. Your victory will be painless. Know the weather and the field. Your victory will be complete.
> *The Art of War*, 10: 5.14–18

LEFT **Sun Tzu, the Chinese military general, strategist and philosopher, who has had a significant impact on Chinese and Asian history and culture. During the nineteenth and twentieth centuries, *The Art of War* grew in popularity and his work continues to influence both Asian and Western culture and politics.**

The key to remaining strong in today's world is to realistically appraise it in all its aspects. In this respect, self-knowledge is as important as knowledge of one's enemies. The US attitude has been coloured by the hubris of empire and the psychology of decline, and as a result the country is at an acute disadvantage compared with the relative clear-sightedness of China.

> You must be creative in your planning. You must adapt to your opportunities and weaknesses. You can use a variety of approaches and still have a consistent result.
> *The Art of War*, 8: 2.1–4

China has sought a range of soft and hard strategies with which to challenge the US, each one gently probing for weaknesses that can be exploited over time to further China's expansion and extend the contraction of US power.

> You can kill the enemy, and frustrate him as well.
> *The Art of War*, 2: 4.6–7

Competition with one's enemy can take many forms, a multitude of which exist prior to exercising the ultimate option of war.

> The trees in the forest move. Expect that the enemy is coming. The tall grasses obstruct your view. Be suspicious.
> *The Art of War*, 9: 4.7–10

> You must make use of war. Do not trust that the enemy isn't coming. Trust your readiness to meet him. Do not trust that the enemy won't attack.
> *The Art of War*, 8: 4.1–5

> You must use surprise for a successful invasion. Surprise is as infinite as the weather and land. Surprise is as inexhaustible as the flow of a river.
> *The Art of War*, 5: 2.4–7

Never in history has an economic power with an expanding trading system failed to militarise its power, and then proceed to use it. These passages from Sun Tzu's text, old as they are, can be read as a clear warning to the West of Chinese intentions.

> You need all five types of spies. No one must discover your methods.
> You will then be able to put together true pictures.
> *The Art of War* 13: 2.7–10

As I shall discuss below, the 'copy and assimilate' process so necessary to China's recent phase of technological catch-up with the West has been driven by extensive espionage, particularly in the area of cyberspace.

> If you are too weak to fight, you must find more men. In this situation,
> you must not act aggressively. You must unite your forces, expect the
> enemy, recruit men and wait. You must be cautious about making
> plans and adjust to the enemy.
> *The Art of War*, 9: 6.1–5

> This fact must make a wise leader cautious. A good general is on guard.
> Your philosophy must be to keep the nation peaceful and the army intact.
> *The Art of War*, 12: 4.18–20

As it declines, the US must protect its resources, cherishing and saving them for use only when and where they can make a decisive difference to its geopolitical standing.

> You must ask: which government has the right philosophy?
> Which commander has the skill?
> Which season and place has the advantage?
> Which method of command works?
> In which group of forces lies strength?
> Which officers and men have the training?
> Which rewards and punishments make sense?
> This tells when you will win and when you will lose.
> *The Art of War*, 1: 2.3–11

Western beliefs about how people should be governed, particularly that democracy is the most evolved political system, fails to recognise that at a time of constricting resources, a centralised government, such as that in China, might actually be the most appropriate system.

> You will find a place where you can win. You cannot first signal
> your intentions.
> *The Art of War*, 1: 4.17

Again, the critical strike must only be launched when the outcome is absolutely certain; patiently waiting out one's time is key. The Chinese are masters of this discipline.

> If you exhaust your wealth, you will quickly hollow out your military.
> *The Art of War*, 2: 3–5

The US's financial overextension and debt structure mean that it will now struggle to support its massive military complex, only tilting the shift in power further to the East.

> Politicians create problems for the military in two different ways.
> Ignorant of the army's inability to advance, they order an advance.
> Ignorant of the army's inability to withdraw, they order a withdrawal.
> We call this tying up the army. Politicians don't understand the army's business. Still, they think they can run an army. This confuses the army's officers.
> *The Art of War*, 3: 4.5–11

The politicisation of the West's armed forces, particularly in the US, has greatly interfered with the way its generals fight wars. The US military has been weakened as a result. China is not likely to fall into such a trap.

> You can fight a war for a long time or you can make your nation strong.
> You cannot do both.
> *The Art of War*, 2: 1.25–26

> Small forces are not powerful. However, large forces cannot catch them.
> *The Art of War*, 3: 3.19–20

As we have seen, the US's wars in Iraq and particularly Afghanistan have had a dramatically weakening effect on it.

> Use a cup of the enemy's food. It is worth twenty of your own.
> Win a bushel of the enemy's feed.
> *The Art of War*, 2: 4.2–5

> Fight for the enemy's supply wagons. Capture his supplies by using overwhelming force. Reward the first who capture them. Then change their banners and flags. Mix them in with your own to increase your supply line. Keep your soldiers strong by providing for them. This is what it means to beat the enemy while you grow more powerful.
> *The Art of War*, 2: 4.8–14

Such strategic considerations underlie China's centralised resource strategy, while the growing scarcity of resources and paucity of advance planning in the West will only strengthen China's hand in the coming decade.

As I have shown throughout this book, since the 1990s the Chinese leadership has demonstrated a high level of strategic thinking that can be said to exemplify the maxims quoted. Assiduous application of these principles (and others found in Sun Tzu's work) has greatly advanced China's strategic position far in excess of its leaders' initial expectations, to say nothing of the expectations of the West.

The Lessons Thus Far

There is no longer a global hegemony regulating the pace of geopolitical change. Although the US, as a dominant if declining superpower, for the time being remains the most technologically and militarily advanced nation in the world, as we have seen, its standing has diminished drastically, and it faces mounting competition from China and elsewhere. As we arrive at a vision of the world as it may become, it is worth recapping the position of the waning American Empire at the end of the first decade of the twenty-first century.

As Communism collapsed, the US matured under George H. W. Bush (Sr) and Bill Clinton. The former presided over a decisive war that showcased US power to the world, and the latter's tenure was characterised by moderate foreign policy. However, neo-Conservatism and Christian fundamentalism were latent, and they flowered as the question of what the US should do with its singular power became more contentious. Under the administration of George W. Bush (Jr), US politics acquired a hubris that rendered unimaginable any notion of a serious challenge to its empire, and the US assumed a proselytising role with regard to the spread of American-style Capitalist democracy in the belief that it would become the new world order.

Meanwhile, China resolved to grow slowly in the shadow of US power. By 11 September 2001, it had already been active for five years in the vacuum left by US foreign policy, securing long-term commodity contracts across the world to ensure an adequate nourishing of what it knew would be no less than a Second Industrial Revolution. China conducted this subtle expansionist strategy for nearly a decade before the US became aware of any potential threat to its interests.

As we have seen, Sun Tzu held as essential the philosophy that opportunities cannot be manufactured but must be awaited, recognised when available and then acted upon. Thus nearly fifteen years after embarking on its visionary strategy to establish secure sources for commodities, China is witnessing the pendulum of power shift from the US with unimaginable speed.

Meanwhile, in the Islamic world, the US must contend with the challenge to its interests from the sectors of society that have undergone radicalisation, and from Iran, with its increasingly expansionist ambitions (greatly aided by American foreign policy miscalculations in the Middle East).

ABOVE A trader on the floor of the New York Stock Exchange tries to absorb the shock of the events of 15 September 2008. During the afternoon's trading, the Dow Jones Industrial Average fell by more than 500 points as news broke that Merrill Lynch was selling itself to Bank of America Corp, Lehman Brothers had filed for Chapter 11 bankruptcy protection and insurance giant American International Group was approved to secure capital from itself.

The response of the US to the 9/11 attacks – namely, the launching of disastrous wars in Afghanistan and Iraq, its involvement (both covert and overt) in distasteful and illegal practices with regard to interrogation and detainment, and its diminishing of domestic civil liberties in the name of 'homeland security', have all caused it the loss of the moral high ground as well as its own self-belief – an essential element in the sustainability of an empire.

Overextended and attempting to leverage expensive wars abroad, the US economy came to depend on squeezing every last drop of growth from the development of a complex credit system that allowed the empire to appear economically powerful well after it had ceased to be so. The global economic crisis that erupted in 2007 brought hardship and humility to a nation formerly driven by bullish confidence. The policies of the Bush administration from 2000 were revealed to be as bankrupt as the financial institutions that failed in 2008, Bush's last year in office.

It seems remarkable – although it should come as no surprise to any student of history – that at a time when multiple threats to its status as the only superpower were developing, a single (albeit large-scale) destructive event in 2001 could set the US on a course of action that would accelerate its inevitable decline over the following decade. In retrospect, the period from the late 1990s to the present has represented a lost

WALTER REED
AMBULANCE
RESCUE 29 QUAD

opportunity for the American Empire – one that could have allowed the country to recognise the patterns well ahead of time and manage the gradual change in power dynamics from a position of relative strength. That being said, without the failures of the Bush years, the US might not have arrived at the conclusion that such a transition (of power from West to East) was inevitable.

Much as a period of rejuvenation follows a forest fire, the US electorate reacted to nearly ten damaging years of neo-Conservative policies, generating the swing of sentiment that powered Barack Obama into the White House. Although the Obama Administration has had success to date as a bearer of high principles and commitment to change, it has not solidified its power base and translated its ascent into unequivocal success, either at home or abroad. Obama understands the possibility of changing the role of the US in the world for the better, using not just 'hard' (i.e. military) power but 'soft' power as well. He understands too that we have to face the challenge of building new kinds of global relationships to avoid the great and numerous dangers rushing towards us. 'People the world over have always been more impressed by the power of our example than by the example of our power,' said Bill Clinton during a speech at the 2008 Democratic National Convention, which saw Obama's presidential campaign enter its final and decisive phase. Indeed, this sentiment neatly sums up the new administration's approach to geopolitics, focusing on regaining moral credibility and global leadership. However, it is still a work in progress and there is much to be done before the US under Obama can establish these new values decisively. Multiple crises present themselves daily, and the national mood as well as that of the wider world remains confused and angry.

Key to the future success of this change in the American Empire will be the response by nations representing a threat to the US: tensions can be avoided as long as they perceive this new approach as one of not weakness but mutual opportunity. Democracies have been underestimated by expansive nations before, usually to the latter's detriment. It is essential, therefore, that the US remains the globe's supreme military power, the better to ensure that the evolution of the new multi-polar world proceeds in relative peace. The US's forty-fourth president would do well to abide by the words of its twenty-sixth, Theodore Roosevelt, 'Speak softly and carry a big stick.'

The 'soft' values that can be expounded by the US would be best linked to no less than a renewed commitment to the values of the country's Founding Fathers, towards not only Americans themselves but also the wider world, declaring freedom, security, tolerance and justice for all. One of the most discouraging messages of the Bush years, articulated most adamantly by former Secretary of Defense Donald Rumsfeld, was that the US did not have any stake in nation building. Today, and for the foreseeable future, the US must become expert in nation building – that is, using its capital resources wisely to help to improve the economies of emerging nations, particularly in Africa. In this way, the bonds between nations evolving within the new geopolitics may be strengthened and secured.

LEFT **The then-Democratic presidential nominee, Senator Barack Obama, speaks during a campaign rally at Veterans Memorial Arena, Jacksonville, Florida, on 3 November 2008 on the eve of election day. The polls correctly showed that he was leading in the race against the Republican nominee, Senator John McCain.**

BOTTOM LEFT **A 200 ft gash in the Pentagon's south side as a result of the crash of American Airlines Flight 77 on 11 September 2001 in Arlington, Virginia. The plane, which took off from Washington, DC's Dulles Airport, was flown into the building by *al-Qa'ida*-affiliated hijackers, resulting in 189 deaths.**

Conflict Management and the Consciousness of Empire Cycle

It is the prevention of conflict, according to Sun Tzu, that is essential to sound military strategy: war itself is a last resort, with high costs attached, although in most cases it is unavoidable. This perspective is particularly critical for older nations with limited national energy that need to deploy maximum resources to prevent conflict. One sound protective measure is to establish strong diplomatic and intelligence corps that are not preoccupied with only one aspect of geopolitics – not, in other words, beholden to a single overarching doctrine, their vision narrowed by the dictates of ideology and unable to see the big picture in its entirety. Sound diplomacy and intelligence watch for, anticipate and adapt to new threats, and they keep a leadership properly informed early in the polarisation process and are primed to develop appropriate coping strategies.

However, should conflict become unavoidable, the wisest response is to wield the full weight of national resources in order to ensure the swiftest possible closure to the conflict with a minimum of casualties. Such a formulation sounds simple but is rarely executed; it requires intelligent planning as well as political and military agreement on the objectives required for success.

The British involvement in Iraq is a good example of how not to fight a war. The country entered the conflict without a clear and honest rationale with which it could justify doing so to the British people. As a result, lacking the widespread sense of righteousness required to fight, national feeling quickly turned against the war. The British treasury consistently failed to provide the military with the proper funding that it needed to prosecute the occupation of southern Iraq, to the point where troops were insufficiently equipped to protect their own lives, let alone the local population. The British army was thus prevented from a quick prosecution of the war and a successful outcome, before the various militias and insurgents gathered momentum. Its position became untenable, ultimately forcing a humiliating withdrawal and leaving Iran in *de facto* control of the south. In short, Britain's leadership did not demonstrate sufficient intent to win the war. The same mistakes are being repeated in Afghanistan, sending signals to potentially aggressive nations and stateless forces that Britain (and Europe by extension) is toothless in geopolitical terms.

Consciousness of the way the empire cycle functions can prevent Armageddon from taking place in the coming decade. Moreover, it is a higher collective awareness of the patterns of empire throughout history that must be developed, not merely within the leadership but also among the general populations of the world.

A transfer of power from the US to a wider group of nations must transpire. This shift can take place either with much pain or cushioned by awareness followed by action. Either way, the outcome is as unavoidable as inevitable old age and death for any individual. Far better, though, to manage the process with conscious, constructive control than to repeat the destructive patterns into which societies typically fall as an empire moves into decline, creating a contested power vacuum that results in yet

ABOVE The official photo at the Commonwealth Heads of Government Meeting at the National Academy of Performing Arts in Port-of-Spain, Trinidad and Tobago, on 27 November 2009. The event is held every two years, bringing together Commonwealth leaders to discuss key global issues, policies and initiatives.

another form of supremacy. The cycle then repeats itself again, the new empire rising and inexorably once more heading towards eventual decline.

The more enlightened process that I am recommending would inevitably be led by the older developed nations, spreading by example to the younger, emerging nations of the world. The objective is nothing less than the development of a new, pan-global civilisation to replace the game of empires.

Lest this concept appears too far-fetched, it should be remembered that behavioural dynamics only make quantum leaps when people are confronted by extreme challenges and forced by circumstances to change. This is often the case for individuals but occurs on the collective level as well. Humanity now finds itself facing just such challenges, wherein we either change our ways and evolve, or suffer the catastrophic consequences of continuing to enact the unconscious patterns of the empire cycle.

By way of analogy, imagine a corporation in which an older executive is challenged by a young, rising individual. The former has more experience but less energy, while the converse is true of the adversary. A difficult economic climate raises the tension and each feels that their job is at stake. Two outcomes are possible. In the first, the pair openly engage in hostile competition, the energy of the younger rival most likely outlasting that of the older. In the second, the older executive, observing the potential for conflict, takes the younger under their wing, sharing the benefit of their experience and values and, most importantly, empowering them. The energy that would have been lost through competition and conflict is instead channelled into positive and constructive cooperation. There is a gradual, but managed, handover of power to the young challenger. In the terminology of game theory, it is a 'win–win' situation.

The US can learn much from the devolution of the British Empire into the Commonwealth of Nations. This process left fledgling former colonies with sufficient infrastructure to become independent, yet offered them a role as part of a greater intergovernmental system. This concept marked a significant departure from most empire collapses throughout history. Although the American Empire has not been based on the colonial model and comprises military and economic spheres of influence, the principles are similar. It is imperative that the US seeks to empower the emerging nations within its sphere, building partnerships between equals as these nations mature.

Once again, it is only with the development of consciousness that such an outcome can come about, in conjunction with the will that allows for radical new approaches. Enough people in both developed and emerging societies must arrive at a new awareness of their times, acknowledge the changes needed and resolve to work together, demanding new, more conscious strategies from their political leadership.

The Eastern Orientation of the New Multi-Polar World

To institute a win–win relationship with regard to the rising power of the China-led ASE, the US must begin with a realistic appraisal of its capabilities and resources alongside an in-depth awareness of the stage of empire in which it finds itself. Unfortunately, recognition that the decline of the American Empire is in progress has come a decade too late. However, although it is has been economically weakened, the US military remains a dominant force that can be used as a bargaining chip with which to facilitate a peaceful power shift to a new, pan-global political system. Next, the US must identify the three main strategic challenges that it faces, and then develop appropriate ways of dealing with Russia, as well as with Islamist elements, allowing it to focus on containment and political integration measures with respect to China.

RUSSIA: LEGACY

Spread over a massive territory with ample resource wealth but poor demographics, historically prone to paranoia and saddled with a poorly redeveloped military, Russia should see the advantages of making peace on its western borders in order to be able to remain steadfast against China in the East, which, as it expands, might find Russian commodities a tempting target.

The role of the North Atlantic Treaty Organisation (NATO) is, as ever, an important element in Western-Russian relations. Formerly a defensive hedge against Soviet power projection into Western Europe, the alliance's objectives post-1991 were rather subverted by US expansionism, as well as by the offers of membership that it extended to nations formerly within the USSR's sphere of influence (and within the USSR itself) but which, in reality, could never have been effectively defended by NATO. Russia understandably regards such manoeuvring as aggression, and the damage that has resulted from NATO expansion needs to be healed in order to reassure the country. Yet the potential integration of Sweden into NATO would significantly strengthen the alliance's northern flank and put pressure on Russia to consider a new relationship with Europe. (Sweden's respected Saab-manufactured Gripen fighter jets, after all, are based only 150 miles from St Petersburg.)

The US must attempt to reintegrate Russia into the Western family, and treat it as a valued ally. In so doing, it must accord the country the political respect that it craves. Most importantly, the US must time such an overture with the commodity-cycle trough that began in 2007. The longer it delays, the greater the inevitability that it will face a belligerent Russia when oil prices rally once again, rendering negotiations impossible. Similarly, Europe needs to reduce its dependence on Russian gas in order to neutralise the monopoly that Russia stands otherwise to acquire, making it more flexible in its view of its relationship with Europe. The growth of China on Russian borders must be viewed with concern by the US. If Russia and the US do not progress beyond sparring with and baiting each another, China will be able to drive a wedge between the two countries. Russia must therefore be courted back into the fold as soon as possible. As unlikely as this prospect may appear to some observers, Russia and the US share common ground as elder nations informed by Christian values – both key players in the WCSE. Analogous to Britain and France at the turn of the twentieth century, these two countries must carefully address all outstanding political issues and resolve them diplomatically. Making Russia an ally would permit the rededication of vital military resources to the containment of Chinese military expansion, and thereby confirm Europe as an uncontested region.

THE MIDDLE EAST: LATE REGIONALISATION

The Middle East represents a vital strategic area for the US on account of the oil resources, at the command of various states, which are needed to feed the US's insatiable energy requirements. Any serious measures taken by the country to transit away from oil dependence by developing new energy sources would immediately alter the strategic value of the Middle East, although the US would no doubt maintain its alliances with the *Sunni* Arab states as well as with Israel, and its relations – cool at best – with Iran. Israel and Iran have made no secret of their mutual hostility, and the *Sunni* Arab states have long regarded Iran as a threat. China, on the other hand, has worked hard to cultivate Iran in return for future oil and gas supplies. The region is therefore at risk of becoming the flashpoint for a future proxy war between the US and China.

The US will need to be seen more aggressively urging its satellite Israel to come to a workable agreement (acceptable to the Arab world) with the Palestinians in the Israeli Occupied Territories. Doing so would go some way towards weakening Islamist polarisation. Israel, with its nuclear arsenal (one of the region's worst-kept secrets), is also vulnerable as US power-projection capabilities ebb around the world – so the quicker it achieves a *rapprochement* with its neighbouring states, the better for all concerned.

Islamist extremism can also be countered to some degree at least by the encouragement of greater political freedoms and a healthy middle class in such states as Egypt and Saudi Arabia, two countries with particularly warm ties to the US. However, the longer the Middle East remains the primary focus of US foreign policy, the more China's rise will continue unabated.

RIGHT **Supporters of the Islamist-rooted AKP wave flags in front of a banner of Turkish Prime Minister Recep Tayyip Erdoğan during an election rally that took place in Istanbul on 22 March 2009. Despite the country's deepening economic crisis, the party received a vote of confidence in local elections.**

TURKEY

Turkey, a prominent US strategic partner since World War Two, must also remain a key focus for the US. Its population is 70–80 percent *Sunni* Muslim, although its political culture has been resolutely secular since the 1920s and this remains intact if under challenge. It is expanding demographically with a youthful population that has driven it economically, and that will continue to do so. For much of the time since 2002, the ruling Justice and Development Party (AKP) has struggled against strong opposition on the grounds that it has maintained a soft Islamist hidden agenda. Turkey makes a natural ally for the US as well as for Europe – its geographical location, for one thing, makes it vital to European Union (EU) security. Nevertheless, the European Community has been reluctant to grant accession to Turkey despite the fact that its application dates back to 1987. This delay could prove to be a serious error, because a more assertive and confident Turkey in the coming years might well choose not to join the EU, looking eastward instead.

During the Cold War, Turkey was a close NATO ally and stood with the US as a southeastern bulwark against the Soviet Union. This special relationship began to erode during the Second Gulf War in 2003, with the US invasion of Iraq. Turkey would not allow the US 4th Infantry Division to land on its soil, thereby thwarting a plan to attack Iraq from the north and keeping this vital and technologically advanced division out of the conflict at a critical stage. Relations since then have been uneven, influenced by Turkey's anxious eye on the close relationship that the US has enjoyed with the Kurdish Regional Government of Iraq. The Iraqi Kurds, long oppressed under the regime of Saddam Hussein, proved themselves to be the one stable ally upon which the US could depend during the war, and Iraqi Kurdistan has developed quickly since 2003, independent in all but name. Turkey, on the other hand, has long struggled with its Kurdish minority. Its Kurdistan Workers' Party (PKK), which is considered by several countries (including the US) to be a terrorist organisation, has used Iraqi territory to launch attacks into Turkey. Along with Syria and Iran, Turkey fears the PKK's irredentist leanings towards a Greater Kurdistan, encompassing Iraqi Kurdistan as its centre together with territory within these other countries' borders.

It should not come as a surprise that Turkey's power base is in the ascendant as US power declines. For the greater part of the past 1,700 years, Istanbul has been the centre of regional power as an imperial seat, Christian or Muslim. Modern Turkey has grown steadily in regional power, driven by rising demographics that have helped it to attain the largest GDP in the Islamic world. The last ninety years, then, should be considered an aberration in Turkish power projection; the country is once more moving to exert regional influence, particularly as US power in the western Middle East declines.

AK PARTİ
AK PARTİ
AK PARTİ
PARTİ

Moreover, Turkey's glorious imperial past endows it with a perspective on the Middle East that far exceeds the limits of its own borders. Thus it could be well placed to balance Iranian expansion in the future. It is also a force for regional stability: the country maintains good ties overall with Iran, the Arab world and Israel, despite occasional periods of greater or lesser neutrality and criticism of the latter. This unique position goes some way towards accounting for the greater regional role that Turkey enjoyed in the Spring of 2010, when, in conjunction with Brazil, it succeeded briefly in making Iran less tractable on the nuclear question. It is highly likely that Turkey will remain an ally to the US and the EU (not least through its membership of NATO), while cultivating influence in Iran, Iraq and elsewhere in the Middle East.

In June 2010, Turkey took the lead in confronting Israel over its Gaza blockade, and about the ill-fated Israeli raid on a ship of Turkish provenance carrying supplies to the Palestinian territory as part of the 'Gaza flotilla', when nine Turkish activists were killed. Israel's conduct during the raid not only polarised Muslim sentiment in Turkey but also boosted the country's leadership position and credibility in the Islamic world.

Much as at the turn of the twentieth century Britain empowered US influence over Latin America, the US would do well to consider performing an equivalent role for Turkey with respect to western Asia as a whole.

IRAN

Iran, on the other hand, represents the most organised Islamic challenge to US power in the Middle East. Its aspirations to extend its influence and rebuild its power base are perfectly understandable, given its imperial history and its impressive size and demographics. While the country was never colonised, serious and repeated intervention in its affairs by Russia, France, Britain and the US in the nineteenth and twentieth centuries defused any possibility of its attaining more than second-rate status in the shadow of the West. Seen in this light, the 1979 Islamic Revolution constituted a reawakening of the Iranian will to self-determination. As such, it was initially supported by many progressive elements in Iranian society who were less attached to the 'Islamic' part of the equation than the 'Revolution'. This marked the first stage of Iranian accession to the status of regional power.

The revolution soon hardened, and moderate voices among the leadership have found themselves shut out or persecuted for the better part of three decades. Iranian paranoia is also acute, and with some justification: the country borders a nuclear-capable Pakistan to the east and is surrounded by US-occupied Iraq and Afghanistan.

Iran has the largest military capability in the Middle East, comprising some 545,000 active troops and 350,000 reservists, and it continues to modernise its forces. As a means of cultivating its power using asymmetric tactics, Iran has also supported Islamist extremism, and has focused both its anti-US and expansionist energies through hostility towards Israel, using *Hizbullah* and Hamas as proxies as well as developing closer ties with Syria. (The high price of oil has played a key role in Iran's new geopolitical leverage. It is a major provider of energy to China, Japan, India and other countries.) In Israel, Iran has chosen the perfect foil; it can claim to be taking an ideological lead

ABOVE Priscilla, a 37 kilotonne atomic bomb, is detonated in the Nevada Desert in 1957 during Operation Plumbbob, the most controversial atomic test series in the US. Some 1,200 pigs were subjected to biomedical experiments and blast-effects studies, and 18,000 members of the US forces participated in exercises during the experiment to ascertain how the military would stand up, physically and psychologically, to the rigours of the nuclear battlefield.

against a regional enemy that is widely despised for its occupation of Palestine, in vivid contrast with the Arab *Sunni* states of the Middle East. Iran's strategy appears to be to render Israel the common enemy of the entire Islamic Middle East, uniting disparate 'tribes' (*Sunni/Shi'i*, Arab/Iranian) against a common enemy, as Genghis Khan did, and to begin its empire accordingly.

Iran's pursuit of nuclear power – another stepping stone in its quest for regional supremacy – represents a serious and urgent threat to regional stability. (It is not difficult to perceive that any political language used to the contrary, including the insistence on nuclear technology for peaceful purposes and the *fatwa* issued in 2003 by Supreme Leader Ali Khamene'i against nuclear weapons as un-Islamic, is simply a smokescreen.)

The standoff that had been building up as a result of US demands for Iran to halt its uranium-enrichment programme, and the threat of sanctions, have been ever present for the past few years.

Should Iran eventually succeed in developing nuclear weapons – a likely scenario without a dramatic political breakthrough by the moderate elements in the Iranian leadership, a persuasive economic aid package or a pre-emptive military strike – the Middle East would then face an arms race between the Arab states and Iran. Moreover, if the US fails to defuse Christian-Islamist polarisation, a new wave of Iranian-led radicalisation could sweep up the poor in the *Sunni* nations into a revolution, despite the deep religious divide between the *Shi'is* and the *Sunnis*. The result would be a redirection of westward-flowing oil eastwards to China.

Iranian nuclear weapons capability lies at the heart of Iran's drive to become a regional power. Thus far it has been successful in playing for time, further developing its nuclear and launch technology while stalling diplomatic attempts to counter its activities. Until 2010, Russia was instrumental in supporting Iranian objectives, seeking to dilute US influence in the region and to form a global natural gas cartel with Iran. China, too, supported Iran by neutralising the call for sanctions through its Security Council veto, coupled with billions of dollars' worth of investment in Iran's oil and gas infrastructure. In addition to the obvious prize of Iranian energy, China's strategy may have considered a newly radicalised Middle East of the future, wherein it would be a massive beneficiary of a foreseen Islamist aversion to selling oil and gas to the West.

Iran toned down its nuclear efforts in 2003 when the US unleashed its military machine on neighbouring Iraq. However, when the US became bogged down in a counter-insurgency war both there and in Afghanistan, Iran applied additional pressure on the US through the *Shi'i* militias in Southern Iraq, which reported to Iran, costing American lives and distracting the US sufficiently to enable Iran to resume its nuclear programme. The greater the attrition in Iraq, the weaker the US appeared. This setback for the US, combined with the majority Iraqi *Shi'is* acquiring greater power, the Chinese backing obtained in return for energy security and what looked to be a blossoming Russian partnership, left Iran further emboldened and expansive. The 2006 war between Israel and *Hizbullah* was one result, amounting (along with *Shi'i* insurgency in Iraq) to a series of pilot wars that set the scene for continued confrontation. Iran's approach, in contrast to the West's ineffectual strategies, has been defiant – and successful.

Israel, which bombed Saddam Hussein's Osirak nuclear test reactor in 1981 (shortly after the start of the Iran–Iraq War), has indicated repeatedly for years that it has plans in reserve, backed by training and simulations, for an aerial assault on Iranian facilities that are capable of producing nuclear weapons. (In 2007, Israel destroyed a facility that it claimed – along with the US – was a cache of nuclear materials in preparation for the construction of a covert reactor. Despite initial scepticism on the part of the International Atomic Energy Agency (IAEA) and elsewhere, IAEA investigations in 2008 and 2009 revealed significant traces of processed uranium from site samples.)

However, there are serious difficulties posed by the number of nuclear facilities maintained by the Iranian programme: the distance between them, the efforts that Iran has gone to in order to conceal and fortify them, and the matter of locating an air corridor through either Iraq or Saudi Arabia through which to successfully attack them. This kind of mission may well be beyond Israel's capabilities, requiring the use of the kind of ground-penetrating nuclear weapons that are better suited to US military action. There are notable targets, such as the Russian-supplied reactor at Bushehr and the Natanz centrifuge production centre, but at least ten other sites – enrichment plants, conversion centres, heavy water plants, reactors, etc – are known.

The situation in Iran has two possible outcomes, neither of them comforting:

1. **IRAN DEVELOPS NUCLEAR WEAPONS AND ATTEMPTS TO USE THEM, PROBABLY AGAINST ISRAEL.** This scenario has, to some extent, been countered by the development of a multilayered ballistic defence system designed to defend Israel, which could intercept a limited missile strike in the mid- and terminal phases of its path. In time, this shield could be bolstered by US Aegis anti-ballistic-equipped warships in the Gulf and the Red Sea. (The US is intent on demonstrating its willingness to defend its allies against Iran: in January 2010, the US sent Patriot defensive missiles to Qatar, the United Arab Emirates, Bahrain and Kuwait, and it has already maintained ships in the Gulf that are capable of intercepting missiles.)

2. **IRAN IS ATTACKED BY ISRAEL WITH GROUND-PENETRATING NUCLEAR WEAPONS, OR BY THE US IN A CONVENTIONAL STRIKE.** Such an attack would engender massive outrage and unease in the Islamic world, so the instigators would need to feel certain that their actions would incapacitate and further delay the advent of Iran's nuclear programme for many years. Balanced against the need for certainty in target acquisition is the risk that Iran's covert programme is much more advanced than current intelligence estimates suggest.

In April 2010, Iran announced the development of its own missile defence system. Although its efficacy has been disputed or dismissed outright by Western observers, it was nevertheless intended as a warning to the West, and specifically to Israel, that any incursions into Iranian territory would be met with force. Moreover, the US and Iraq would need to brace themselves against devastating retaliation by *Shi'i* forces in Iraq controlled by Iran, and Israel would need to stand ready to endure not only retaliatory strikes on its cities by Iran directly but also hostilities launched by *Hizbullah* and Hamas.

AFGHANISTAN

As precarious as the situation in Iraq is at present, in Afghanistan it is downright unstable and devastating setbacks there remain a possibility. The threat from the *Taliban* must be neutralised and the best means of doing so is not an increase in military force but real improvement in the lives of the traumatised local population.

It must be clearly understood that Afghanistan is an American pilot war, and the stakes are very high – not only in the context of any threat by Islamic extremists but also in terms of how China perceives the US's intentions, as well as the US's ability to inhibit its own expansionist aspirations.

The 2001 invasion of Afghanistan was initially a defensive war designed to destroy *al-Qa'ida*'s base of operation, then under *Taliban* protection. However, the US did not retain a sufficient number of troops on the ground to stabilise the state once the *Taliban* had been routed. More importantly, it failed to devote sufficient resources to the enormous investment required to lift the country from three decades of war into a new way of life. A huge opportunity was lost, at a time when the Afghans could have been helped onto a path of sustainability and self-determination.

The *Taliban* was therefore enabled to return, renewed, and it has so far resisted all attempts at being dislodged, particularly through its strategic use of western Pakistan's lawless tribal zones to launch attacks. The US response has been to strike in Pakistan's territory, which only further inflames Islamic fundamentalism in that country – a trend that, should it continue, risks complete destabilisation of the region and Pakistan becoming a nuclear power controlled by extremists. A radical Pakistan would then be likely to engage India in war, which in turn could suit China (the spectre of a nuclear exchange notwithstanding), because its biggest regional competitor and valuable member of the US/Asian alliance (with Japan) would be neutralised.

RIGHT Burqa-clad Afghan women beg in Kabul, Afghanistan, in February 2009. It is thought that up to 80 percent of economic activity across the country is undertaken by those people who have neither adequate incentives nor suitable mechanisms to formalise their business activities. Excluding the multi-billion dollars that are made from opium production, it is estimated that the private sector contributes approximately 89 percent of the total GDP.

The seriousness of this position cannot be overstated. The US must urgently employ maximum military and political resources to bring this conflict to a successful conclusion. There are several key imperatives to achieve US success in Afghanistan:

1. The US must show strong political will to win the war, whatever it takes and however long it takes. If it fails to do this unambiguously, the *Taliban* will simply wait it out. The US must therefore deploy the maximum number of available forces with the prime objective of restoring the life of the population to normal parameters, providing it with security, access to medical care and a viable economic framework. The Afghan people must be treated with the greatest respect, and every measure must be taken to ensure the minimum loss of innocent civilian life. Doing so is essential for a successful hearts-and-minds campaign.

2. Troops should be equipped to the highest protective standard so as to reduce the rate of casualties to an absolute minimum.

3. Cultural-sensitivity training should be given to all soldiers operating in the Afghan theatre, so as to avoid alienating the population and maximise the military's effectiveness in identifying suspects.

4. The insurgents must be effectively isolated from their supply sources.

5. The country's international borders must be sealed and monitored.

6. Control zones must be created to limit movement across countries.

7. Where possible, the population must be isolated from the insurgents in order to prevent resource resupply and intelligence gathering.

KBL

8 Each zone must be completely under control before spreading outwards, thus forcing insurgents into smaller areas where they can be neutralised or forced to surrender more easily.

9 Joint deployments must be undertaken with indigenous forces so that the latter can support the occupying forces and eventually take over.

10 The Afghan government must be supported and strengthened to ensure appropriate and sufficient support of the population.

CHINA: ASCENSION TO EMPIRE

The People's Republic of China is the one nation in today's world in the ascension stage of empire. It is expanding economically, politically and militarily at an accelerating rate. The rest of the world can only struggle with the strategic implications of such a rapid rise. As I have endeavoured to show throughout this book, without proper grounding in the momentum behind China's growth, global leaders will find themselves in a perpetual struggle with this change to the world order. China's ascension to empire is inevitable, and, just as the American imperial incarnation differed markedly from Britain's, so we should expect China's imperial status to be manifested in its own unique way. The country has the capacity to drastically alter the way of life for many people across the world, and this power must be taken very seriously indeed, with every possible measure undertaken to acquire a deeper understanding of Chinese culture.

The insatiable appetite for commodities required by China to feed its growth cannot be underestimated. As we have seen, the Chinese strategy of securing resources from nations adrift in the vacuum of US foreign policy has been very successful for over a decade. However, where ten years ago the world had little experience of Chinese values in commerce, today the Chinese meet with increasing resistance from nations that are reluctant to do business with them. (The collapse of a deal in June 2009 between the Chinese state-owned Chinalco and Australia's Rio Tinto mining company is a good mainstream example of the negative perception that China faces in the global corporate world. Shareholders and the Australian population were wary of increased Chinese involvement in the company and country, so much so that they rejected the $19.5 billion that China would have injected into the debt-ridden corporation, which also incurred $200 million in fees for walking away.) Yet, should China become frustrated by its inability to continue amassing the resources that it needs through peaceful means, there is a risk that, like many empires in a similar position throughout history, it will become more aggressive in its acquisition strategies. China's navy already behaves as though the East China Sea is under its jurisdiction, as opposed to being international waters. It is vital that the relationship between the political and military branches of Chinese leadership is understood better in the West – particularly the different mindsets that pervade each one. The People's Liberation Army Navy may well harbour even more aggressive and expansive ambitions than the *Politburo*.

Competition between the US and China is a contest for supremacy not only between national entities but also between two state systems: Western-style democracy and the centralised leadership by which China has always been governed in one form or another. These are two Capitalist contexts in which the state plays either a minimal or a heavy role. As the era in which natural resources were readily available now draws to a close, the new paradigm could favour Chinese centralised resource acquisition rather than free-market dynamics that are not driven by strategic (state) imperatives. Western democracies should counter this by providing substantial tax breaks for specific strategic-investment activities, such as offshore oil exploration and production.

The US has for years been adept at exporting its culture and values across the world, in arts, commerce, education, etc. Its projection of such 'soft power' is perhaps without parallel. China has not previously enjoyed any comparable intimate familiarity on the part of the world at large, and particularly in the US. Insularity, like centralised rule, is another feature of Chinese governance across the millennia. The People's Republic has effectively existed in relative isolation until fairly recently. Yet in terms of longevity and historical influence on the world, Chinese culture dwarfs most others: it carries traditions formed by more than 4,000 years of continuous civilisation. It is now

BELOW **A reclaimer works in Yandicoogina stockyard (owned by the mining and exploration giant Rio Tinto) in Western Australia's Pilbara region, loading high-grade iron ore onto a conveyor.**

in its sixth empire cycle. Critically, Chinese political culture has acclimatised itself to the governance of a massive population over a very great time span, mostly within an imperial context. Moreover, as a natural consequence, the culture of individuality is less pronounced than it is in the West.

Patience is another trait that has served Chinese empires well, from the First Emperor through Mao Zedong and to the People's Republic in the present day. As a result, China is eminently capable of managing its expansion despite the present global US military dominance, until the day arrives when the transfer of wealth has resulted in the creation of a military power befitting a twenty-first century empire. To effectively inhibit this course of events, the US must implement an eight-stage containment policy, as follows:

LEFT The Mutianyu area of the Great Wall of China. Several walls collectively referred to as the Great Wall have been constructed and maintained between the fifth and sixteenth centuries BC. The most comprehensive archaeological survey, using advanced technology, has determined that the entire structure, including all of its branches, stretches a distance of 8,851.8 km (5,500.3 miles).

1 LEARN TO RESPECT CHINESE HISTORY AND CULTURE. Empires in decline have, throughout history, hubristically failed to respect or appreciate the magnitude and capabilities of a rising challenger. The US must not repeat this error. It is vital to 'know one's enemy', the better to contain threats and lay the groundwork for friendship and mutually beneficial partnership. The dramatic contrast between the two cultures – that is, between uncomplicated American straightforwardness and complex Chinese hierarchical operation – may make mutual understanding difficult. Nevertheless, the two civilisations have enjoyed open contact with one another during the last decade that was undreamed of during the height of the Communist era, and there is now greater mutual familiarity than ever before.

2 STABILISE THE US FINANCIAL CRISIS. Although the Obama administration has managed to wrest a certain amount of control over the US economy as the financial crisis progresses, the US remains dangerously overextended. It risks a rapid decline and resultant economic vacuum similar to that suffered by Britain after the Suez fiasco. As I noted earlier, the parallels between Britain after Suez and the US in the present day extend to the crucial fact that, much like the US's relationship with Britain during Suez, China is now one of the US's main creditors as well as its greatest challenger. The point at which this dynamic becomes a weapon of significant power will occur when the Chinese centralised economy becomes large enough to no longer need to depend on the American consumer for its growth. However, before that point, the US and China will continue to be locked in a mutual consumer/producer embrace, China will not risk accelerating the decline of the US, its most profitable market, but will remain a reluctant supporter of the US economy, even as it grows its military capabilities and expands its foreign influence. This relationship can buy the US a small amount of time, which it must use wisely.

ABOVE Stacks of $100 bills pass through a circulator at Washington, DC's Bureau of Engraving and Printing on 14 October 2009. The US has been printing money at an increasing rate during its overextension and decline.

Empires do not die easily, and the US will use all conventional means at its disposal to arrest its decline. The Federal Reserve will do its utmost to maintain the stability of American equity markets, including the maintenance of current equity-to-debt ratios, which perpetuate the illusion that the country's economy is still not just solvent but a powerhouse. The US Mint will continue to print money to shore up the system, expanding the national debt. In turn, the dollar will lose significantly more value at some stage. None of these measures will slow down the decline of the US. To radically change the course of the US economy, courageous, innovative action must be taken. If the country were a corporation with the real debt-to-equity as high as it is, the only practical way to prevent the debt burden from strangling its future growth would be to convert as much of the debt to equity in a swap. In the case of the US, such an equity–debt swap would have to take place across the spectrum, from national to company and household debts. Such a solution would require huge commitment from politicians, as well as difficult decisions regarding how different creditors would need to be treated. Ideally, the majority of the write-off would only take place domestically, leaving international creditors relatively unscathed. There is massive stigma associated with such restructuring, which until now only emerging-market nations, like Russia and Argentina, have had to endure. It would be instructive to recall, however, that the International Monetary Fund had to bail out the UK after the collapse of the British Empire. Were the US to follow a similar path, it would not, therefore, be unprecedented. The removal of the US debt burden would then facilitate new radical tax policies like a flat tax rate of 25%, that would re-energize the American entrepreneurial spirit and economy.

The choice is clear: the US muddles on in a state of denial, leading to gradual economic decline, or it faces the hard decisions and implements radical action to promote future growth, in turn ensuring the stability of the world's financial system. When the time comes for the dollar's inevitable steep decline, economic alliances will again become vital. The support of Japan and Europe to counter the sale of US debt and dollars by China will be necessary in order to prevent global financial trauma.

3 BE PREPARED FOR A WAVE OF CHINESE INNOVATION. Having moved into the stage of ascension to empire, China will replace its strategy of 'copy and assimilate' with one of astounding innovation that will surprise the world and put an end to a common perception of the Chinese as good implementers lacking in creativity. While the US should do all that it can to continue encouraging its own creativity, the once-powerful force of American ingenuity will no longer suffice to compete with China. The US must therefore adopt its own copy-and-assimilate strategy. In order to do so, it must relinquish its hubristic self-image, associated with its past glory and primacy in innovation in technology.

The new phase of Chinese innovation will see many of the global standards in technology and commerce once set by the US being determined by China. A good example is global credit ratings, which were the domain of US companies until 2010, when the Dagong Global Credit Rating Company was formed. Naturally, this new operation views the world from a very different perspective. It downgraded the US and France a notch, and raised China by a similar amount. Such an adjustment is reasonable, given the relative economies of these nations, and it corrects American delusions typical of the decline stage. In addition, the use of Chinese as a global language can be expected to increase, much as the use of English expanded along with the British and American Empires.

4 FOCUS RESOURCES ON CONTAINING CHINA. Depressurising the effects of Middle East polarisation and considering such critical aspects of strategic realignment as bringing Russia back into the Western fold will remove the substantial distractions that have tended to draw US attention away from the rise of China.

5 **DEVELOP A COMMODITY STRATEGY.** As I have already indicated, Chinese strategy has for more than a decade been focused on cornering commodities to feed the nascent empire's industrial growth. This requirement has forced China to augment its presence in the outside world, from Latin America to Africa, the Middle East and Asia. On the other side of the coin, the Chinese commodity grab might be slowed down if China's commodity sources were to be constricted in areas such as Africa. The US must therefore allow the current free-market Capitalist system to evolve on that continent in order to compete with the Chinese policy of long-term resource acquisition. The US has an opportunity to dramatically increase its engagement with Africa, on terms that provide African states with a competitive alternative to dealing with China. Such a strategy would also include developing new, reliable sources of energy, including nuclear power.

6 **COUNTER CHINA'S PUSH-BACK STRATEGY.** China has been able to push back the US's spheres of influence using a strategy of subtle intervention in states that have complicated or hostile relations with the US. This approach has included encouraging North Korea to build nuclear weapons; providing military and technological support to Pakistan; and, until 2010, opposing sanctions against Iran. Direct military confrontation has not been a feature of this policy, and the geopolitical consequences have also acted as a distraction to the US, providing a smokescreen for further Chinese expansion. For the US to counter this strategy, it first needs to recognise that it is taking place.

7 **PAY CLOSER ATTENTION TO AFRICA.** With a current population of 800 million, which is expected to double by 2040, and with thirty countries sustaining growth of over 4 percent, this continent holds the key to the future of the world. The West must recognise the extent to which China has been actively seeking to infiltrate Africa, acquiring the resources to feed its rising empire by offering financing and aid to capital-starved states on exceptional terms. The continent is an excellent investment opportunity for the West in its own right, but more importantly it is a critical strategic arena. If the West could offer financial and economic aid packages that present an alternative to those of China, it would deny China the resources that it craves. An ascending empire without the resources that it requires for continued growth, or the military with which to appropriate them, is a stillborn power. Intervention in Africa could slow Chinese growth, and mitigate the friction of expansion past the critical 2025 watershed in the commodity cycle and Chinese demographics.

8 **FORM NEW ALLIANCES TO COUNTER DIRECT CHINESE MILITARY CHALLENGES.** China's construction of a military complex is proceeding at a significant pace. The US needs to build stronger links with India, Australia, Japan and even Russia in order to limit any military options China might elect to use in the next twenty-five years. A NATO-like organisation is required, repeating that body's success in maintaining peace throughout the Cold War. Notably, in the theatres of war of the future, more emphasis would fall on sea than land power. US objectives should focus on blocking any avenue for Chinese naval advancement. It needs to encourage the integration of China into a new global order. The US must maintain its military supremacy as far as its finances allow, transferring technology to its Pacific and South Asian allies.

9 **EXPOUND A POLITICALLY INCLUSIVE POLICY.** The US must invite China to be an equal stakeholder of power before the latter acquires the ability to take that power by economic and military force. This would entail such initiatives as backing China's integration into multinational bodies like the G8. (China currently sits on what is often called the 'G8+5', the five representing the other leading emerging economies of Brazil, India, Mexico and South Africa.) The US should work hard to make China an equal partner in resolving global issues, such as climate change (China was a late signatory to the 2010 Copenhagen Accord, along with India), disease and food supply, encouraging the signing of bilateral or multilateral agreements. Embarking on such a course could well include accepting the inevitable fall of the dollar from primacy in the global economy as it is replaced by a collection of reserve currencies – a bitter pill for the US to swallow, to be sure. Yet if these two great powers could work together on such challenges as climate change and new energy sources, their combined resources would change the world.

Alliances Old and New

ALLIANCES OF THE TWENTIETH CENTURY

With the world populated by a number of competing powers of similar magnitude, it is only natural that new alliances will emerge in an attempt to establish a balance of power (see Figure 63). We can only hope that their formation will not be the precursor to major conflict, as inevitably it has been in the past (e.g. World Wars One and Two).

In 1890, Britain still regarded itself as the world's only significant naval power, and its principle interest in Europe was to ensure that no single dominant power arose. The Triple Alliance of Germany, Austria–Hungary and Italy was formed to counter any aggression from Britain, France or Russia.

In order to compensate for the rapid rise of the new global powers of the US and Japan and a resurgent Germany, Britain sought a series of strategic redeployments from 1898 to 1907. First it made a pact with Japan that was renewed in 1905 and 1911 and that held firm during World War One, which was designed to balance France and Russia in the Far East. In 1905, anxious about Germany's ascendance, Britain and France put aside their historic and geopolitical differences by systematically working through their points of friction. This Anglo-French *Entente Cordiale* would expand to include Russia – tenuously at first, then later cemented during World War One – because the British needed to secure their oil supply from Iran for its new navy. In 1907, Britain ceded control of the Americas to the US, putting aside almost a century of enmity since the War of 1812 to become allies for the following century.

This strategic realignment of the globe meant that Britain could redeploy its navy to areas that it considered under the greatest threat (i.e. where Germany was strong). However, in the process, it reduced the total number of cruisers and smaller ships it had that were available to visit foreign ports and show the flag. The beginning of the end of the British Empire was thus inaugurated.

This salutary tale is pertinent to the US's developing alliances with India and Japan. Among any three allies, internal power plays will always exist. In the current scenario, India – the youngest – seeks to rise above its partners as its power grows, much as the US did with Britain from the 1890s to the Suez Crisis in 1956, when it finally became the dominant partner.

NEW ALLIANCES OF THE TWENTY-FIRST CENTURY

Geopolitical power in the run-up to the twenty-second century will be split into three key blocs: the centralised powers, the democracies and the neutrals.

THE CENTRALISED POWERS

What I call the centralised powers are those states that seek to revise the current world order and are the motivators of change. China is at the heart of this trend, steered by a centralised government, economy and military. It is driven by expansion and, as we have seen, is able to summon huge economic funds with which to secure access to natural resources from commodity-producing countries. Its recruitment strategy can be summed up by the well-worn phrase 'the enemy of my enemy is my friend'. In this case, the principle perceived enemy is the US.

The road to conflict starts when states turn their backs on each other as allies. Since the collapse of the Berlin Wall, the West has taken two diplomatic courses in relation to Russia. The first, spearheaded by the US, continued to press its advantage and expand NATO's hold on the borders of Russia, triggering historical insecurities and wounding Russian pride. The second, led by Germany, sought to integrate Russia into a European system for fear that the consequence of snubbing it would be to force it eastward. Germany has worked hard to engineer strong links between Russia and the EU, but its efforts have been overshadowed by the actions of the US. Russia is turning against the West in a significant step backward towards the Cold War.

Alliances for the New World Order *(fig. 63)*

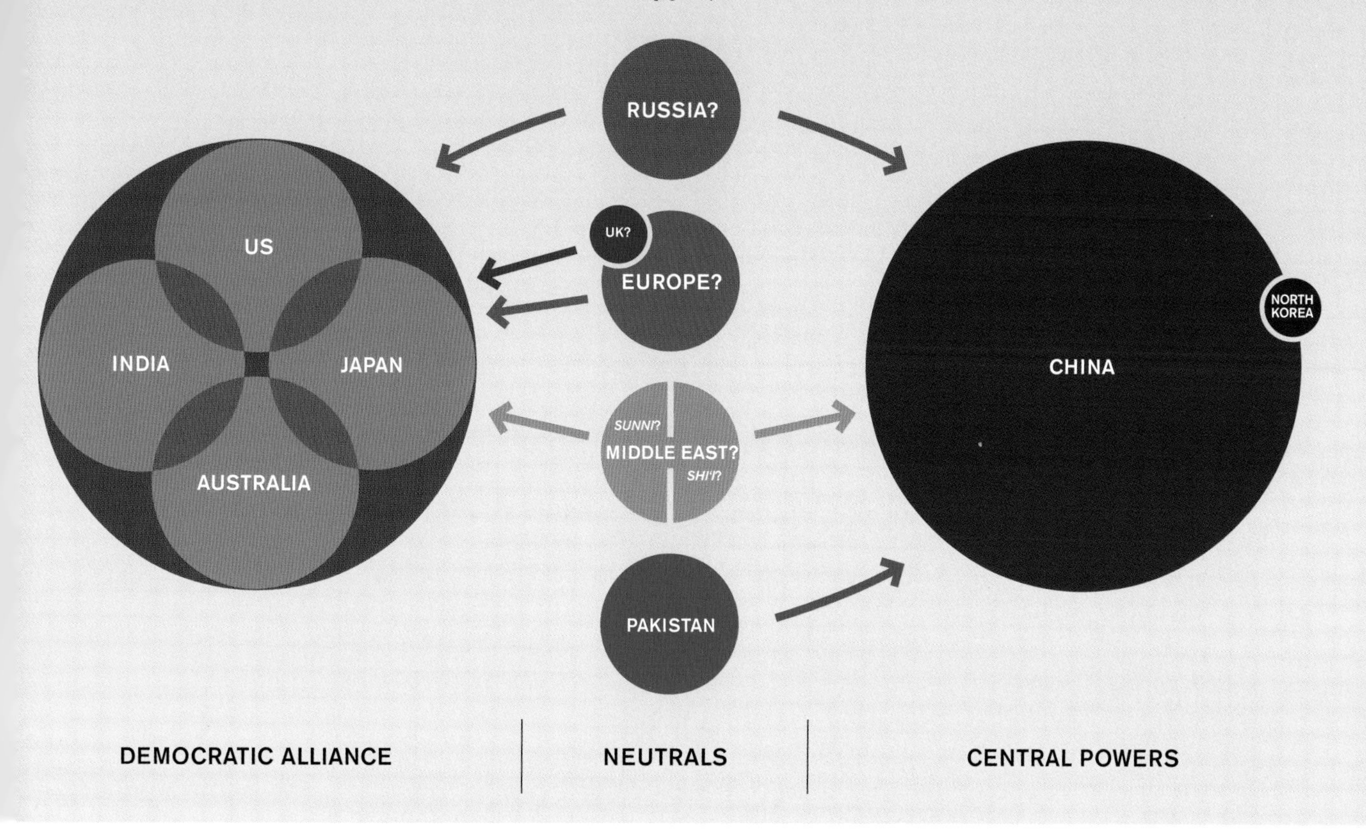

As Russia moves further from alignment with Europe and actively against the US, it will be drawn to the prospect of stronger ties with China. Although there is great mutual distrust between these two states, and while initially it may not suit China to show its hand too early, the countries do have complementary profiles: one is long on people and short on resources; the other the precise opposite. Both countries also have centralised governments and a history of strong political links. Only a growing Russian relationship with Germany might counter the lure of China for Russia.

China's secondary threat is India – the only country that matches its manpower resources. China has thus fostered ties with Pakistan that will only become stronger as that state's relations with India and the US worsen in the face of mounting radicalisation and fallout from the war in Afghanistan. China might attempt to encourage the growth of Islamic fundamentalism directed towards the West, and also the growth of anti-Indian sentiment. Such a strategy would encourage the removal of US influence from Pakistan. The global financial crisis has had a profound influence on Pakistan's economy, and it is China, not the US, that is looking to shore up Pakistan (no doubt at a price). China has also sought to ensure that Pakistan's nuclear arsenal balances that of India, and it is currently supporting the Pakistani nuclear programme in its development of plutonium-based weapons that are smaller and better suited to a variety of missile-launch vehicles.

A centralised alliance would therefore consist of China, Russia and Pakistan. Given that these states are all are linked by overland communications networks, it would be predominantly a land-based alliance, although it could be expected to pose a significant naval challenge along the lines of Germany's in 1914.

Rounding out the group is a host of states that stand in opposition to the US and are thus potential allies of either Russia or China, such as Cuba, Syria and Venezuela.

THE DEMOCRACIES

The democratic states, although led by the US in decline, will work to translate their technological military advantage and political influence into the structuring of an alliance. They share common values in international policy that are grounded on the freedom of the individual, and they command enormous financial resources backed up by US military technology and power.

Japan is something of a US offspring, with the second-biggest GDP in the world and an equal level of technical sophistication. Their common cause is the containment of Chinese power in the region, and the importance of Japan to the US is demonstrated by the anti-ballistic missile defence technology that the two countries share.

Both countries are forging strong diplomatic links with India, which is the third democracy in the alliance. Japan's diplomatic overtures to India began in 2004, when it provided it with substantial economic aid. The US followed in 2005 under the Bush administration, somewhat in the tradition of Richard Nixon's engagement with China in 1971–72, which brought to an end two decades of mutual estrangement. (Similarly, and most critically, it also further isolated Russia, redefining the Cold War map.) Although Bush's visit was hardly noticed outside India, the mission was a great success, drawing the vast country away from its historic military links with Russia.

The US recognised that India's billion-plus population and growing economic power would make it a formidable ally, and its achievement of deeper ties came at the price of disregarding India's non-compliance with the NPT, as well as its promise to strengthen political and military cooperation. It also paved the way for the US to replace France and Russia as India's major source of arms.

A democratic alliance would consist of the US, Japan and India. As a commodity-rich country identifying as Western, Australia would link itself to this axis, in principle as well as for protection against an expanding China. Canberra's ties with Washington are strong, but in time it would ally itself more firmly with Tokyo and New Delhi.

THE NEUTRALS

Europe is still in a regionalisation stage, redefining itself within a new and larger political structure. As such, it can be expected to be neutral with respect to projecting power outside its borders. Europe does not currently have the military infrastructure, capability or will to involve itself in the Asian sphere of influence. All it can do is defend itself against any future direct attack. Only in the event that Europe faced attack from a remilitarised Russia would this dynamic see an exception, demanding the rejuvenation of NATO. Such a scenario, however, is unlikely in Russia's legacy stage. The role of Britain, on the other hand, is not determined. Although a longstanding member of the EU, it has strongly allied itself to the US throughout the twentieth century. Under a disastrous labour government, Britain suffered a significant economic setback, but its geopolitical view of the world and its enduring Commonwealth ties could make it a key part of the new democratic alliance, particularly with respect to India and Australia. Indeed, the decline of Britain's special relationship with the US will arguably enable British foreign policy to be more balanced and objective, and thus effective, in the coming decades.

DECLINING ALLIANCES

As an empire falls into decline, so does its web of alliances, which have been created to strengthen its borders. The larger partners may be able to defend themselves, but the smaller ones become prey to the new order. With respect to the US, three diplomatic relationships are worthy of discussion in this regard: NATO, South Korea and Israel.

NATO

Formed at the end of World War Two to counter the Soviet threat, NATO achieved its primary mandate with outstanding success until the end of the Cold War. Since 1990, it has struggled for a sense of purpose and has become involved in war zones (like Afghanistan) that dangerously displace it from its original principles. At the centre of the NATO rationale was the need for the US to remain linked to Europe for the latter's defence, a perspective that derived legitimacy from the US's successful intervention in the two world wars. NATO's participation in the US's overseas wars (in which very few of its member states contributed armed combat forces) threatens the future of the alliance. Furthermore, the US has used NATO to relentlessly extend its influence to Russia's borders, inflaming Moscow's insecurities even as NATO's perimeter extends way beyond any serious defensive structure, affecting its very credibility.

With the decline of its power, the US can be expected to withdraw from Europe as it faces the challenge from China, leaving the former to defend itself. In response, the EU, gaining traction as a centralised entity and wishing to take its dream of a 'United States of Europe' to the next level, will have to organise its own military force. The creation of such an institution would be the beginning of NATO's replacement over the next two decades. NATO's current perimeter would then contract to a more defensible size, and a buffer zone might also be created with the old NATO states that do not participate in the EU's new alliance.

SOUTH KOREA

The US's alliance with the Republic of Korea was the first to come under pressure as the new world began to unfold. China covertly encouraged North Korea to develop nuclear weapons and confront the US directly with its flagrant proliferation. In the process, South Korea, fearing it could become a nuclear battleground, moved away from the US towards a position of acceptance of the new balance of power. As a result, although US soldiers remain along the demilitarised zone, their numbers have been reduced and the relationship between the two countries has not yet regained its former solidity.

ISRAEL

Israel's statehood in 1948 was by and large a US creation, placing the US's stamp of power over the Middle East and the declining British Empire. Although today Israel has both a nuclear deterrent and a highly capable military, it remains a client state of the US, which funds the country in many overt and subtle ways through both government and private finance mechanisms. Its dependence is exemplified by the development of anti-ballistic missile technology, which is now becoming so essential to Israel's survival. If the Jewish state's benefactor declines in power, particularly within the context of rising Islamic extremism, the risks to Israel will multiply.

LEFT **NATO defence ministers meet at the organisation's headquarters in Brussels on 11 June 2010. Secretary General Anders Fogh Rasmussen set out austerity principles for the defence alliance with an economic straitjacket to hit defence budgets for years to come.**

BELOW LEFT **A barbed-wire fence that runs across the landscape in the demilitarised zone (DMZ) serves to separate North and South Korea. The barrier effectively cuts the Korean Peninsula roughly in half, crossing the 38th parallel on an angle, with the west end of the DMZ lying south of the parallel and the east end lying north of it. The fence is 250 km long and approximately 4 km wide, and it ranks as the most heavily militarised border in the world.**

Empires and Exogenous Quantum Shocks

No discussion of future geopolitics would be complete without at least a brief overview of the probabilities of various 'shocks to the system'. It must be stressed that expanding empires have a greater ability to recover from such unexpected paradigm shifts than those that are in contraction, much as the young heal faster and more thoroughly than do the elderly.

GEOPHYSICAL RISKS

Some 75 percent of the world's population lives in areas affected by natural disasters of geophysical origin. Since 1998, 500,000 people have died from earthquakes alone. Of particular relevance is the so-called 'ring of fire', where tectonic plates collide. Asia is encompassed by this ring, but the growth patterns of the ASE are unlikely to alter – in the worst-case scenario they might be slowed down temporarily.

A massive solar flare is another event that could devastate the world by, say, disabling global communications systems. Countries with more advanced networks would suffer most (i.e. the Western state). Such a disaster would speed the shift of power from the West to the East, perhaps even providing an opportunity for aggressive action.

The 'million-dollar unknown' might come in the form of an asteroid strike, but the size and location would determine the economic and social impact of such an event.

THE SLOWING OF AGING

A refrain throughout this book has been the fact that demographics drive empires. Declining death rates (in conjunction with rising birth rates) would ensure that productivity continues into later life as well as staying constant, and these two factors could prolong the life of an empire. Of course, aggression and risk are the domain of the young, but healthy populations with higher average ages and a high quality of life could possess greater wisdom and thus maintain their edge, thus changing the normal cycle of human affairs and allowing us to escape the unconscious collective patterns that underlie empires. This process is taking place already in developed countries, as health care improves and cures are being found for degenerative diseases.

NEW TECHNOLOGY

As already noted, rising powers are more likely to develop revolutionary new technologies, yet older ones that continue to innovate can extend their own lives. Areas of potential interest include the wide-scale development of nanotechnology, new code-breaking technologies, advanced robotics and biotechnology applied to weapons systems. The latter generally has enormous potential, although avenues of development (e.g. cloning) are often stunted by controversy in the developed world on account of ethical considerations, even as they are embraced in the emerging world, which does not focus on this aspect as much, preferring to consider the advantages instead.

RIGHT **A mosque in the centre of the wrecked town of Banda Aceh in Indonesia, one of very few buildings to survive the devastating tsunami of 26 December 2004. The long-running civil war in the region hampered international efforts to help survivors. It is estimated that Indonesia alone suffered losses of 110,000 people during the disaster.**

BELOW RIGHT **Thousands of people flock to look at the devastated town of Beichuan in Sichuan, China, which was destroyed by the 12 May 2008 earthquake. It is estimated that this natural disaster caused 69,000 deaths and left 4.8 million people homeless.**

Chapter Twelve

Conflict, Disease and Climate Change

Resource Scarcity

As the Five Stages of Empire model clearly demonstrates, wars are fought for the singular purpose of resource acquisition, cloaked by various ideological or religious rationales. The constriction of access to resources has, therefore, historically been the most significant catalyst for conflict.

Population growth in the developing and industrialising world is now exerting intolerable pressure on the world's resources, and major conflict is all but inevitable as we approach the next Kondratiev peak in 2025. Attempting to ignore this build-up to war is unjustifiable and inadvisable.

The East–West Balance of Power and Commodity Prices *(fig. 63)*

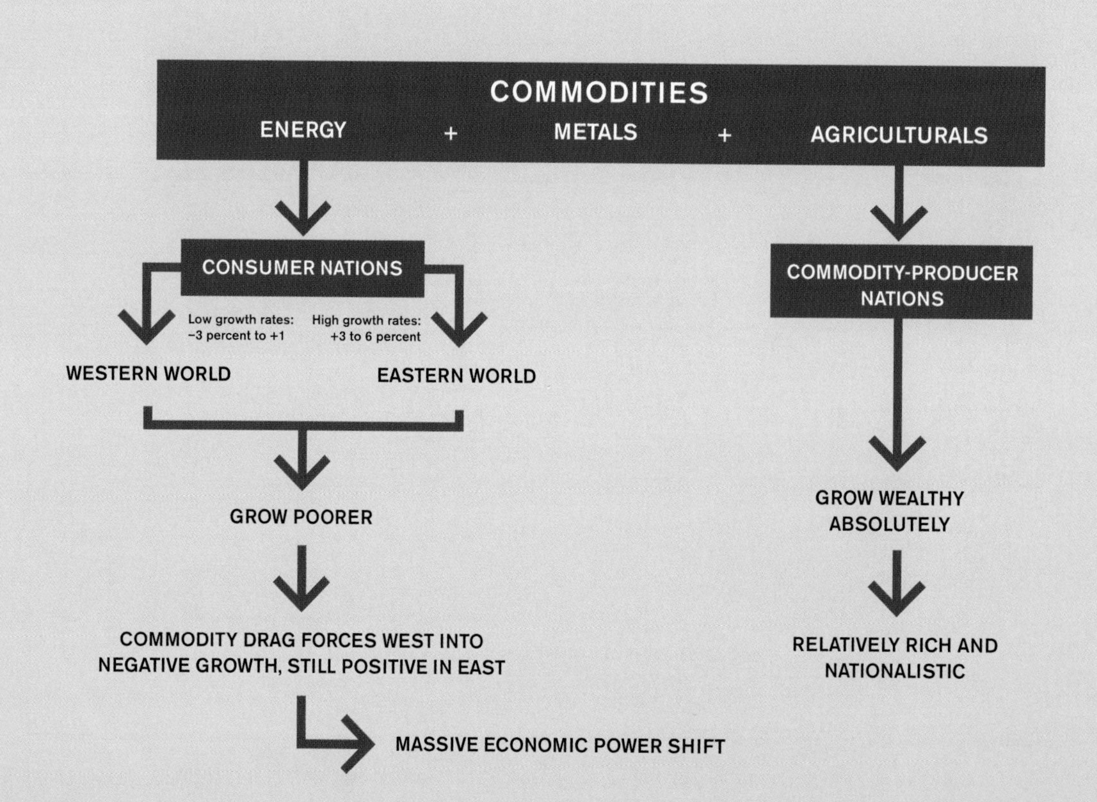

COMMODITY INFLATION: THE NEMESIS OF THE US

As the US and Europe continue to stagnate economically while China grows rapidly, pressure will be maintained on commodity prices feeding through to the continuation of high input prices and the dynamics of inflation in China. However, with no growth, the West will suffer the ignominy of stagflation – the worst of all worlds economically. In this environment, China will hold its own with moderate real growth, while the West contracts with negative growth and endures the social consequences that come with such decline. Meanwhile, commodity-producing countries (in particular those with the rare resource of oil) will continue to prosper as long as they are well managed economically, with balanced budgets based on realistic oil prices. Strategically, this is the time for countries to learn from the price spikes of the second decade of the twenty-first century, and to invest in resource acquisition to prevent the consequences of the new spike. However, the West is in decline and relatively impoverished; only China and perhaps the Arab Gulf states will think in this way. With the East in strong positive growth and the West in negative real growth, the decade from 2010 to 2020 will see the former overtake the latter at triple the traditional speed.

Prior to 2000, the rise of today's wealthy oil- and gas-exporting states would have been inconceivable. However, the majority of these countries became somewhat more belligerent when oil reached its peak in 2008. Unless energy-consuming nations implement significant preventative measures, they should expect to be confronted by similar attitudes in the future.

The developed world must take the initiative to seek solutions, preventing such tensions from mounting and effecting positive geopolitical change. Alternative energy sources must be developed as part of emergency civilian programmes, and given the highest priority – for time, too, is another resource of which we have a limited supply. The following are key principles of this proposed redress:

1. **EFFICIENCY.** Natural resources must be used with minimal wastage and maximum recycling, buying us time to make more fundamental changes to our resource demands. Price increases will be a prime driver of this mechanism, but, without government incentives and planning, the time lag between the deployment of new methods and rapid price appreciation may not be sufficient to reduce the competitive dynamics.

2. **SUSTAINABILITY.** Carbon dioxide (CO_2) emissions should be offset wherever possible, and such techniques as carbon sequestration, which captures and stores CO_2 in subsurface reservoirs and other forms of 'carbon sink', should be encouraged and shared among states. The most efficient way to achieve these ends is to institute an effective global carbon-trading scheme. The EU leads the field in this regard, and it can only be hoped that such a programme will act as the cornerstone for its expansion into the wider world.

3 SUBSTITUTION. Technological solutions driven by research, development and innovation must be prioritised in order to maximise untapped resources (like oil shale and undersea oil exploration and exploitation) using the most efficient methods possible. Substitution must also occur for base metals, as Asian industrialisation continues to push up prices for this commodity. (The scale of construction taking place in China – and the Gulf – is comparable to that in the heyday of British and American industrialisation, although today the West's demands remain static because its infrastructure is already in place.) Typically, the drive for substitution would take place during just such a period of high demand, and new methods and materials sought with which to build more cost-effectively. However, base-metal acquisition friction is expected to continue well into 2025, and this will colour geopolitics as many commodity-rich countries strike agreements with China (which will continue to pursue higher stakes in mining companies across the world, even after the end of the financial crisis).

LEFT Solar modules of a new 100 megawatt photovoltaic on-grid power project in July 2010 in Dunhuang, in China's northwest Gansu Province. The Chinese government has set a target to install 20 gigawatts (GW) of solar energy capacity and 100 GW of wind power by 2020. The gigawatt is 1 billion watts. The US's total wind power in 2009 was 35.2 GW.

BELOW LEFT Wind turbines at E.ON AG's Scroby Sands installation near Great Yarmouth, Britain, in June 2010. The company is among the utilities leading a worldwide push to develop offshore wind power, overcoming a lack of work ships, stormy seas and higher costs to make almost twice the profit that it would on land.

4 PRAGMATISM. A realistic outlook is called for with respect to alternative energy sources. Development can be painfully slow, and initially promising technologies are frequently shown to be inefficient and our expectations of them overblown (wind power, for example, is sometimes spoken of as being as viable as nuclear power in fulfilling our energy requirements but sadly this is a delusion).

5 SHARING SOLUTIONS. Once the developed world has begun solving resource challenges, it must share its discoveries with the rest of the globe.

6 RECONSIDERING ECONOMIC HABITS. In a world dominated by Capitalism, where commodities trade freely on global markets at prices that are set according to supply and demand, it needs to be understood that China's centralised resource acquisition system bucks the trend because its government is prepared to pay premium prices now to ensure future growth over the long term. A commodity squeeze would affect and possibly undo US-style Capitalism, which is unable to buy for the present at a loss in anticipation of future gain. This mismatch in patterns of purchasing commodities at source requires careful consideration by Western governments and a new scheme of tax incentives. If this does not happen, inevitably the developed world, with low-to-zero growth, will be swamped by imported inflation from China that will choke off any meaningful recovery from the financial crisis.

NUCLEAR POWER

As of February 2010, there were 436 reactors online in thirty-one countries, with a further 55 under construction and several dozens more planned or proposed. Failing effective, affordable carbon-free coal burning, the only other reliable energy source with which developed countries can drive their national grid networks is nuclear power. It is vital that developed and emerging countries alike embark on a massive emergency programme of nuclear power plant construction to replace coal, oil and gas. The safety of modern nuclear reactors is above reproach; the only matter of concern regards the disposal of their waste products. One sensible idea that addresses this challenge is the construction of deep geological repositories, such as containment facilities within salt domes in stable regions, like those that are located beneath the North Sea. (The crystalline structures of the salt domes re-form following any tectonic movement, thereby maintaining a seal around the facility.) Nuclear waste products from fission reactors, however, are also raw materials for nuclear weapons. This power source is, therefore, one that the developed world does not currently encourage in emerging countries. Yet if the developed world were to become predominantly powered by nuclear fission, carbon emissions would be reduced considerably, along with friction resulting from its energy dependence on Russia and the Middle East.

Nuclear fusion could potentially provide a next-generation energy solution, emulating the process of the sun – itself a fusion reactor. Fusion is the act of joining two atoms, commonly the hydrogen isotopes tritium and deuterium, to form a larger, unstable molecule that then breaks down. The result is a stable helium atom, together with a large amount of energy that is released during the transformation of an unstable state to a stable one. Tritium and deuterium are available in near-endless quantities, and fusion reactions would leave no waste products (though the reactors themselves would become radioactive for five decades – an insignificant amount of time compared with fission reactors). However, the process requires an enormously high temperature to operate, equivalent to that of the sun's surface. The challenge is thus to devise a means of containing this high-energy plasma long enough to release more energy than that required to heat it in the first place.

Many incremental advances have been made, but funding for research into viable fusion reactors has not been commensurate with its potential to solve the world's energy crisis. Intensive cooperation across the developed world to solve the fusion question must be made an urgent priority. The Joint European Torus (JET) project, the largest-scale research programme in existence that is concentrating on the problem of plasma-containment physics, became operational in 1983 and has as its aim to contribute to the European grid with fusion power within three decades. The EU, the US, India, South Korea, Japan, China and Russia are all participants. That JET exists and is working well is encouraging, but, unless there is a dramatic increase in investment, fusion will not be viable in time to avoid the 2025 commodity squeeze and its consequences.

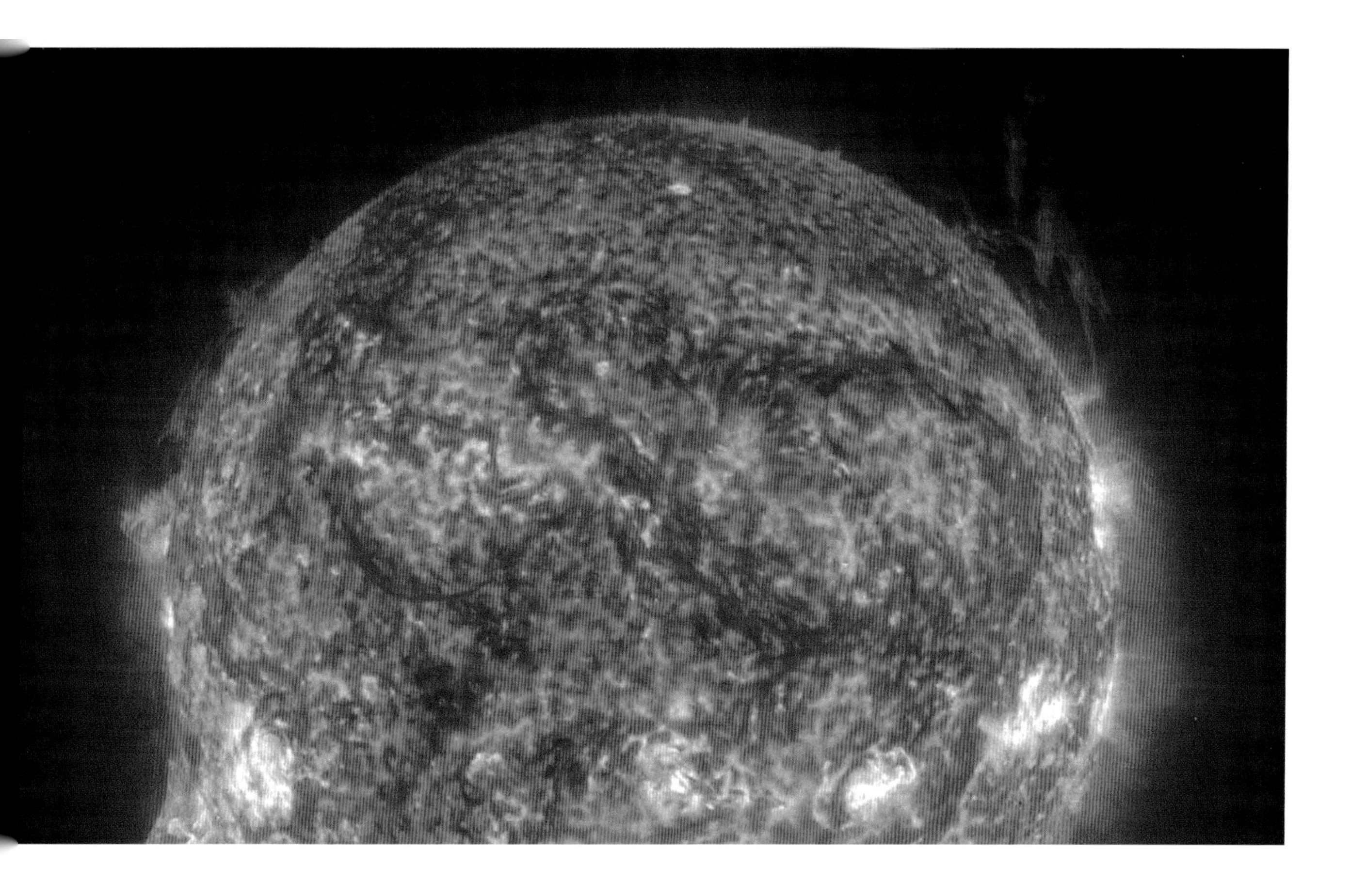

ABOVE A solar eruption from the sun imaged by NASA's SOHO satellite on 1 July 2002. Scientists call the event a prominence – a huge cloud of relatively cool, dense plasma suspended in the sun's corona with magnetic fields building up enormous forces that propel particles out beyond the sun's surface. These occur about every eleven years for three to four days and can affect the Earth.

NEW FUEL TECHNOLOGY

Gasoline engines must ultimately be replaced and, in the short to medium term, at least improved. The high price of oil in recent years has at least encouraged investment in new fuel technologies. There are many potential avenues of innovation, although not all of them address the issues of either energy efficiency or the reduction of carbon emissions. The South African energy corporation Sasol, for example, is a leader in petrol substitutes derived by liquefying coal and natural gas. These technologies will be in increasing demand to make up for the shortfall in oil, but they exact a high price by doing nothing to alleviate carbon-accelerated climate change.

Biomass (organic material) fuels, such as ethanol and biodiesel, will become more valuable, although the land required to produce them may put pressure on the food chain and cause agricultural commodities to increase in price over the next decade. Battery technology is also growing in use, and there is an expectation that the ability to charge and recharge vehicles from the national grid will be available in the developed world before long, but this will require cheap power.

Solar power is another potentially groundbreaking power source that may become even more viable with better power-transmission technology, minimising the huge power loss that occurs in the transformation of source energy to a usable form.

COMPETITION FOR METALS

The industrialisation of Asia is the greatest driver for the rise in price of metals, which will become increasingly difficult for the West to access. The massive new infrastructure of the current Chinese economy demands large quantities of metals, and rare-earth metals are used at the highest levels of technology. The scale of building currently taking place in China (as well as in parts of the Middle East) is comparable to the industrialisation of Britain and the US at their heights. On the other hand, the West's requirements remain static because its infrastructure has already been built.

Typically, during such a period of high demand, we should expect to see a process of substitution implemented, and new methods and materials innovated with which to build more cost-effectively. However, the friction caused by the issue of base-metal acquisition is expected to continue well into the next decade, defining the politics of many commodity-rich countries. New approaches by the West will be required to maintain good diplomatic relations all round. Indeed, China continues to acquire stakes in mining companies across the world in the aftermath of the financial crisis in order to ensure its future access to the necessities of empire building.

RIGHT **An aerial view of crops in Gauteng Province, South Africa, showing a massive centre pivot irrigation system. This rotates and causes a circular area centred on the pivot to be irrigated, creating a circular pattern in crops that is visible when viewed from above.**

BELOW RIGHT **The Hoover Dam at Lake Mead, Arizona, in July 2007. The white 'bathtub ring' on the rocks is formed from mineral deposits left by higher water levels. A seven-year drought and greater water demand spurred by huge population growth in the southwest has caused the water level at the lake, which supplies Las Vegas, Arizona and Southern California, to drop by more than 100 ft to its lowest level since the 1960s.**

FOOD AND WATER

Population growth, climate change, water shortages and the development of biomass energy sources will all place considerable pressure on agricultural prices. In areas where climate change has already reduced the amount of rainfall, such as the African tropics, traditional non-irrigated farming methods are very near failing to provide for the populations that they once served. The only remedy will be to replace these with massive irrigated farming projects. This is no light undertaking, although it is actually one of the cheaper solutions to our commodity problems – particularly if the developed world participates in the investment process, as this would provide food security and a channel for developing states to build viable, sustainable economies.

However, as agriculture accounts for more than 70 percent of global freshwater usage, increases in irrigated farming will exacerbate an already potentially dangerous problem with water supplies. Some 2.5 billion people now live in regions suffering from water scarcity, and of these a further 900 million lack access to freshwater supplies. As a consequence, 5 million people die every year.

The sea has traditionally provided humans with limitless supplies of protein. However, we have long been extracting its resources at an unsustainable rate. Current fish stocks are 95 percent down on 1900 levels, particularly in the case of larger fish, and they have suffered a complete collapse in the case of other species. We must rethink our means of garnering food from the oceans as 15 percent of all of our protein derives from them. (Asia is particularly dependent on fish consumption.) One obvious solution is the development of a new and expansive aquaculture industry.

Globally, the reduction in the water supply on account of melting glaciers and changing rainfall patterns, as well as the effects of forest clearance and subsequent

desertification, will make water security one of the most inflammatory issues likely to spark conflict. Governments worldwide should anticipate a rapidly altering environment and assume that change – for the worse – will take place in a quarter of the timeframe currently envisioned. They should take measures to ensure that food and water stores meet demand. The issue of water and food security has come of age. We shall see a return to the most basic concerns of our ancestors, unless the vision of desalination plants powered prolifically, cleanly and cheaply by nuclear and fusion energy across the globe can quickly become a reality.

COMMODITIES AND THE EAST–WEST BALANCE OF POWER

CHINESE SEA LANES

There are three essential elements to expanding an empire: the development of a thriving economy that produces goods that can be traded; the existence of trading partners; and the securing of trade routes to facilitate commerce. The prevention of trade-route interference from foreign powers and pirates is the responsibility of the army on the land and the navy on the sea.

In the twenty-first century, the US navy and what is left of the British and French navies have been reduced to fewer than 350 high-value ships to patrol the world's sea lanes. Each is highly combat-ready but, of course, is unable to be in more than one place at a time. The problem of effectively covering such a huge area is compounded by the fact that the majority of these vessels are designed to operate together in task forces. There is currently no provision for a fleet of smaller coastal craft that could secure the littoral waters of vital conflict zones. The upshot is that the West no longer controls the sea lanes as it once did. The resurgence of piracy off the Horn of Africa, which has reached endemic levels, is a grave sign that Western trade-route security can no longer be guaranteed. As China grows in economic stature, and as its empire expands, it will build a blue-water navy with which to secure its trade routes, taking advantage of the declining naval power of the West. Since the US currently enjoys a technological lead, this development may take a decade or so more – but it is inevitable. China will need to protect the strategic choke points that govern the flow of imports as opposed to exports, as it must be assumed that in the event of conflict it is imports that will be essential to feed the Chinese industrial machine. These routes across land and sea have not changed much since the time of the Tang Dynasty, with the exception that resources from West Africa will have to sail round the southern tip of Africa, and those from Brazil will either have to navigate Cape Horn or pass through the Panama Canal.

The flow of resources from the Middle East and Africa will have to skirt round Sri Lanka, where they will be vulnerable to the growing naval power of India that is now being supported and encouraged by the US as it seeks to redistribute its declining naval resources. Once safely past India, the route passes through the narrow Strait of Malacca, dominated by Singapore. Some 80 percent of all China's imports take this route.

Such is the difficulty associated with controlling this strait that the Chinese are working hard to develop ports in Myanmar, as well as direct road links, to avoid this area. However, continuing along the sea lanes into the South China Sea, vessels will then be

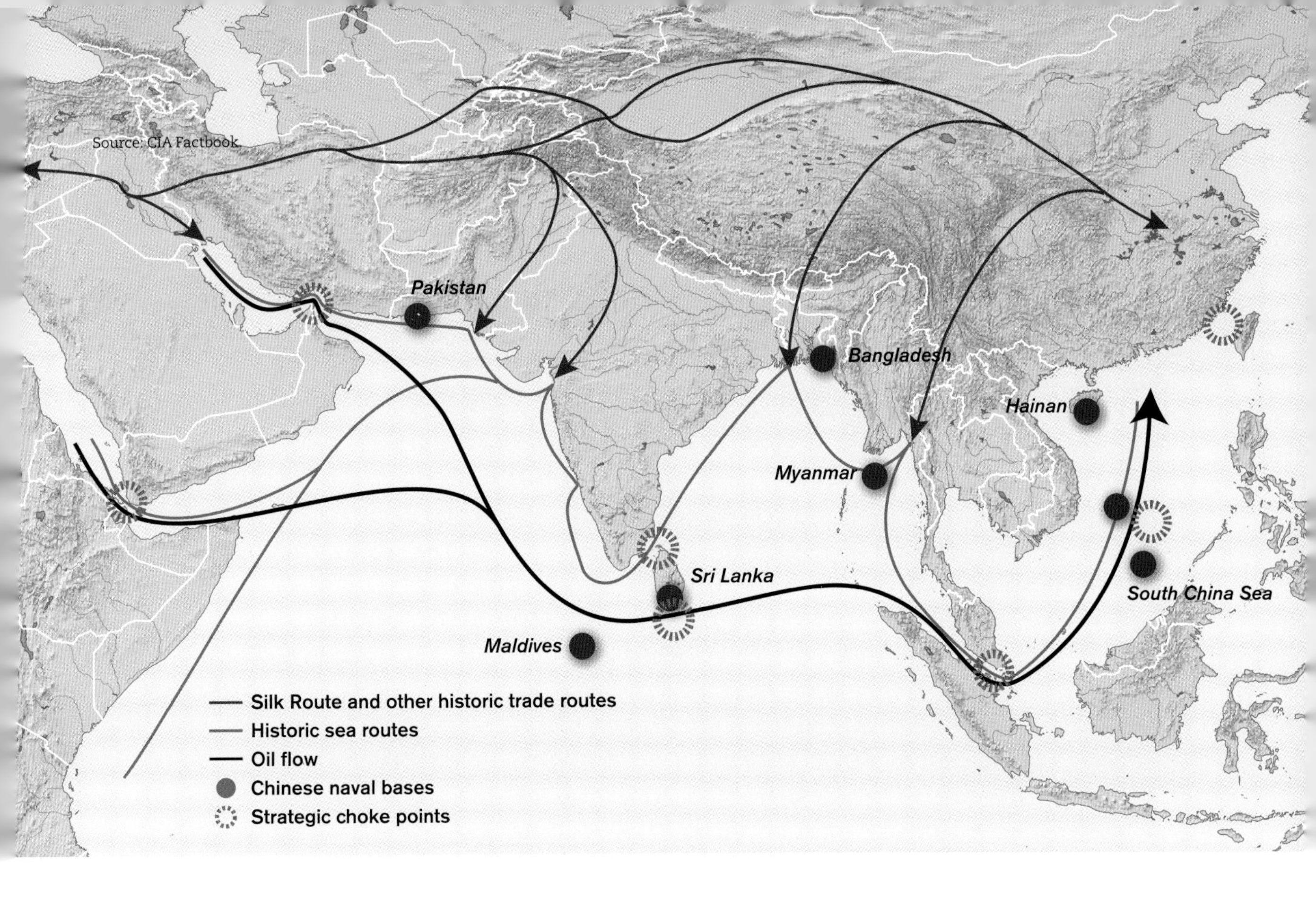

ABOVE Map of China's land and sea routes, which have not changed since the Tang Dynasty, along with the strategic choke points that are sure to become a focus for future Chinese military power projection.

required to travel along the Chinese coast to their port of choice. Once again, therefore, the importance of securing the Taiwan Strait becomes clear.

The Chinese string of pearls system of naval bases along this east–west trade route has fed China's ruthless courtship of governments that are able to assist with this objective. One example is the support offered by the People's Republic to Sri Lanka in 2009 in the form of arms shipments, which enabled Colombo to finally defeat the Liberation Tigers of Tamil Eelam resistance group in a long-standing civil war. In return, China has secured a significant strategic advantage with the vital port at Hambantota.

Chinese maritime interests necessitate a significant naval presence across the Indian and Pacific Oceans, as well as on related inshore seas. The US has been able to achieve this; it is just a matter of time before China does as well.

The development of overland connections across Pakistan and Myanmar to China have produced strong political relationships with these countries, and the future firming up of ties with Indonesia, Malaysia and Vietnam will also take strategic priority.

One major area of contention is that Japan shares trade routes with China. It is only natural that a naval arms race will ensue between these two countries as the polarisation factors increase. Indeed, the China-vs-Japan-and-India competition for commodities during the third phase of the current Kondratiev wave is reminiscent of the sixteenth-to-eighteenth-century struggle in Europe for sea-lane control. These lines of friction will become the most likely focus for a war, which, as has been said, has a very high probability of breaking out at the next expected Kondratiev peak in around 2025.

Moreover, what happens to the current global trading network of oil-powered ships when the price of oil moves into its phase of terminal price appreciation? At what stage does the price of oil affect the cost of transportation of commodities to the point where globalisation slows down? The development of super-sized nuclear-powered bulk transport is one answer to this future challenge in the sphere of global commodities, with all of the issues of security that this implies.

Polarisation

POTENTIAL CONFLICT ZONES BY REGION

Before looking at some of the potential solutions to polarisation, it will be useful to identify those regions where conflict could erupt during the next decades.

NORTH AND SOUTH AMERICA

The Americas are a relatively stable zone, with the north controlled by the US with good relations with Canada. From a military perspective, the US remains invulnerable to attack from all but a full-scale nuclear assault and, as it develops its ballistic defence systems, it should remain well protected against attack from the rising nuclear powers.

To the south, there is a small risk of a Mexican civil war if the drug lords continue their ascendancy to power. Further south, if in the future energy prices rise and remain high, it is possible that Venezuela will attempt to challenge Brazilian power, resulting in minor, low-level regional conflict.

EUROPE

Europe is in legacy and reformation and so is no threat to its neighbours, given that it is unable to generate sufficient political coherence to project military power outside its immediate area. Russia is in legacy, its economic malaise supported by its commodity wealth, so, despite tough talk, it is unlikely to provoke conflict unless oil prices rise higher. Increased wealth could mean that it will once more become bellicose. For now, neither Europe nor Russia has the resources for a new arms race, and neither army can project power into the territory of the other, making the zone stable. But the risk remains that if Russia and China become allies, Russia could be drawn into a Chinese-driven conflict with the West.

THE MIDDLE EAST

The Middle East, on the other hand, is a prime focus for war, with Iran at the centre of three major potential areas of conflict. The first is a regional challenge from the Iranian *Shi'is* to the *Sunni* powers, driven by Iranian aspirations to be the dominant power in the Middle East. Direct intervention within the affairs of Iraq is one example of this process, and,

The Nature of Conflict as a Function of the Five Stages of Empire *(fig. 65)*

PHASE OF CHALLENGER	PHASE OF DEFENDER: EXPANSIVE	PHASE OF DEFENDER: PEAK	PHASE OF DEFENDER: CONTRACTION
EXPANSIVE	Can be initiated by either side, and becomes a long, drawn-out war of attrition that only ends with decisive victory by one party. For example, Rome vs Carthage, Britain vs France, Russia vs Germany.	A war that can be relatively fast, depending on the power differential between the two sides. The mature empire can show surprising resilience. For example, Germany vs Britain 1914.	If the challenger has timed and planned the attack well, the war can lead to a rapid collapse of the old empire, which is then subsumed by the new empire. For example, China vs West 2017?
PEAK	The only way this type of war might unfold is if the expansive nation is only at a regional level, and the power differential of the hegemony allows it to attack pre-emptively without risk. For example, colonial expansion at the end of the nineteenth-century, the US in Iraq and Afghanistan.	Very unlikely as each side will not have the national energy and risk:reward ratios to attack the other.	Very unlikely as a peak mature empire is unlikely to choose war over peace.
CONTRACTION	Not an option for the old empire. With financial limitations and a sheer lack of will to project its power, it is more concerned with keeping what it has.	Not an option for the old empire. With financial limitations and a sheer lack of will to project its power, it is more concerned with keeping what it has.	Not an option for the old empire. With financial limitations and a sheer lack of will to project its power, it is more concerned with keeping what it has.

ABOVE **The nature of the conflict between the challenger and defender as a function of their respective stages of empire.**

if Iran becomes a nuclear power, it is fairly certain that the Saudis will square up to them, using their connections with Pakistan. The second is a challenge to Israel by proxy through *Hizbullah* and Hamas, but, if Iran becomes a nuclear power, also potentially through a nuclear stand-off. The third is an implacable desire to remove American influence from the region, wherever present. This started in Iraq and has spread to Afghanistan.

INDIA AND PAKISTAN

India and Pakistan represent one of the world's most potentially dangerous areas as the force of Islamist expansionism in Pakistan could drive one nuclear nation to provoke another, more powerful one to the edge of nuclear war. The US has already intervened once – in 2003 – to prevent this from happening, but with Pakistan moving closer to becoming a fundamentalist Islamic state, and if the current Indian Sikh prime minister,

Manmohan Singh, is replaced by a Hindu, the risk of conflict could escalate considerably. The nuclear cold war between these two nations is running a similar risk profile to that of the WCSE's Cold War that preceded it. Only once each side has built its nuclear arsenal and mutual assured destruction (MAD) becomes a reality will nuclear tension recede. However, over the next decade the risk of a nuclear conflict remains high. Not only would such a conflict be a disaster for the world but strategically it would knock out the only real competitor that China has in terms of demographic potential.

AFRICA

The greatest risk to Africa is the north–south, Islamic–Christian resource conflict that could over the next decade become a significant problem. Uganda, Kenya and Somalia could well be in the forefront of this process, as Islamic expansionism and resource scarcity force the desert peoples south, and into conflict. The Somali Islamist jihadists group, al-Shabaab, is at the forefront of this movement, and to counter it and encourage stability, Uganda has some 6,000 men, part of the peace-keeping African Union Mission in Somalia (AMISOM), stationed in that country. In time, it will undoubtedly be forced to increase its commitment as it is forced into ongoing pre-emptive action.

It is also possible that as China continues to gain influence in Africa, it will intervene in the Islamic–Christian resource conflict in the form of proxy wars to further its resource plundering of the continent.

ASIA

As has been discussed, China is the power to watch in this area. Although it is currently no match for US forces, if it translates its growing economic power into military power, the capability gap could close – and over the next decade. Even before then, there are risks that should be considered. However, since Japan is a mature power, any regional conflict will be as a result of a conscious decision by China to precipitate conflict.

North Korea is close to economic failure, has a massive military complex and is led by an autocratic leader with all of the risks that this implies of dramatic and ill-considered action being taken when the country is faced with economic collapse. The situation is worsened in that, as China grows, the US will be less confident in projecting power into the region, giving North Korea an exaggerated sense of security. Conversely, China will not want to see a peaceful failure of North Korea that results in its reunification with South Korea and a new regional power on its doorstep.

Taiwan is also a key risk area because, until the island has been brought under its direct control, China will not consider its civil war over. Certainly its military is configured to be able to straddle the Taiwan Strait in a lightning strike that would be over before American forces could be mobilised. As the power projection of the US weakens in the area, and it is distracted by events in the Middle East, the risks of Chinese invasion could grow. Conversely, as Taiwan's perception of China as a world power increases, it is likely to pragmatically choose closer ties with its neighbour and loosen those with the West, thus depolarising this hot spot. However, the growth of China's amphibious warfare capability, developed initially to focus on Taiwain, will no doubt be the basis of an enlarged capability

ABOVE Pakistani and foreign students of religion sit their final examination at Karachi's Jamia Binoria *madrasa* on 18 July 2009. According to government records, there are at least 15,148 Islamic seminaries in Pakistan, which teach more than 2 million students – some 5 percent of the 34 million children in formal education.

that ultimately will make the Indian Ocean, China Sea and Pacific a focus for future conflicts. The last potential flash point is between Japan and China as the two nations share the same sea lanes and compete to extract resources from the East China Sea.

DEPOLARISATION: AVERTING CONFLICT

CASE STUDY: SAUDI ARABIA

The Kingdom of Saudi Arabia is a firm ally of the West. Paradoxically, it is also a significant centre for the growth of Islamist extremism, fermented by the severely conservative form of *Sunni* Islam known as Wahhabism that is the country's official religion. The Saudi government has attempted to control the expansion of radical Islamism in its midst, implementing a programme of anti-extremism, counter-terrorism and de-radicalisation strategies. The kingdom boasts several distinct advantages in such initiatives:

1. OIL WEALTH allows it to spend freely on combating nascent terrorism.

2. RELIGIOUS MONOPOLY also provides Riyadh with another tool against the jihadist elements within its society.

3 **THE TRIBAL STRUCTURE** of Saudi Arabia encourages obedience to authority and can be used in the state's favour to quell dissent.

4 **SOCIAL AND STRUCTURAL REFORMS** that balance the equability of Saudi society are slowly being introduced, much as the British did in the latter part of the eighteenth century prior to the French Revolution.

5 **EFFECTIVE REHABILITATION** of captured jihadists is an important component of the Saudi programme, so that those who are apprehended can eventually rejoin the very society that they once rejected.

These strategies have been exported, with Saudi financial support, to Yemen, Pakistan and Afghanistan, although the administrative circumstances in these countries are varied and quite different from the Saudi context, so success will be more difficult to achieve. Nevertheless, Saudi depolarisation strategies should be studied by all countries containing extremist elements that wish to build a wider support base in the population.

DEPOLARISING TENSIONS ACROSS THE WORLD

As has been discussed, competition for resources is more intense than ever among the growing number of people in our shrinking world, most of whom quite naturally wish for the kind of high standard of life that those in the West have traditionally enjoyed. Increased polarisation between nations is consequently all but inevitable. Growth in population, shifts in empire and resource hunger are all drivers of polarisation, and, as long as those forces are present in global society, conflict will remain a pervasive danger. Increased polarisation is, as was demonstrated in PRESENT, the beginning of the road to war – a gradual build-up of tension before the eruption of (apparently sudden) conflict.

However, as much as the polarisation process unfolds in a series of ever-greater catalytic steps, it also works the same way in reverse, if a depolarising sequence can be devised to lower tension levels. To develop an effective depolarising strategy, it is critical to understand that younger, challenging nations exhibit primary polarisation, inducing, in turn, secondary polarisation in older, more established nations.

The US, in the role of established empire, can choose how it responds to the challenging primary polarising energy from China and the Islamic world. The second Bush administration unwisely chose to respond to Islamist primary polarisation with a similar aggressive secondary energy that only added fuel to the fire, diminishing any claim to the moral superiority of American values.

A more carefully considered response would have emphasised respect for mainstream Islam and common cultural values, and called on moderate elements in the Islamic world to contain their radicals – a 'coalition of the moderates' should have been conceived, not a 'coalition of the willing', a phrase that George W. Bush and his administration used copiously to describe the invasion of Iraq in 2003. The key question over an older nation is whether it responds in kind to a primary polarisation or yields, flexes and minimises the effects of the opposing expansive energy to prevent it from gaining traction. To borrow a concept from the deceptively lethal martial art of *tai chi chuan*, it is best to 'invest in the loss': that is, by yielding, a centred defender can use an attacker's outward-directed force against him, expanding in turn as he overextends and becomes vulnerable to the slightest input that effortlessly redirects the aggressive energy.

Democracies have a unique capability of changing their leadership at critical moments of collective need. The US, having suffered the repercussions of the neo-Conservative Bush regime, did exactly that in 2008, electing a leader of diametrically opposite ideas. Obama has a new range of policy options available to him and he appears to understand the need to depolarise friction, where possible, with significant changes in foreign policy. The trend is potentially encouraging, but genuine threats remain and are increasing in urgency.

BORDER DISPUTES

In recognition of the fact that tensions will rise globally in the next decade and that, historically, these tend to come into focus over long-running border disputes, the world leaders of G20 (the Group of Twenty) would do well to immediately begin campaigning to resolve such issues before they become catalysts for conflict. A special global task force should be set up to implement radical negotiation strategies in these cases. In areas rife with ethnic differences, inequality of wealth, historical antagonism and resource-access issues, achieving meaningful outcomes will be extremely challenging, but the alternative is much worse. Examples from the late twentieth century include the Falklands War (still simmering) and Iraq's invasion of Kuwait. Today, attempts to resolve key land conflicts need to encompass Sub-Saharan Africa, the Middle East and the Caucasus, as well as border disputes between India, Pakistan, Bangladesh and China. Priority should be given to those disputes that border the interests of expanding powers, such as China, as they carry the highest risk of being catalysts for future major conflict.

With resources in such great demand, ownership of the seabed is, and will continue to be, a major focus for international tensions. Areas of dispute often include the overlaps created by coastal features and islands on the continental shelves. The need for resolution is most frequently driven by questions of oil exploration and fishing rights. Some of the most contentious areas are in the South China Sea, and these should be addressed before they balloon into major conflict.

Border Disputes Between Sovereign States *(fig. 66)*

Source: CIA Factbook.

This map covers a variety of terrestrial and maritime disputes, from militarised disputes and unilateral claims to managed situations and dormant claims.

Border disputes on the boundaries of expansive powers have the greatest potential risk of conflict.

Potential Future Conflict Zones *(fig. 67)*

CONFLICTS DRIVEN BY NATIONS IN REGIONALISATION

Region	Aggressor	Defender	Relative Empire Status	Comment Plus Risk Factor	Risk of US Intervention
Africa	Islamic countries	Christian tribal countries	**Both parties are in the same regional state**	The competition for food and water resources causes a migration of the northern population southwards, creating conflict along the Sub-Saharan border. Somalia and Kenya are prime examples.	Low
Mexico			**Internal**	This could be a very localised war driven by the growing power of the drug lords that would draw in US support and resources.	High
Latin America	Venezuela	Brazil	**Both sides are in regional state, but Brazil is clearly more powerful**	This might occur when the price of oil increases and as Venezuela seeks to exert control in its small sphere of influence against the growing economic power of Brazil. However, neither side has a military infrastructure. Colombia/Venezuela recently came to the edge of conflict over Ecuador and disputed borders as a guise for proposed expansion.	Low
Middle East	Iran	Israel	**Both sides are in regional state, but Iran is more expansive**	War is currently being fought by proxy rocket attacks fired by Hamas and *Hizbullah* in Gaza, the West Bank and Lebanon, and could expand to an Israeli bombing campaign against nuclear targets or an Iranian attack on Israeli homeland.	High
	Iran	US	**Regional power of Iran seeks to expand against declining empire**	Potential pre-emptive US strike.	High
	Pakistan	India	**Both sides are in regional state, but Pakistan is more aggresive, India more powerful**	Expansive Islamic policies seeking to challenge the stronger regional power of India. In this case, India is the defender as its Hindu values are not expansive. The legacy of partition and two wars still make this conflict highly probable. The greatest risk factor is the radicalisation of Pakistan.	Diplomatic Intervention

CONFLICTS DRIVEN BY NATIONS IN ASCENSION

Region	Aggressor	Defender	Relative Empire Status	Comment Plus Risk Factor	Risk of US Intervention
Asia	China	Taiwan	China ascends to empire and annexes the regional power of Taiwan.	China demonstrates its new growing power and the weakening power of the US by annexing Taiwan, a military operation that its forces are fully focused on achieving. Politically, this would be viewed as the end of the civil war between the Communist and the Nationalist forces.	High, possibly with Japan
Asia	China	Japan	China, as it ascends to empire, challenges the mature power of Japan.	Both nations exist in the same region, have long-term historical grievances and seek resources like those under the East China Sea.	High
Africa	China	Africa	China, as it ascends to empire, seeks the resources of Africa.	This might take the form of proxy wars, whereby China supports Islamic African nations against southern Christian nations.	Medium, may not have the resources

CONFLICTS DRIVEN BY NATIONS IN MATURITY AND DECLINE

Region	Aggressor	Defender	Relative Empire Status	Comment Plus Risk Factor	Risks of US Intervention
Europe	Russia	Europe/US	Mature Russia challenges legacy Europe and declining US	A conflict between two arms of the declining WCSE, both in contrasting phase, is very unlikely, particularly as America has the military upper hand and Russia, due to its demographics, will not in the foreseeable future be able to build an economy to finance a new arms race.	High

DEPOLARISING TENSIONS BETWEEN THE US AND THE ISLAMIC WORLD

The common heritage of Islam and the Judeo-Christian beliefs of most of the US is an underused resource that should be brought to bear in relations with the Muslim world. Many Americans are ignorant of this – even, for example, that Muslims worship the same God ('Allah' is Arabic for 'God') or consider Jesus among their most important prophets, and are just as capable of living secular, modern, agnostic or atheist lives as their counterparts in the West. For their part, Muslims should not reflexively associate the US with hostile policies. (For example, it is often not recognised that the pro-Israel policies of the US are at odds with many Americans – Jewish and Christian alike.)

BELOW Palestinians in a shop in the West Bank city of Hebron watch Barack Obama's televised speech at Cairo University on 4 June 2009. The US president reitererated his country's support for a Palestinian state, coexisting in peace with Israel. He called on Palestinians to renounce violence and on Israelis to put an end to settlements.

Emphasis should be on the similarities as opposed to the differences. Obama's Cairo speech of 4 June 2009 was historic, and was justifiably hailed as a landmark. It brought great relief and hope to average Muslims to hear an American president speak about the shared aspects of Christian and Muslim civilisation, recognise the legacy of the great Islamic empires, communicate respect and the need for understanding, and – very unusually – acknowledge some responsibility for an act of aggression in the Islamic world (i.e. the reference to the US's part in overthrowing Iran's

democratically elected and reformist prime minister, Mohammad Mossadegh, in 1953). The impact of his words ought not to be understated, although, quite reasonably, media commentators across the world, particularly the Islamic world, opined that words were a good start, but were to be trusted only when backed up by action and change.

Making further progress will require a strategy involving three crucial elements:

1 GENUINE STEPS TOWARDS RESOLVING THE ISRAEL–PALESTINE CONFLICT. Gradually, the powerful pro-Israel lobby in the US is recognising that it cannot win every argument, and critical voices (Jewish, Christian, Muslim, etc) are being given wider currency. Israel is so firmly lodged in the US's sphere of influence that the US will be held responsible for injustices committed by the Jewish state. Israel must be given credit where its due for its democratic system, however flawed, and for the rights of its citizens generally to dissent – a unique characteristic in Middle Eastern polity. However, in recent years under an increasingly hawkish leadership, it has repeatedly followed the example of its imperial protector with regard to the ill-judged use of indiscriminate force that has impacted on civilians. Examples include the attacks on Lebanon in 2006 and Gaza in 2009, and the raid on the Gaza aid flotilla in May 2010. World opinion has come down strongly against Israel, and by association has inflicted damage on the US's reputation under Obama. The US must, now more than ever, treat Israel more objectively, particularly with regard to the illegal settlements in the Occupied Territories – a perennial sore point that, given increasing religious polarisation among militant Orthodox Jews (including soldiers), will remain a difficult issue, even for Israeli policymakers who wish to see it resolved. The US must do all it can to prevent disproportionate Israeli aggression and repressive Israeli policies, which are self-defeating for Israel and also hurt the US as its major financial backer. In short, the US needs to get tough with Israel and set the road to resolving the Palestine question once and for all on a more equitable basis. Moreover, it must encourage moderate elements in Israeli politics, as hardline politicians are elected when Israeli citizens feel threatened. Iran's nuclear aspirations are thus directly linked to the reduction of polarisation in the Middle East (though unofficial confirmation in 2010 of the fact that Israel is currently the region's only possessor of nuclear weapons will make the situation with regard to Iran much more intractable).

2 THE US MUST PROVE THAT WHEN IT INTERVENES MILITARILY IN THE AFFAIRS OF A COUNTRY, THAT COUNTRY'S INHABITANTS BENEFIT IN THE END WITH IMPROVED PHYSICAL AND ECONOMIC SECURITY. Iraq is now increasingly in the hands of its own government, but an acceleration of economic aid to help to

rebuild the country could go some way towards compensating for the US's gross errors of occupation for some years after invading in 2003. Careful, steady attention must be paid to giving the Iraqi people a sense of confidence that their country is now, or will soon be, a better place than under Saddam Hussein. Similarly, the US must bring all of its resources to bear in rebuilding Afghanistan's economy and infrastructure as rapidly as possible. (Afghanistan is now the linchpin of future US success as an interventionist state)

ABOVE **Israelis barricaded on the rooftop of a religious school throw food at border guards who are trying to take their position on 23 August 2005 during the evacuation of the northern West Bank settlement of Homesh. In spite of international condemnation of the founding of such settlements in the Occupied Territories, the illegal practice continues.**

3 IN ALL OF ITS INTERACTIONS WITH THE PEOPLES OF THE COUNTRY THAT IT IS TRYING TO 'LIBERATE', THE US MILITARY MUST DEMONSTRATE RESPECT FOR LOCAL VALUES AND CUSTOMS. The US must win the hearts and minds of those whose fate it determines, even as it holds true to the best of American values. Its unbiased treatment of its own soldiers, from whom the very highest moral conduct must be expected, is an essential component of such thinking. The notion that US soldiers are immune to prosecution from war crimes must be abolished, and the rule of law as applied to enemy

combatants under international treaties to which the US is signatory is also of endless importance. The current programming of soldiers for total war must give way to the post-Vietnam emphasis on moral integrity.

DEPOLARISING TENSIONS BETWEEN THE US AND CHINA

In any scenario that has it entering into serious hostilities against China, the US will have no military options for resolution short of a full-scale war. Were this to happen, it would likely occur between 2017 and 2025, as the current commodity cycle reaches its peak. Interestingly, 2025 is also when China's demographics are due to stabilise according to current trends. As such, its drivers for expansion will decrease considerably. In effect, then, over the next decade and a half, the US must accommodate itself to the idea of yielding where it can to China – buying time while simultaneously neutralising the growing power of the People's Republic. Such a balance can best be struck by integrating China into the world order, ultimately as an equal. Once again, the key to this relationship is respect, mutual understanding and recognition that the Chinese experience is borne of a history as the world's oldest continuous civilisation. The US must appeal to this nation's inner wisdom, even as it blocks its avenues for military expansion. In addition, ordinary Americans as well as those in government should be encouraged to study Chinese language, history and culture as an essential part of a process aimed at reducing fear and suspicion – of depolarisation, in other words.

A threat grave enough to prompt the US and China to combine forces would provide invaluable grounds for developing mutual trust and positive partnership. The urgency of adapting to and minimising climate change has, to date, failed to bring these two powers together, and has instead only highlighted their competition. However, the challenge of devising new energy sources and fighting globally threatening epidemics could at last provide such an opportunity. American leaders since John F. Kennedy have often repeated his imprecise 1959 rhetorical usage, in a speech, of the Chinese word *weiji* ('crisis'), which is widely believed to contain the characters for 'danger' and 'opportunity'. In fact, it is a false extrapolation, as this noun in Chinese cannot be broken down so simply. However, the sentiment is an inspiring one, and perhaps the error's popularity among politicians is a hopeful sign – if its meaning is heeded.

THE CRITICAL MASS OF POLARISATION: DANGER SIGNS

Although one party in a polarisation process might elect to attempt depolarisation, the other might prove resistant to its effects: the process might have progressed too far already, or the revolutionary elements of the nation in primary polarisation might have gained too much control over the population. In such a case, the nation in secondary polarisation must recognise the point at which its overtures no longer have any chance of success, and duly prepare for the inevitable conflict. If it has sincerely sought peace but is nevertheless forced to go to war, it will have a just cause for which to fight and its resolve will be strengthened in the war that it tried to prevent.

One key feature of advanced primary polarisation is the level of militarisation undertaken by a state, not just in terms of its offensive strike capability but also with respect to its government: when the institutions of the military and the political leadership become closely entwined, the danger rises precipitously. The final catalyst for war occurs when the state believes that it can advance its interests through military action with a high probability of success. Its envisioned superiority combines with the relative military advantage that it may enjoy, and an assessment of how prepared the enemy is to react. In the case of two empires, pilot wars often serve as a crucible, persuading the nation in primary polarisation that the older one can be defeated easily.

Returning to the example of Iran in the present day, we can see these factors in operation. Despite its large, moderate and Western-friendly electorate, the so-called 'hardliners' there firmly control their society. Despite the yearning of Iranians for a change of leadership and their attempts to make it happen, the conservative element will remain in power and possibly draw nearer to the acquisition of nuclear weapons. Unless diplomatic efforts and/or the imposition of sanctions prove effective, pre-emptive military action by Israel and/or the US appear all but inevitable.

A second example is China, which commands a military force that grows in its capabilities daily, although it still lags behind the US by at least a decade. Yet the military appears to wield strong influence or outright power over the government. Hawkish military action (e.g. satellite shoot-down exercises and interception of US reconnaissance aircraft and ships) suggests that the Chinese political establishment is not in full control of its military machine. This is a serious danger that will require constant monitoring.

Lastly, revolution is the extreme result of polarisation within a single nation, and certain conditions are more fertile than others for this – relative wealth being one. The ratio of high-, middle- and low-income people across the world is 1:3:2, although this could change to 1:3:5 by 2050 as a result of the future growth of emerging economies. The trend is towards increased inequality. It is worth noting that revolutions are not likely to originate with the so-called 'bottom billion' (they are busy simply surviving), but with those who are secure enough to choose to rebel and improve their situation.

The Global Balance of Military Power

THE EMERGENCE OF NEW TECHNOLOGIES AND MILITARY TRENDS

Any examination of military power in the future must take account of new technologies and military trends. Key among these are:

1. NANOTECHNOLOGY. The ability to manipulate matter at both the atomic and the molecular levels at sizes of 100 nm (a nanometre is equal to one-billionth of a metre) or less will lead to advances across the spectrum of modern science, as well as a new generation of miniaturised weapons and swarmed autonomous systems.

2. BIOWEAPONS. Weapons with genetic capabilities risk disturbing the world's power balance, focusing on different genetic markers between

the populations of the East and the West

3 **ANTI-BALLISTIC MISSILE TECHNOLOGY.** The nature of MAD, which held sway in the twentieth century, will change. The issue of nuclear-weapons proliferation may emerge more clearly as a result.

4 **ROBOTICS.** Employed effectively, mechanised combatants could well reduce casualties while at the same time, ironically, increasing the destructive effects of conventional combat.

5 **DIRECTED-ENERGY WEAPONS.** Electromagnetic, light, particle and acoustic weapons are already a reality and their near-instantaneous transmission adds up to a revolution in targeting solutions for fast-moving objects. Such weaponry will become widespread in time.

6 **CODE BREAKING.** The capability of the Allies between 1939 and 1945 to break the codes of the Germans and Japanese was critical to their victory, enabling them to deploy their forces where they would be most effective and maintaining the element of surprise on the battlefield. Techniques for code breaking will be just as important in the coming decades. The implementation of quantum key inception will guarantee secure communications and, conversely, the advent of quantum information processing will provide its users with the capability of breaking any code it wishes, thereby enhancing cyber war capabilities

7 **WEATHER MANIPULATION.** The battlefields of the future may be reshaped to optimise conditions, affecting both morale and mobility.

8 **ARMS CONSORTIUMS.** The prohibitive cost of modern weapons has resulted in only the richest countries (e.g. the US and Europe) owning advanced arms-manufacturing facilities and becoming net exporters of arms to other states. Soon we should expect China, and India, to adopt a similar role. As they do, their geopolitical influence will also increase. This is likely to become apparent by 2020. As the price of sophisticated, large, single-ticket weapons systems rises (the F22 has a programme cost of $65 billion and a unit cost of $150 million), the only alternative will be for allied states to form arms consortiums, with all the risks of compromise this entails, effectively reducing the ability of a state to be autonomous in its defence policy.

9 **INCREASING EQUALISATION.** Rising military power in the emerging world and declining US power will narrow the military gap. US military intervention across the world will decrease as a result, its ability to wage war with impunity having been diminished. The Western paradigm of

war will change completely. As technology becomes more equal, sheer mass will again become a deciding factor – a variable that favours China with its enormous population. Evenly matched opponents and escalating casualties are distinct disadvantages for declining as opposed to rising empires.

ABOVE **The Sea-Based X-Band Radar, which is a key component of the US Missile Defense Agency's Ground-Based Midcourse Defense System, completes sea trial testing in the Gulf of Mexico in July 2005. Currently based on Adak Island in Alaska, it can roam over the Pacific Ocean to detect incoming missiles.**

10 CULTURAL STUDIES. The cross-cultural nature of any East–West conflict will require a massive improvement in the understanding by the West of its potential enemies, with great emphasis on cultural backgrounds and collective mindsets. Because of the West's long dominance, the converse is not true: other countries understand the West very well. One example is in intelligence – Western agencies favour specific channels of focus while the Chinese use a more holistic, 'mosaic' approach to information gathering.

THE NECESSITY FOR AMERICAN MILITARY DOMINANCE

Over the next two decades, the world will continue to experience one of the most consequential power shifts in history. As we have seen, essential to the peaceful birth of a new geopolitics is the smooth integration of emerging economic powers into the world order – foremost among them China. Power vacuums subsequent to the decline of an empire are habitually chaotic, but in an age of rapidly proliferating nuclear weapons, this tendency represents a great danger. If the US retains its relative power advantage, either singularly or later, during decline, through alliances with India, Japan, Australia and even Russia, then this vacuum will not hold the potential of a devastating war. At this stage of its empire cycle, the US must be alert to new technological innovations that could rapidly tip the power balance (like the dreadnought did prior to World War One).

However, the decision by a rising power to challenge the dominant power of its time is about not only the balance of military hardware but also the perception of national will. The US therefore needs to be very mindful of its limited resources, to maximise political leverage and scale down all non-essential military activities that are not focused on the containment of China. Doing so requires a strategic rewrite of the geopolitical map as well as a realistic appreciation of the US's position within its empire cycle: the empire must perceive its limits. For example, the First Gulf War was one of expansion conducted by choice, under the hubristic misapprehension that the US was anything but in overextension. Although the war has resulted in a freer Iraq (albeit one shattered, exhausted and living on the precarious edge of civil war), the price for the US has been high: in addition to the enormous cost of the war, Iran has seen a boost in its power and influence, and North Korea used the US's preoccupation with Iraq to fully develop its nuclear weapons capacity. The message for the US is that it cannot be in multiple places at the same time; new alliances are needed. Counter-insurgency needs to be de-emphasised and the US navy re-emphasised.

The only wars that the US should consider are defensive wars, which have a very different set of strategic drivers than those of expansion, and only unfold after failed pilot wars encourage a challenger to believe that military action will suit its strategic aims. Over the next two decades, the US must stay focused on preventing China from believing that a military option for expansion in any direction could have a chance of success. It must be exceptionally alert to potential regional hot spots, into which it might be drawn to fight a war. Once again, a knowledge and understanding of its enemies can provide the best defence. In this regard, intelligence must be up to date. If, despite the country's best efforts, a pilot war should break out, it must be prosecuted with the utmost attention and an appropriate commitment of forces to bring about a swift resolution. American political will must be resolute, and the blurring between political and military leadership in the US needs to undo itself so that the military can execute a chosen course of action without constant political intervention that risks failure. (The Chinese, after all, would not suffer from such an encumbrance.)

NUCLEAR PROLIFERATION

The acquisition of nuclear capability is particularly desirable for states that oppose US power, not least as a statement of defiance. Proliferation acts in the same manner as a pilot war, threatening US global dominance. For example, North Korea's success was the US's failure, and it further emboldened China. The Chinese supported Iranian nuclear efforts until the Spring of 2010, when the country agreed to back sanctions. Yet Iran is very close to the breakout point, and Iranian nuclearisation would certainly mark another milestone in the decline of US power.

Proliferation is inevitable. Once a new technology is conceived and implemented, it is only a matter of time before it spreads across the globe. However, the immense destructive power of nuclear weapons places them in a unique category, and as such their management and control are vital to the world's stability. Clear insight into Europe's survival throughout the Cold War, caught in a standoff between the USSR and the US, is important in preventing nuclear exchanges in anger. Unfortunately, the same factors that kept the two superpowers in check are not present in the current context. Non-proliferation treaties are fundamentally flawed, as they allow for a very rapid leap from the civilian use of nuclear technology to a military application, giving the UN Security Council, or a country like the US, only a small window for pre-emptive action. The combination of access to the technology and the will to develop it further has proved very difficult to control, particularly as democratic powers do not generally favour the use of military strikes to destroy nuclear programmes in their infancy.

A nuclear state in the making requires an effective conventional deterrent as well, which could be employed in the case of a strike against its nuclear facilities. North Korea has Seoul in its sights, while Iran has the ability to launch conventional missiles against Israel, to block the Strait of Hormuz and deploy its proxies, *Hizbullah* and Hamas.

The MAD mechanism that has kept the world safe from nuclear warfare since 1945 is no longer the most workable paradigm for defence against the nuclear threat. US-developed missile shield systems have come a long way since their genesis in the Star Wars programme, and the US will be able to prevent a small-scale nuclear missile strike against it. In time, shield technology will become even more capable of making the US relatively invulnerable for the first time since 1949. The control of space will become a vital cornerstone of this new strategy.

The looming question will then concern with whom the US chooses to share the technology, as well as how long before the shield concept blossoms into a system that covers the globe and is capable of intercepting any missiles at the boost phase of launch. Current capabilities and rate of innovation suggest that these goals will be reached over the next decade, if not sooner.

Anti-ballistic missile defence will not, of course, guard against the detonation of a nuclear weapon by terrorists in a major city, or at any kind of critical installation or facility. It is current US policy to track the hallmark signatures of radioactive products from the world's known reactors, allowing it to attribute a source to a nuclear event. The state that produced the fissile material would then have to answer to the US, UN, etc, and risk some kind of retaliation.

THE ACHILLES' HEEL OF THE US

Although the US, with the most powerful military force in the world, is currently all but unassailable, power is an ever-changing continuum in geopolitics. In 1588, as the Spanish Armada set out to conquer England, Spain did not foresee that its huge, unwieldy galleons would fall prey to the smaller, more nimble English ships with their revolutionary cannon technology. Nor could the Royal Navy, in 1897, have anticipated that forty years of investment in the world's most powerful fleet would, on the launch of the dreadnought, be negated in a stroke by a new and debilitating arms race. France's Maginot Line had Europe's biggest army behind it, dug in to prevent German encroachment. Did it expect to surrender only six weeks after the outbreak of hostilities? And would the USSR in 1980 have imagined that, within one decade, its fearsome military machine – unmatched in size and designed to attack the heartlands of Europe – would be outclassed by US technology and left adrift by the collapse of the political system that underpinned it?

These great powers were presumed invincible, but in actuality at the edge of defeat, all had one feature in common: Spain, Britain, France and the USSR all faced economic decline at these key junctures, just as the US does today. Meanwhile the challengers (England, Germany and the US) were, like China today, experiencing a surging economy. What, then, is the US's Achilles' heel?

The key weakness of the US is its declining prosperity, which will only get worse as its empire fails. The contraction of an empire forces profound changes in foreign policy, engendering withdrawal and isolation. National energy diminishes, and there is a reduced readiness to project power. Thus the challenge for the US in the coming years will be to summon sufficient political will to deploy its forces effectively – both in the right place and at the right time.

As part of this process, China would be expected to copy and assimilate US technology (primarily through espionage), and later to innovate, finding ways to negate American advantage. Any resultant arms/technology race will favour the Chinese, who will more readily be able to afford such an undertaking, while the US could be broken economically by such a challenge. The US may even be reluctant to see the potential of new military technology, committing itself to older forms in which it has invested – its Maginot Line, in other words.

NEW FORCE STRUCTURES FOR FUTURE WARS

A certainty with respect to dominant military powers is that they are mentally invested in the technology and strategy of their last war, not the next one. As such, the military structure of the US is still a legacy of the Cold War in which the navy played a secondary role to the air and particularly the land components of US defence policy, because, at the time, the US required a large army to protect Europe. If the US is to adapt to the threat of Chinese expansion, it will need to reconfigure its military power to emphasise its space, air, naval and amphibious capabilities over its conventional army. Just as in the decades before 1914 and 1939, we can expect to see the current naval arms race in the Pacific region accelerate and the US forced to match the pace of development and naval innovation set by China. There will be a pressing need for more US aircraft carriers and

R08
ROYAL NAVY

1000

surface ships, but most importantly hunter-killer submarines that will provide the front line in any military blockade of a war with China. Pearl Harbor should serve as an enduring image of what can happen when a nation becomes complacent and fails to understand the intentions of an aspiring empire.

CYBER WARFARE

Although China is no match for the US in terms of technology, its massive population means that it can deploy a large number of very capable people in other capacities; for example, the legions of computer hackers presumed to be in the employ of the government, who access both industrial and military technology, and other networks around the world. One 2009 study by the Canadian group Information Warfare Monitor claims that a Chinese 'ghost' network has hacked into government and private information systems in 103 countries, obtaining sensitive and classified documents from computers belonging to NATO, foreign ministries of the countries in question (including Iran, India, Taiwan, Germany and Pakistan) and financial institutions. In 2007, British intelligence warned UK businesses and government that Chinese hackers had been raiding them. China naturally denies all such charges, and no hard evidence has been produced to demonstrate that Beijing controls an army of hackers, but the attacks have been confirmed as being Chinese in origin and are consistent with current trends in their military strategic theory. Countries around the world now find themselves defending against similar raids on a daily basis, in a sustained and largely unpublicised cyber war.

One potential area of weakness, therefore, lies in the degree to which public, commercial and military infrastructure in developed countries has been committed to electronic processes. Sabotage from hackers could shut down US military networks from the inside, thereby, for example, reducing the army, navy and air force from a highly integrated system to one of individual units. Power generation and telecommunications on a vast scale are also susceptible. Western intelligence services, overwhelmed by the hacker threat and preoccupied instead with security against the perceived Islamist danger, need to form dedicated agencies to protect against Chinese cyber incursions, preventing further leaching of state or corporate secrets and defending against the possibility of a remote-controlled shutdown of vital national infrastructure.

Networks could also be disabled from the outside through electromagnetic pulses (EMP), which would render much of the US weapons systems inoperative. EMPs were first developed as a derivation of nuclear blasts – such energy channelled into electromagnetic radiation can then fry any electronic components that it encounters. The detonation of such a weapon in the upper atmosphere could shut down immense areas. The first use of nuclear weapons would elicit dire retaliatory consequences, of course, and the US might even manage to intercept such a missile attack prior to detonation. However, if China developed an EMP that used a conventional power source, like super-cooled magnets, it could inflict serious damage without going above the nuclear-response threshold. (This area of warfare has advanced greatly in recent years.)

LEFT **An artist's rendering of the flight deck of one of two new aircraft carriers, HMS *Queen Elizabeth* and HMS *Prince Of Wales*. Due to enter service in 2014 and 2016, respectively, they will be the biggest, most powerful surface warships ever constructed for the Royal Navy and a key part of Britain's improved expeditionary capabilities needed to confront the diverse range of threats in the current security environment.**

BELOW LEFT **An artist's rendering of the Zumwalt destroyer (DDG-1000), a new class of multi-mission surface combatant ship designed to operate as part of a joint maritime fleet, assisting marine strike forces ashore as well as performing littoral, air and subsurface warfare. These ships will deploy revolutionary technology, including the new rail gun, which, using electromagnetic repulsion, is able to fire a shell to a 5 m accuracy over 200 miles at ten rounds a minute. The US navy has commissioned three, each costing nearly $3.3 billion. The first is to be in use by March 2015.**

THE NEW SPACE RACE

Humans, and American ones at that, last left the moon on 14 December 1972. In the ensuing four decades, no craft has travelled beyond a couple of hundred miles from the Earth. Notably, this apex in the first space race matched the peak of the USSR's power, coincident with the 1975 Kondratiev commodity peak – demonstrating that even when the Soviets were at their height, they were unable to afford to keep up in the enormously expensive race to space. As the USSR's economic flex declined, so did its space programme, reinforcing the point that economic power and space races go hand in hand. The US lost interest in the moon when the USSR cancelled its manned landing, and it was able to continue developing its reusable space shuttle programme, gaining valuable access to, and experience in, near-space orbit.

RIGHT **Mars: an image from NASA's Planetary Photo Journal Collection.**

FAR RIGHT **An artist's impression of one of NASA's two Mars Exploration Rovers (MERs). The MER robotic space mission is an ongoing project to explore the planet's surface. The total cost of building, launching, landing and operating the rovers for the initial ninety-day mission was $820 million. Five mission extensions have cost an additional $124 million.**

BELOW RIGHT **An artist's impression of potential manned exploration on Mars. After driving a short distance from their Ganges Chasma landing site, two explorers stop to inspect a robotic lander and its small rover. The traverse crew also check out the life-support systems of their rover and space suits while still within walking distance of their base. (Artwork created for NASA by Science Applications International Corporation's Pat Rawlings.)**

The relative cost of space exploration has since declined dramatically, opening up the competition to multiple emerging countries, ushering in a second race that could ultimately facilitate the migration of humanity into space. The US and Russia have no wish to see their hard-fought leadership in space technology diminish, and so they have announced numerous ambitious programmes. Yet the question for the US is whether, as an empire in decline, it can sustain the political will and economic resources to remain ahead of the pack in the next decade. The Russians, for their part, are better placed in relative economic terms, but their technology has lagged behind significantly. Its national will to commit to a new race is strong, but a country in its state of demographic decline is unlikely to produce the technological creativity and institutional organisation required for great advances in this context.

Ultimately, leadership in space will almost inevitably go to the new ASE. A space race will develop, with China, Japan and India as the main competitors (and in that same order of supremacy). India's ambitions echo those of China, but the country will remain a decade and a half behind its neighbour on the empire curve. Its space programme is unlikely to rival the frontrunners in the near term, but in time it could overtake Russia. India might not be a likely partner in any Japanese-US space alliance, as its technology lags behind those two countries, but its 2008 lunar probe mission was a success, fulfilling 95 percent of its mission objectives (including the detection of water on the moon and the planting of the Indian flag). India proposes further space missions, targeting Mars and near-Earth objects. It is also planning a manned mission by 2015.

Spurred by competition from China, Japan has become a new contender in space. Its plans are ambitious, and it possesses sufficient technological capabilities to achieve a moon landing by 2020 and a manned lunar base by 2030, as planned. Japanese technology is as capable as that of the US, if not as proven, and is therefore more advanced than that of China. However, the country's continuing demographic crisis and declining regional-power status risk interfering with its space-programme funding, particularly given the demands of the concurrent arms race with China. A natural solution to this problem would be an extension of the current defence partnership between Japan and the US, although this too, as noted earlier, is troubled. A significantly

more promising response to the challenge of a space race would be to structure an even more elaborate partnership to include the European Space Agency.

China clearly occupies the dominant role in the Asian space race, and it could potentially dominate world space exploration and applications in the next decade. The China National Space Administration has outlined a far-reaching lunar exploration programme that could see a moon base established by 2030, and it has taken a serious interest in exploring the moon for metal-mining potential, particularly with regard to the extraction of helium-3 for fusion energy. In 2007, China launched its first lunar probe, the *Chang'e* 1, which mapped the moon in greater detail than had been achieved previously. The craft's mission was extended to 2009, whereupon it was deliberately crashed into the lunar surface. The *Chang'e* 2 is scheduled for launch in the Autumn of 2010 and will continue the work of its predecessor as it prepares for a soft landing by the lunar rover/explorer *Chang'e* 3, which is slated for 2013, and *Chang'e* 4, an automated lunar sample return mission that is scheduled for 2017. The *Tiangong*-1 space station, meanwhile, will be placed into orbit in 2011, to be followed by the space laboratories *Tiangong*-2 and *Tiangong*-3 by 2015–16. The *Tiangong* orbiting modules will be compatible with China's *Shenzhou* range of spacecraft, of which the last three, *Shenzhous* 5–7, launched in 2003, 2005 and 2008, respectively, have been manned. (All *Shenzhou* missions, beginning with the first in 1999, have been successful.) Further to these operations is the *Yinghuo*-1 Mars probe, which is due in 2011.

China has worked hard in the copy-and-assimilate phase to catch up with US space technology, using all means at its disposal to narrow the technology gap. For now, it is still some way behind, but it will not remain so for much longer. Notwithstanding its hopes to exploit space resources, Beijing's drive to stake out space is fuelled by the same desire to prove global leadership that the US and USSR displayed during the Cold War, and Chinese *taikonauts* will undoubtedly become the next humans to walk on the moon.

Rocket technology has evolved only by small increments since the launch of Germany's V-2 in 1944. In the twenty-first-century space race, we might well observe quantum leaps, endowing the inventor nation with an unassailable lead. Based on the accelerated creativity of an empire in the ascension stage, it is mostly likely that China will take this role, leading mankind further into the solar system. (It has declared an intention to embark on a manned Mars mission by 2040.) As previous empires were able to gather resources from locations considered inaccessible by their competitors, so China could confirm its imperial primacy in forthcoming decades by being the first nation to repeat this in space.

Disease

Disease impacts differently on developed and emerging countries. The developing world suffers from higher disease levels associated with the social stress of high population densities, poverty and poor access to medical care, compounded by the low standards of the care available. Consequently, lower average lifespans and greater child mortality feature more frequently. Inhabitants of the developed world, however, enjoy extended lifespans with better health care and access, with many of the diseases that affect their counterparts in the developing world either absent or limited in presence. Although the average inhabitants of wealthy countries do not often pause to consider the state of health in poor ones, such nonchalance is ill advised and possibly deadly. The globalised world of the foreseeable future is criss-crossed with travel routes; new opportunistic diseases incubating in one country can be rapidly delivered to the rest of the world. Two key examples are HIV and drug-resistant tuberculosis.

The modern world has also witnessed the birth of new viruses in areas like China (SARS) and Mexico (H1N1, or 'swine flu'), both of which were cause for pandemic scares in 2002–2003 and 2009, respectively. High population density and the proximity of animals in poor city and rural regions are high-risk incubation zones for new viruses. Pandemics can ravage an expanding human population, and easily spread viruses with high mortality rates that have had decimating effects many times throughout history – the fourteenth-century Black Death and the 1918 Spanish flu being just two obvious instances. If, as I have discussed, empires are vulnerable to disease as they enter the stage of decline, then the US could soon be particularly susceptible to a viral pandemic. The 1918 flu affected 28 percent of Americans and killed some 500,000–675,000; the 1957 Asian Flu pandemic killed around 70,000 Americans. Influenza caused an average of 36,000 deaths annually in the US during the 1990s, out of a population of over 250 million. The annual death toll from seasonal flu has since risen. A pandemic flu with high lethality as well as high communicability could easily infect one in two people and cause four deaths per thousand – a frequency that would kill 1.0–1.5 million Americans and 70–150 million people worldwide. Conservatively, the US economy would lose $500–750 billion, and the world, $4–5 trillion (approximately 5 percent of global GDP).

With such high potential costs at stake, the developed world should commit resources to ensuring that disease reservoirs in emerging nations can be contained. Financial and medical aid could reduce the frequency and severity of infectious diseases in the poorest regions. Given that poverty is linked to disease, more equitable economic growth could further reduce the spread of disease – one more reason to encourage emerging economies. The formation of a global health fund with a mandate to channel aid from developed nations into disease-fighting programmes would be a good first step. Such a group could coordinate its efforts with organisations like the Bill and Melinda Gates Foundation, which has, in effect, recognised the urgency of addressing global health concerns ahead of the world's governments.

The reduction of infectious disease in developing countries might also, over one or two generations, slow down birth rates in the emerging world. However, the irony is

that the very technology that will allow man to overcome naturally occurring diseases has and will increase our ability to construct biological weapons, which by intention or accident could be released to cause a new global pandemic. Currently, fewer than a handful of states have offensive biological weapons, but that number will surely increase with the advent of widespread genetic technology. Accordingly, there is as urgent a need for a non-proliferation treaty for biological weapons as there is for nuclear weapons.

LEFT Emergency workers in hazardous-material suits walk among mock rubble during Top Officials 3, a terrorism response exercise coordinated by the US Department of Homeland Security, in New London, Connecticut, in April 2005. Participants responded there to a simulated chemical attack and in New Jersey to a simulated biological attack in a drill that lasted for a week.

Climate Change

Change to the Earth's climate will continue to present an urgent and dramatic challenge and can only exacerbate any existing social, political and resource stress, consequently lowering the threshold risk for global conflict. On a micro-level, Darfur is an example of how the effect of climate change on basic food and water supplies causes a densely populated area to break down into conflict.

With the world's industrialised states still competing with each other, solutions will be delayed or prevented. Until the most powerful countries truly understand that everyone's children will suffer the effects of climate change, the ingrained collective behaviour that has governed humanity for so long will continue pushing us towards disaster, making it unlikely that we will change in time.

Climate change is a function of the empire cycle, as we have seen, with the ASE rising to supplant the WCSE as it declines. There is simply no room on the planet for both industrial super-empires to coexist in their current manifestations. The US, throughout the 2000s, missed a historic opportunity to use the twilight years of American power to channel wealth into climate-change policies, thereby modifying the way China and India grew their industrial power bases. Now, overextended and economically turbulent, the US is less able than ever to divert resources to helping the countries of the ASE to develop clean technologies; it does not even have the resources to lead by example by effecting such changes in its own economy.

China may come to determine that climate change is a drag on the global economy, but, owing to the momentum behind its own growth, it may also consider that it will be able to counter climate change as its empire expands – particularly into Africa. Yet the effects of climate change on China may become more devastating and at a faster rate than the country can mitigate through expansion.

If the US appealed to Chinese pragmatism, agreed to reduce its own sphere of influence and worked with the country to develop a groundbreaking climate-change agreement, then India would surely follow. The crossover between the issues of energy friction and climate change could both be tackled under the same policies, representing a once-in-a-lifetime opportunity for politicians from the West and the East to ease the worsening pressures that daily menace us all. However, political agreements in a crisis context need to be enacted in a meaningful way before it becomes too late to effect change. One solution with which to achieve real results would be the creation of a world body, invested with the authority to conduct the necessary research and coordinate efforts into coherent and binding plans of action.

One daunting problem is that there are just too many people in the world, all seeking the benefits of industrialisation. Carbon allowances would need to take into account population growth as well as industrialisation rates; population controls would thus become an integral part of climate-change policy. Reconciliation must take place between developed countries with flat demographics and benefiting from a high standard of living (from having long been industrialised), and which have already generated massive amounts of climate-changing carbon, and the world's emerging countries with swelling demographics, which are now following the cheapest and most polluting paths to industrialisation. Key to this discussion is the pollution rate per capita, which is highest in the West and lowest in the developing world. By the time the latter is approaching the same level of industrialisation as the West, per capita the Earth will have quickly and inevitably passed the 5°C raised-temperature point, and then the biosphere will exact its own severe penalties on human civilisation.

The US and China are essential to resolving the problem. The US has the highest per-capita emissions rate in the world. China comes in at only 40 percent lower, but, because it is more than three times the size of the US, its total emissions levels are very close to its Western counterpart. The predominance of coal-fired power stations in China exacerbates the situation. Unfortunately, these two nations are engaged in a power struggle, linking climate change to the evolving multi-polar world and the cycle of empires. However, assuming that the US and China can come to an agreement with respect to the sharing of power, the steps towards mitigating climate change might proceed according to the following principles:

1 FOCUS ON THE IMMEDIATE SOLUTION. The most urgent step is to dramatically reduce per-capita greenhouse emissions. The West must make available whatever emissions-reducing technology it has to the developing world, so that the latter can stop at the same emissions ceiling. The developed world, particularly the US, must lead the way. Being the first to take action could count strongly towards depolarisation.

2 IMPLEMENT A CARBON CREDIT SCHEME. This is paramount. As a means of self-regulating emissions, carbon credit is the best way to balance commercial and environmental demands. Structuring such schemes is complex and commitment is necessary from governments worldwide to create a loophole-free system. It will be imperfect to begin with but can be improved with time. The credit scheme can also be linked to environmental taxes, which should penalise polluters and benefit responsible companies and individuals. After all, new technology always costs money, requiring state support to implement on a wide scale in order to effect the modernisation of national infrastructure.

Programmes such as the World Bank's Forest Carbon Partnership Facility supply carbon credits to regions with rainforests in danger of decimation, in exchange for help to preserve them, not only offsetting emissions in these sensitive regions but actively supporting their rejuvenation.

3 **COMMIT TO DEVELOPING NEW ENERGY SOURCES.** Novel means of generating power are needed in order to halt the high rate of carbon emissions as a result of conventional energy consumption. Growing energy friction will act as an additional impetus to institute carbon sequestering and construct new nuclear reactors (initially exploiting fission, and later, development permitting, clean-burning fusion).

4 **REWARD INNOVATION AND SHARE TECHNOLOGY.** Economic rewards for new and effective climate-change solutions must be a regular and generous by-product of effective carbon-trading schemes. Again, developed countries must encourage the transfer of new technology to the developing world – nothing short of a global environmental revolution should be attempted. New technologies should not be limited to emissions-reduction solutions but also encompass such fields as agriculture and bioengineering, such as planting fast-growing willows; changing crop colours and generating oceanic water-vapour clouds to increase levels of ultraviolet radiation reflection; and cultivating plankton farms to have a carbon-sequestering effect.

5 **ENCOURAGE AND IMPLEMENT A SEA CHANGE IN SOCIAL ATTITUDES.** In many ways, ancient societies understood their environments far better than we do at present – people lived in tune with the seasons, and with their ecosystems, even when they harnessed the natural world for their own benefit when hunting, farming or diverting water from rivers. Respect for the natural world was present to a far greater degree. First industrialisation withered this attentive approach to the environment; then complacency and a preference for convenience set in, along with the perpetuation of a new ignorance of the natural world down through successive generations. We now need to recognise beyond doubt that we must live in balance with our habitat, and that unless we rediscover the means of doing so, our negligence will cost us dearly. Only genuine sustainability can provide our children with a future and give the biosphere the protection that it requires. Only thus can our civilisations continue to prosper.

Empire Earth

Returning to the concept of fractals, discussed in PAST, I should like to consider the position of the entire human race according to the Five Stages of Empire model. Geopolitics takes place at the level of individual states competing for resources, with the result that some, driven by demographics, become empires. Super-empires emerge as a collective made up of empires, or regional powers led by an empire, as with the WCSE and the ASE, respectively. The greater fractal would take in the planet as a whole, positing the Earth as an empire in its own right. Where is the world as a whole on the empire cycle? There are several key indicators providing insight into this question:

1. The planetary biosphere is past the point of being able to sustain an endlessly proliferating human population that is undergoing a second wave of industrialisation. We are living at odds with our environment and consuming at a greater rate than allows for replenishment.

2. The global social structure is two-tiered. The WCSE, with its neutral demographics, is steadily relinquishing power to the emerging states, consistent with the transfer of power that begins just prior to an empire's decline, when traditional elites give way to those formerly subjugated.

3. Increased social stress, which is the result of competition for increasingly scarce energy, water and food resources, prepares the path for war and disease to devastate the population.

These signs all suggest that the human race is balancing at the frontier between overextension and decline in a regional planetary cycle, with our population curve verging on plague levels. To prevent potentially devastating consequences for the planet's billions of inhabitants, it is imperative that we become aware of currently unconscious behaviour patterns, and realistically confront and resolve the huge problem of population. Voluntarily lessening demographic pressures by reducing populations is an option, although as a remedial measure it is not politically acceptable virtually anywhere in the world. It may only be viable once a majority has been persuaded by overwhelming evidence, and by then it would simply be too late. However, confronting this issue must become part of a coordinated foreign policy among the developed countries, and would best be integrated with the issues of disease and climate change. Recruiting China and India to this perspective would be a major advance, although realistically, assuming that these countries well understand the relationship between demographics and power, they would have no motive for subscribing to any measures that would limit their progress towards empire.

At present, the developed countries are barely able to maintain replacement ratios of two children per couple – perhaps on account of the cost of child rearing

Humanity's Inflection Point: Regional Planetary Decline, or Ascension to a Planetary Empire *(fig. 68)*

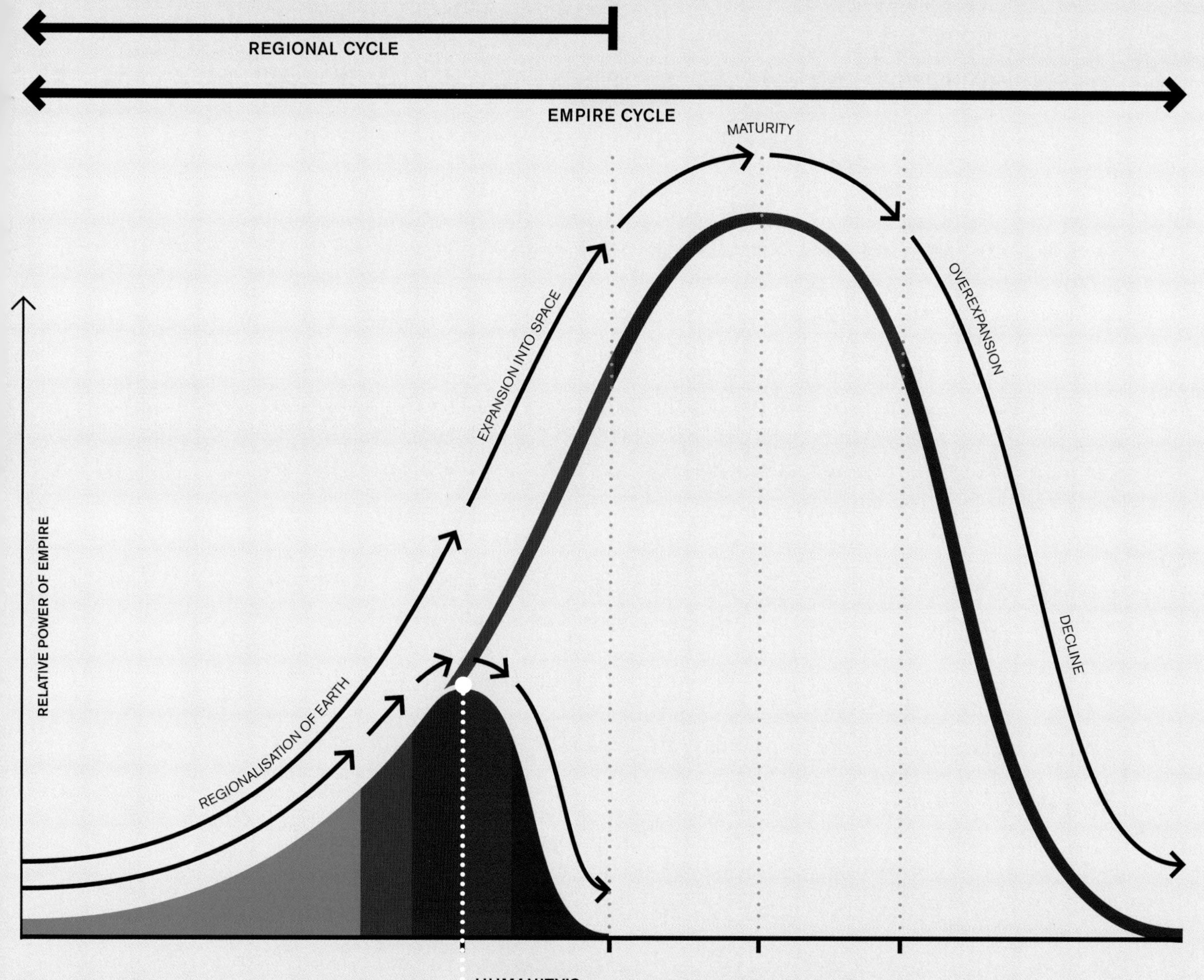

HUMANITY'S INFLECTION POINT

Will humanity evolve to create a global intergrated society that can then expand to the stars, or will it fail through conflict and enter decline for this cycle?

balanced against the high probability of survival, which favours lower birth rates. Moreover, the more urbanised a country becomes, the more birth rates decline, suggesting that population density and social stress are contributory factors too. (Countries with formerly high birth rates, such as Brazil, are experiencing noticeable declines in rates among their increasingly urban, as opposed to rural, population.) Urbanisation, therefore, has the potential to moderate birth rates. As we have seen, the global population has only just passed the 50 percent-urban mark. As the urban-dwelling majority pass from being a phenomenon to being the norm, the moderating effect of a predominantly urbanised human race on population growth will become more pronounced. Working against this possibility is the process of increasing industrialisation, whereby more resources are needed to sustain each urban dweller than their agrarian counterpart.

RIGHT **The Space X Falcon 9 test rocket lifts off Pad 40 at Cape Canaveral Air Force Station, Florida, on 4 June 2010. This was a test flight for the craft, which is scheduled to undertake supply missions to international space stations in the future.**

If voluntary population control has no chance of succeeding as an option, other self-regulatory solutions to rescue humanity from decline will need to be sought. Such a plan will inevitably surface, but likely not before we reach the critical point of 2017–25, encompassing both China's demographic downturn and the peak of the next commodity cycle. Perhaps hope can be found in the notion that, if we have avoided conflict by 2025, China's stationary demographics combined with those of the West will be sufficient to induce an evolution in thinking that could allow humanity to seek a new, sustainable approach to living, until we have the technology to undo the damage that we have wrought or escape our closed system and dwell among the stars.

(One possible hope just emerging on the horizon may come in the form of advanced robotics and artificial intelligence (AI). Although automatically associated with science fiction and perhaps often dismissed accordingly, the field of robotics in the late twentieth and early twenty-first centuries has seen astonishing progress, not least in Japan, which, tellingly, is undergoing one of the worst demographic crises in the world. The rise of once-unimaginably sophisticated robots as substitutes for labour and military manpower, and perhaps even for nurturing in place of children, is no longer an outlandish concept. Downward demographic pressure coupled with exponential leaps in robotics and AI technology may provide a solution for the absent bodies.)

To sum up, in the context of the biggest empire cycle possible, the human civilisations of the Earth are currently at the stage of maturity and are at risk of heading into decline, after reaching the limits of the planet's ability to sustain human hunger for resources. Whether or not we survive this inflection point depends on whether or not the Earth's last two great powers can resolve the competition-and-domination cycle. If this resolution requires a war, the consequences could be so damaging that the downward spiral of Empire Earth could proceed beyond hope of any meaningful survival. However, if we can summon the will to use our reason and consciousness instead, we can address our differences peacefully and inaugurate a new geopolitical paradigm. The next stage along the planetary empire cycle could then see the ascension of the human race to a new, stellar empire.

SPACEX

Afterword
The Story of Easter Island

Around ten or more centuries ago, in the southwestern Pacific Ocean, the population of one or another group of Polynesian islands (most likely the ones now called the Marquesas) set out in canoes and catamarans seeking new land to settle. They arrived at what we know today as Easter Island, or *Rapa Nui* in the local language. Under the command of their gods and their *ariki* (king), the people proliferated and prospered within a feudal system on the island, which measures about 25 km by 12 km.

The inhabitants of *Rapa Nui* committed tremendous energy into sculpting and erecting the huge, iconic and ultimately enigmatic stone figures for which the island is so well known: the *moai*. Crafted between AD 1250 and 1500, these were positioned along the coastline and required intensive resources for quarrying, workmanship, transport and erecting. The precise methods employed by the islanders to move these statues, some of which could weigh several tons each, are as yet undetermined, but log rollers were probably used. Evidence indicates that when the island became completely deforested by 1650, the statues stopped being produced.

Further inefficiencies in resource consumption attended the *Rapa Nui* culture. A warrior class seized power and instituted the *tangata manu* (cult of the 'bird-man'). Annually, a contest would be held between warrior clans to swim out to a nearby islet and secure the first of the season's eggs laid by a local species of tern. The victor was awarded complete control of the island's resources for the rest of the year, and this naturally led to the weakening of the island's ecosystem as well as widespread malnutrition and death. Conflicts arose between different factions on the island, leading to, among other things, the toppling of many of the *moai*. The seventeenth and eighteenth centuries were a time of devastation for the island's physical, psychological and spiritual health. By the time European slavers arrived in the 1860s to kill and carry away *Rapa Nui* captives for labour in the mines and plantations of colonial Peru, the island had been so battered by self-inflicted strife that its people could not resist effectively. More of them succumbed to diseases introduced by the Europeans, such as smallpox and tuberculosis. By 1877, on account of death as well as deportation and emigration, only 111 islanders remained, of which only thirty-six had descendants – the low point from which the cycle of population expansion began again. The legacy of their predecessors, who might have numbered as many as 15,000 in previous centuries, was a barren island, guarded by a around 900 giant stone statues amid the wreckage of the past.

The story of Easter Island contains all of the elements of the rise and fall of a society over centuries: the commodity cycle, competition for resources, greed and short-sighted resource management, religious polarisation, warfare, disease and climate change. The similarities to our world of the present day are remarkable, though reasonable, according to fractal theory. The lessons of history have been clearly and vividly spelled out; the only question now is whether or not the hubris of humanity will allow them to be heeded.

OVERLEAF ***Moai*** **on a hillside on Easter Island (*Rapa Nui*), Chile. There are about 900 such statues scattered along the coastline of the island. Carved from soft volcanic rock, they stand 3–12 m high. Their exact origin is unknown, but most experts believe that they were carved by early Polynesian settlers.**

All this happened before and will happen again, unless we choose to consciously change our collective patterns of empire.

Acknowledgements

Writing a book of such magnitude is the product of a life's journey. Along that journey there have been key people whom I would like to thank:

My gifted physics professor, Roy Sambles, who taught me the power of intellectual rigor and the ability to formulate powerful models from seemingly simple observations and variables.

The Papua New Guinea population of the Sepik, who thankfully chose not to eat me and unknowingly showed me that the apparent superiority of Western culture is but a veneer and an illusion.

Tony Plumber, whose works were the first I encountered that showed that others believed that collective human behavior was a reality, and who continues to be a good friend.

R. J. Elliot, whose study of market patterns in my opinion is without compare and whose insights in predicting price moves and the associated market psychology remains breathtaking.

T. K. Kalaris, my boss at J. P. Morgan, who had the confidence to believe in my theory that markets and human behaviour are based not on logic but on collective emotions, and that the resulting sequence of price patterns are predictable. And for giving me the chance to evolve my observations and theories in my own department.

My *tai chi* teacher, Rose Lei, for revealing to me the ancient wisdom of the practice and of Chinese culture, and my good friend Quincy Rabot, who continued my education in these traditions.

The members of the investment community who have supported Emergent, facilitating an environment that allowed me to put into practice the concepts found in this book, and whose enthusiasm for my ideas expressed in my very many speaking engagements over the years encouraged me to overcome my dyslexia and write this book.

________The dedicated, talented and good-humoured team of market professionals at Emergent.

________My book production team, all of whom put in the most enormous dedication and effort into this project and to whom I owe huge thanks for making this book a reality.

________My children, whose presence inspired me to start the task of writing this book to ensure that they, like all the other children of the world, might have a future.

________Lastly, but most importantly, those dear to me whose love has allowed my creativity to manifest itself and for me to walk a personal path of transformation in the process.

Index

Picture Credits

BAE Systems/Aircraft Carrier Alliance: 442t
CSIRO/Science Image: 315
Imperial War Museum, London: 261b, 264, 271, 278
NASA: 27, 104; 140/GSFC/Craig Mayhew & Robert Simmon; 372–373/JPL-Caltech, Susan Stolovy; 445tl/Goddard Space Flight Center Scientific Visualization Studio; 445tr/JPL/Cornell University; 445b/Pat Rawlings, SAIC
Reuters/Goran Tomasevic: 298
US Navy: 442b

All other images courtesy of Getty Images, including those below, which have further attributions:

Agence France Presse: 11, 17, 51t, 93, 101t, 119, 122r, 135, 137b, 146, 161, 169, 180, 186, 189, 193, 202, 207t, 225, 227, 234, 238, 240t, 242, 244, 249, 256, 262b, 288, 330r, 333, 335, 337, 342, 371, 389, 395, 397, 408t, 411, 425, 432, 434, 448
Bloomberg: 123, 124, 127, 129, 182, 198t, 400, 414b
Botanica: 368
Bridgeman Art Library: 14–15, 28, 40, 73, 376
ChinaFotoPress: 102–103, 108b, 122l
De Agostini: 31, 309t
Eco Images/Universal Images Group: 361t
Flickr: 218
Gallo Images: 419t
Kate Geraghty: 250
Mike Goldwater: 207b
Imagemore Co. Ltd: 111
Imagno: 96r, 214
Kallista Images: 330l
Kaveh Kazemi: 233, 241b
Bruno Morandi: 43r
National Geographic: 74, 348, 361b, 362
Yawar Nazir: 352
New York Daily News: 247
Popperfoto: 216b

Zack Seckler: 198b
SSPL: 37, 310
Stocktrek Images: 290, 438
Tim Graham Picture Library: 20b
Time & Life Pictures: 48, 49l, 51b, 52, 72, 142, 143, 156, 158, 215, 272, 281, 303, 311, 391, 408b
Travel Ink/Gallo Images: 44t
Veronique de Viguerie: 176

Many thanks to Mapping Worlds (www.mappingworlds.com) for providing the maps on pp. 112, 168, 171t, 193, 195, 203, 219, 233b, 318, 328, 332, 371t, 421, 428–429

Additionally, all figures are courtesy of David Murrin, except those listed below:

149 (fig. 25) ©1996 Roy Beck, The Case Against Immigration; 163 (fig. 27) ©Federal Reserve Board. Cycle analysis by Tony Plumber; 166 (fig. 29) © BP, NE, Umicore, USGS; 174 (fig. 32) © Emergent 2010; 326 (fig. 52) © WHO 2009. All rights reserved; 354 (fig. 59) © Robert Simmon, NASA. Minor modifications by Robert A. Rohde.

pp. 375 to 379 extracts from *The Art of War* by Gary Gagliardi

About the Author

AUTHOR DAVID MURRIN has twenty-three years' experience in the directional trading of markets. He trained as a geophysicist with an honours degree in geophysics from Exeter University. In 1984, he joined a global seismic exploration company and was posted to Papua New Guinea. For three years he lived and worked with local tribes in the Sepik River Basin, and was inspired to begin formulating his theories on collective emotional behavioral patterns.

In 1986, he joined J. P. Morgan in London, trading on the major bond, interest rate, bullion, foreign exchange and equity index markets. His responsibilities included acting as strategic advisor to the head of trading using his price and behavioral-based analysis techniques. In 1991, he founded and managed J.P. Morgan's highly successful European Market Analysis Group, which was responsible for developed and emerging markets, among other areas. In 1993, he established Apollo Analysis Ltd to advise several bulge-bracket banks in taking directional risk in global and emerging markets.

In 1997, David co-founded UK-based Emergent Asset Management, which initially specialized in the management of macro/emerging market hedge funds, and of which he is chief investment officer. As well as macro-trading, with experience in special situations in oil exploration and sixth-generation oil rigs, he is responsible for the firm's real asset business in Africa. He is chairman of the fund's dedicated Pretoria-based management company, EmVest, which focuses on commercial-scale farming across Sub-Saharan Africa.

David speaks widely on financial markets, appearing frequently as a keynote speaker and on television. He is also a keen military historian of long standing. This personal passion has fuelled years of research and has been instrumental in helping him to construct a macro view of geopolitics in the coming decades – from which this book takes its themes.